Agricultural Innovation for Societal Change

Towards Sustainability

_____

Previous publications by Bo M. I. Bengtsson on sustainability:

Agricultural Research at the Crossroads: Revisited Resource-poor Farmers and the Millennium Development Goals. Science Publishers, Enfield, NH, 2007. ISBN 9781578085149.

Agriculture, Climate and the Future (together with Susanne Gäre in Swedish). Ekerlids, Stockholm, 2019. ISBN 9789188849434.

At the Crossroads. Sustainable Policy for the Future (in Swedish). Ekerlids, Stockholm, 2022. ISBN 9789189323681.

Agricultural Innovation for Societal Change

Towards Sustainability

_____

Bo M. I. Bengtsson

CABI

CABI is a trading name of CAB International

CABI	CABI
Nosworthy Way	200 Portland Street
Wallingford	Boston
Oxfordshire OX10 8DE	MA 02114
UK	USA

Tel: +44 (0)1491 832111

T: +1 (617)682-9015

E-mail: info@cabi.org

E-mail: cabi-nao@cabi.org

Website: www.cabi.org

The views expressed in this publication are those of the author(s) and do not necessarily represent those of, and should not be attributed to, CAB International (CABI). Any images, figures and tables not otherwise attributed are the author(s)' own. References to internet websites (URLs) were accurate at the time of writing.

CAB International and, where different, the copyright owner shall not be liable for technical or other errors or omissions contained herein. The information is supplied without obligation and on the understanding that any person who acts upon it, or otherwise changes their position in reliance thereon, does so entirely at their own risk. Information supplied is neither intended nor implied to be a substitute for professional advice. The reader/user accepts all risks and responsibility for losses, damages, costs and other consequences resulting directly or indirectly from using this information.

CABI's Terms and Conditions, including its full disclaimer, may be found at https://www.cabidigital-library.org/terms-and-conditions.

A catalogue record for this book is available from the British Library, London, UK.

Library of Congress Cataloging-in-Publication Data has been applied for.

ISBN-13: 9781800627789 (hardback)
9781800627796 (ePDF)
9781800627802 (ePub)

DOI: 10.1079/9781800627802.0000

Commissioning Editor: David Hemming
Editorial Assistant: Helen Elliott
Production Editor: Rosie Hayden

Typeset by Exeter Premedia Services Pvt Ltd, Chennai, India

Contents

Preface

Agriculture has developed over centuries. After World War II, it became modernized and more advanced with technical and chemical innovations. They have contributed to raise agricultural productivity, although some people still suffer from food shortages. Agricultural innovations have also had some negative consequences for the environment, both nationally and globally.

I have been strongly associated with agriculture and rural development since I grew up on a small farm in southern Sweden in the late 1940s. After that I spent more than half a century of engagement in various agricultural research and development activities with an international focus. After 2000, several major developments have taken place that have been significant for future agricultural research and food production.

In the late 1980s, the Brundtland Commission presented the concept of sustainability to be used in guiding development to ensure a good life for future generations in harmony with nature. Centuries ago, the Inca rulers recognized our ultimate dependence on Mother Earth. The concept was also accepted by world leaders and later governments agreed to the United Nations Millennium Development Goals, now replaced by the Sustainable Development Goals to be reached by 2030. The question is whether policy makers and the scientific establishment have given substantive attention to the concept for its practical implementation. Agriculture is still given less political priority and managed as though it was business as usual. Natural resources are chiefly considered as a source for exploitation, hardly a sustainable approach over a long-term perspective.

I owe a great deal of gratitude to many colleagues, partners and farmers, and I would like to express my sincere thanks for their guidance and advice during numerous constructive dialogues. Finally, I wish to express my gratitude to my publisher and all members of the CABI team for giving me excellent guidance and support during the whole process. In addition, special thanks to my copy editor for accurate and careful work in a highly cooperative manner and my wife, Christina, for patience with my writings.

Bo M. I. Bengtsson, March 2024

1 Outlook

Abstract

Global demand for food increased up to 2000 in relative harmony with global population growth. Since the global economic crisis in 2008, food shortage is still a problem for too many people. Famines occurred in 2021 and a healthy diet is unaffordable for more than 3 billion people. The Russian invasion of Ukraine jeopardized part of the world wheat trade. Business as usual has been the solution for those with cash in the globalized and industrialized world of food production. This has torn apart the local context in rural societies and depends on technologies that cause growing environmental degradation. In the future, agriculture needs to tackle many challenges, calling for food sovereignty with sustainability, and setting biology as a priority for both political and research establishments. Their actions have been slow in giving prominence to the concept of sustainability recommended by the Brundtland Commission, which world leaders accepted in 1992 at the Earth Summit in Rio de Janeiro.

For quite some time, political actions and innovative technology have contributed to modern societal development in the industrialized countries. This includes globalization of agrifood systems with significant nutritional implications. Global trade and technological change in agriculture have also improved food security for many countries. This involves changing the supply-chain structures, the rise of modern retailing and new food quality standards. Emerging high-value supply chains may also gradually contribute to some income growth in the small farm sector in low-income countries. However, supermarkets there may contribute to obesity among the wealthier consumers.

Globalized and industrialized food production prefers few and cheap staple crops, such as wheat, rice and maize. It has torn apart the local context in rural societies. Globalization has changed the traditional role and competitiveness of small- and medium-sized enterprises. In addition, the world food market can change swiftly, especially in times of crisis.

Up to 2000, global demand for food increased. This took place in relative harmony with global population growth, modern technologies and good harvests, leading to improvements in incomes and diversified diets. In 2004, prices for most grains began to rise but production could not keep pace with a stronger growth in demand. This led to depleted food stocks, followed by diminishing food supply due to lower harvests in major food-producing countries. Rapid increases in oil prices caused higher fertilizer prices and other costs of food production. Food-exporting countries introduced export restrictions and large importers began purchasing grains at any price to maintain domestic supplies. Then, the global economic crisis in 2008–2009 undermined food security in many countries. Business was seen as the solution, but it has an uncontrollable power over food in low-income countries.

The number of people who are short of food in the world is too high. Estimates indicate that 691–783 million people lacked sufficient food in 2022. This was affecting 9.2% of the global population compared with 7.9% in 2019, the year before the COVID-19 pandemic began (WFP, 2023). Some 122 million more people have faced hunger in the world since 2019 (FAO, 2023). In 2022, about 8% of the global population lived in extreme poverty, subsisting on less than US$2.15 per day, whereas half of the global population lived on less than US$6.85 poverty line per person per day (Schooh *et al.*, 2022). Thus, the world is far from achieving Sustainable Development Goal 2: zero hunger by 2030 (UN, n.d). In fact, only 16% of the SDG targets are on track to be met globally by 2030 (Sachs *et al.*, 2024).

Estimates show that 670 million people will be facing hunger in 2030, equivalent to 8% of the world population (UN, 2023b). This is the same figure as in 2015 when the United Nations (UN) launched the 2030 Agenda. Around 2.3 billion people in the world were moderately or severely food insecure in 2021, and 11.7% of the global population faced food insecurity at severe levels (FAO, n.d.).

In 2021, there were famines in Yemen, Nigeria, Burkina Faso and South Sudan according to the World Food Programme (WFP). It warned that 41 million people in 43 countries were close to famine: an increase from 27 million 2 years ago (Al Jazeera, 2021). In 2023, nearly 282 million people in 59 countries/territories faced acute food insecurity (GRFC, 2024). In fact, famines have always been part of the history of Earth and many famines have occurred, even in the last 100 years.

The Russian invasion of Ukraine in 2022, led not only to a tragic war between two of the world's largest agricultural producers. It also jeopardized more than 33% of the world wheat trade, 17% of the world maize trade and almost 75% of the world sunflower oil trade. Within a week of the invasion, prices of future wheat had jumped by almost 60% and those of maize and soybeans by almost as much. Ukraine's ports on the Black Sea were blockaded, limiting crop exports, despite a temporary agreement with Russia to allow them. Moreover, the planting and harvesting of new crops was disrupted (Glauber, 2023).

The Russian invasion of Ukraine was not the first attempt to make use of one of the latter's strategic and natural resources. In the summer and autumn of 1943, German SS Commando units had invaded Ukraine to gather and steal crop seeds and plant material from the region, and seed collections held in the Soviet research stations.

The reduced production of food crops in Ukraine is significant for Africa and China. Countries such as Togo, Burkina Faso, Senegal and Mauritania already import more than half of their wheat needs from Russia and Ukraine. Prior to the Russian invasion, oil prices had increased and rising inflation in early 2023 made the cost of imports much more expensive for these countries. There have been significant price increases, for example, for diesel, fertilizers and electricity.

So far, the international sanctions on Russia have resulted in some reduction of trade, especially with European countries. The effects within Russia are not yet noticeable. In fact, both Turkey and Saudi Arabia, long-term oil exporters to the European Union (EU) market, increased their imports of oil from Russia in 2023 according to Bloomberg (2023a). Overall, this development also has effects on European countries.

A food crisis cannot be excluded in the near future. Worldwide support for food and agriculture accounted for almost US$630 billion/year on average during 2013–2018. The major part is targeted at individual farmers through trade and market policies, and fiscal subsidies. The support is tied to production or unconstrained use of variable production inputs. It concentrates on staple foods, dairy and other protein-rich foods from animal sources, especially in high- and upper-middle-income countries. Rice, sugar and several types of meat are the foods most incentivized worldwide, while fruit and vegetables are less supported overall, especially in low-income countries. Much of this support is market-distorting and may not reach many farmers.

For some time, the Food and Agriculture Organization (FAO) has claimed that global food production must double and do so in a sustainable way – a great challenge. On the other hand, current global food production would be sufficient for all people if evenly distributed across the globe and people on low incomes could afford to buy it. This would call for more trade and globalization, but above all cash for people on low incomes. Regretfully, this is not likely, and globalization may not benefit low-income countries. Thus, there is a need for novel approaches. The current focus on globalization and Western-type, large-scale agricultural production may not be the best solution in the long term, say up to 2070.

In most low-income countries, agricultural policy is weak, despite a large agrarian sector. Also, aid donors have neglected to support the agrarian sector. Agriculture is vital for the economy, jobs, livelihoods and even survival in low-income nations. Future production of nutritious foods is required for healthy diets and income generation. The challenge is to produce food in a sustainable manner, which is far from producing an industrial product. This raises the question of how future livelihoods are to be managed and what future sustainable societies would look like. Biology must be at the forefront.

Globally, there are a few major countries in control of and producing most of the world's food. Good natural resources and land conditions provide a good basis for the United States to be one of the world's leading agricultural producers and suppliers. As in all industrialized countries, since the 1970s the number of people involved in US farming has fallen from about 25% to 1–2% and so have the number of farms. There are farms of many sizes. In 2001, 2 million farms had an average size of some 220 ha, highly mechanized (USDA, 2007). Farm land consolidation has been an ongoing process. Between 1997 and 2017, the number of farms fell by about 200,000 with some 22 million ha farmland lost (Douglas, 2024). In 2023, the average farm size was 187 ha and there were some 1.89 million farms (USDA, 2024). Interestingly, there has been a recent increase in the number of smallholdings in both the United States and Brazil, although the share of cropped land controlled by the large farms has increased (Lowder *et al.*, 2021). Farm ownership has been concentrated in large factory farms, especially those producing eggs and beef. Agricultural exports from the United States were valued at US$196.4 billion in 2022 but were expected to decrease to US$184.5 billion by 2023 (Statista, 2023).

India ranks first in the world for the highest net cropped area followed by the United States and China. It ranks second worldwide for farm output (Bhardwaj *et al.*, 2021). Farmers are important to politicians in democratic low-income countries, as demonstrated when Prime Minister Narendra Modi recently was forced to repeal three recently introduced laws. They were supposed to liberalize the pricing and sale of agricultural products, leading to the disappearance of the state-guaranteed minimum price. In early 2024, German farmers protested EU rules and their approval by the German government, using 6000 tractors and items of agricultural machinery to block the main streets in Berlin. The issue concerned the price of fuel after government subsidies were abolished.

Later, farmers also protested in France and were joined by other farmers approaching the EU in Brussels.

About 90 million out of about 115 million Indian agricultural enterprises have 1 ha or less of land at their disposal according to the recent Agriculture Census of 2015–2016 (Government of India, 2016). Of the country's official 150 million ha under cultivation, about 80 million ha require rehabilitation. In the first high-production arable areas of the Green Revolution, in the 1960s farm productivity has slowed down. Pastures have almost disappeared, and the groundwater is sinking. Soils are becoming degraded from high inputs of agrochemicals. Also, the decreased area of forestry has had negative impact on biodiversity.

India's population may reach 1750 million by 2050, larger than China's and a rapid increase from about 990 million inhabitants in 1999 (United Nations, 2023a). Despite this population growth, India has managed to produce enough food so far. However, the economic contribution of agriculture to India's gross domestic product (GDP) is steadily declining. India is a large agricultural exporter, mostly to low-income and least-developed countries. In 2023, India decided to stop its rice exports due to increasing national food prices (Jadhav *et al.*, 2023).

In China, food production must supply one-fifth of the world's population on one-tenth of the world's cultivated land. This covers about 140 million ha but about 37 million ha has been contaminated by arsenic and cadmium, among others, according to official statistics, or set aside for restoration. Reports indicate that Chinese agriculture is the country's largest environmental destroyer of water, larger than industry. About one-quarter of the country is desert and there are many low-producing areas.

There are diverse types of Chinese agriculture but most food production takes place in a small-scale and terraced agricultural landscape. This requires manual labour, both for production and maintenance. In addition, there are large industrial meat factories and dairies, sustainability-oriented 'high-tech' agriculture, large greenhouses and some organic farms. Over time, there has been a loss of productive farmland due to areas required for new infrastructure, industries and urban development.

Approximately 60% of the Chinese population lives in urban settings, up from 25% some 30 years ago (United Nations, 2023b). This urbanization process will continue. Three-quarters of Chinese urban areas have surface water unsuitable for drinking and fishing. With a consumption-driven economy, meat consumption has tripled since 1990. Industrially produced foods are increasingly consumed by more people. Nonetheless, more than three-quarters of Chinese consumers are concerned that such food could be harmful, according to a study by McKinsey & Company (Zipser and Po, 2021).

In the Soviet Union, large collective and state farms were common and benefitted from state-guaranteed marketing and supply channels prior to 1991. After the break-up of the Soviet Union, the hectares under grain production diminished (Buzdalov, 2015). The agricultural sector had to transform to a market-oriented system. This led to reductions in arable land in several regions (Isaev, 2018). After the devaluation of the rouble in 2014, domestic production increased, enabling Russia to become a larger agricultural power (Serova, 2020). There are some 280 million ha of planted cropland in 2020. Meat production has increased during the last few years and Russia became the world's largest exporter of wheat in 2016, and remains in this position in 2020 according to Statista (2022). The war with Ukraine has changed Russia's agricultural production and the country has challenges for its future agriculture (Serova, 2020).

Another large agricultural actor globally is the EU. It is a network comprising 27 nations with different biological conditions for agriculture. The EU can only stand for an overall policy beyond the current principled system of one account for the payment of membership fees and another account for the payment of national agricultural aid. In the EU, some 157 million ha of land were used for agricultural production in 2020 according to Eurostat (2022). This was 38% of the total land area. There were 9.1 million agricultural holdings: two-thirds of which were less than 5 ha in size.

The number of farms in the EU has declined, but the amount of land used for production has remained steady. Between 2002 and 2021, the EU trade in agricultural products more than doubled according to Eurostat (2022).

Agricultural exports grew faster (5.4%) than imports (4.2%). In terms of exports, the United Kingdom was the main partner (21%) followed by the United States, China, Switzerland, Japan and Russia (Eurostat, 2023). Imports from outside the EU originated mostly from Brazil (9%), the UK, the United States, Norway, China and Ukraine (Eurostat, 2023).

For future global food security, there are at least ten challenges. The first is to give political priority to agriculture as a long-term perspective. In low-income countries, the political objective should be to secure that people always have access to food of excellent quality from sustainable production. This calls for food sovereignty.

Future food and agricultural production should be sustainable, a huge global challenge. In 1992, the concept of sustainability has already been accepted by the world's politicians at the Earth Summit in Rio de Janeiro. This meant 'a development that meets the needs of today, without compromising the ability of future generations to meet their needs'. Both politicians and the scientific community have been slow in giving the concept practical content for implementation. Current food production is not sustainable and not following the United Nations Sustainable Development Goals (SDGs), despite minor exceptions. This second challenge will require dramatic changes in how future food is produced.

The third challenge is to reach and engage all farmers in working towards sustainable production, particularly small farmers. Globally, most farms are small. Nine out of ten of the world's 570 million farms are family farms and produce around 80% of the world's food (FAO, 2014). There are, however, great variations, even between low-income countries. Among Nigeria's farmers, around 88% are considered as owning small family farms of 0.5 ha of land on average. There are also larger farms, many of which have a high share of income from non-agricultural wages (FAO, 2014). The average small farm in most of Latin America is about 1–2 ha but Brazil is different. Brazil is the world's largest producer of sugarcane for export of both sugar and ethanol production, soybeans and the world's third largest producer of maize.

The fourth challenge is to produce food from a biological perspective rather than continued dependence on chemicals and fossil energy.

Environmental degradation and pollution of various kinds must be given higher priority, minimized and preferably avoided completely. Novel crops and cropping systems are necessary to increase biodiversity and improve long-term soil fertility.

The fifth challenge is access to new arable land. So far, global food production has been able to keep pace with population growth. Harvesting levels of major crops are, however, not increasing at the same rate as before. Sustainable agriculture may imply less intensification and so more land will be necessary, an issue seldom highlighted. In fact, current development destroys farmland. There is, however, little new arable land area available to develop. Siberia's large wide grassland may have potential for future large-scale food production when increased global temperature may ease arability (Lustgarten, 2020). This would require effective management of internal and international migration flows and people willing to work in agriculture in that region. North-east China may be of political interest to both Russia and China. Otherwise, there are few other large, unpopulated areas on the globe suited for food production.

The sixth challenge is land grabbing. In 2008, the king of Saudi Arabia launched his Initiative for Saudi Agricultural Investment Abroad. It urged Saudi Arabian companies and wealthy private investors to go overseas and buy land to ensure food security. China is another major actor encouraging its companies and private investors to follow this path. The objective of the Communist Party is self-sufficiency in food; 'our rice bowls should be mainly loaded with Chinese food'. Land grabbing has had negative effects for many people, such as losing their land and being displaced. It also takes place with expanding large-scale plantations for oil palms. These concerns have prompted the International Criminal Court (ICC) in The Hague to include land grabbing as a crime against humanity within its mandate since 2016.

The seventh challenge is that former food crops in industrialized countries are grown for fuel instead of food. About 25% of the maize and 20% of the soybean crops are cultivated for ethanol and biofuel in the United States (United Nations, 2023b). This is not a new phenomenon because alcohol was already produced from maize in 1906 as an addition to gasoline in the United States (Shahbandeh, 2023). Then, no farmer was interested, considering it an option only in tough times. Now, maize and wheat may be sources for fuel. Within the EU, rapeseed cultivation accounts for just over 60% of its biodiesel industry (United Nations, 2023b). In tropical climates, there is an increase of palm oil production which may result in deforestation in South-east Asia and in parts of Africa and Latin America. This problem is on the rise in view of the global need for more energy.

The eighth challenge is whether the current approach to growing globalized food production is the best way to assist national agricultural development towards food sovereignty in low-income countries. The agro-industrial model means a focus on an industrially intensive agriculture for short-term profit. A basic question is: To what extent does this approach, in the short term, help the world's farmers produce food crops in low-income countries?

Another question is: Do current globalized transnational corporations (TNCs) provide the best avenue to sustainable agricultural development in the long term and at what social and economic price? If agriculture develops under the full control of TNCs, using their current approach, there might also be a very rapid shift from subsistence to commercial farming in low-income countries. This may accelerate the reduction in farm employment, albeit raising earnings for some. It may, however, lead to growing unemployment if other job alternatives fail during such a rapid modernization process. There is a need for balanced development to avoid social tension and conflicts.

The ninth challenge relates to agricultural research. So far, most of it has been developed – and designed – for industrialized societies with large-scale, mechanical, agricultural production methods that make extensive use of agricultural chemicals and digital technology. This may not be easily adopted in small-scale agricultural communities in most low-income countries. It cannot be classified as sustainable. What alternatives are available or being developed by different actors who deal with global agricultural research towards sustainability, including those in the private sector?

The tenth challenge applies to most current global food production. In general, low-salary workers or small-scale producers produce the food. In particular, large numbers of migrant workers are used as a low-wage labour force in wealthy nations. In Europe, migrants from various continents make up the workforce for low-paid work. In the United States, illegal migrants from Central and South America are often employed to produce fruit and vegetables. In the Netherlands, farmers are protesting against increasing EU rules and regulations that affect national agricultural policy, which they find to be obstacles to their future engagement in agriculture. In Sweden, thousands of migrants arrive annually to pick berries of various kinds.

The opportunities of urban centres attract young people who wish to leave the countryside. However, youth unemployment in China's cities had risen to a record high of 20% by mid-2023, according to Bloomberg (2023b). Moreover, the Chinese government has requested that certain minority groups to do agrarian work at lower salaries, partly because of the industrialization process (Hoshino, 2019). Fewer Chinese women are willing to work in a labour-intensive agriculture. Cheap and reliable labour disappears when women leave rural life: a phenomenon in countries going through an industrialization process.

In India, about 40% of farmers want to stop farming if they find better jobs. At the same time, half of the population is under the age of 21, which places great demands on employment (Hasell and Roser, 2013). In rich countries, farmers are getting older and wish to retire. Likewise, the Chinese farming population is getting older. Young Swedish farmers find it difficult to set themselves up since it is extremely expensive to buy a farm. According to the US National Young Farmers Coalition, land access is one of the most difficult obstacles for young American farmers (Rippon-Butler, 2023). This negative trend will seriously threaten the agricultural sector in the next few decades. Solutions to this challenge may gradually force governments to intervene using state regulations or by increasing food prices.

Since 2000, a range of efforts has been made to meet the new challenges and problems and to develop agriculture and people's lives in increasingly insecure times. In 1992, the report of the Brundtland Commission was accepted by the world leaders at the Earth Summit in Rio de Janeiro, where they agreed that governments should work towards sustainability (WCED, 1987). Although governments later have accepted the UN SDGs in theory, the approach to solve emerging agricultural problems has been to do better what has been done well before. The question is: Is 'business as usual' the best approach and would it ensure future generations a good life?

The political establishment has been slow in giving prominence to the concept of sustainability and scientific society seems to have been reluctant to focus on it and find practical solutions. Since it is a difficult and controversial dilemma it has not been popular with policymakers. The implementation of the concept requires political and substantive practical solutions and societal changes. Now, it is urgent to find solutions and make changes.

The following chapters will focus on recent agricultural changes in relation to developments since the 1960s with prospects towards future decades rather than the next few years. What agriculture and society would we have, say in 2050, with a focus on sustainability?

References

Al Jazeera (2021) UN food agency: 45 million people on the edge of famine. Available at: https://www.aljazeera.com/news/2021/11/8/un-warns-of-sharp-rise-in-number-of-people-facing-famine

Bhardwaj, T., Deshpande, P., Murke, T., Deshpande, S. and Deshpande, K. (2021) Farmer-assistive chatbot in Indian context using learning techniques. In: Mahalle, P.N., Shinde, G.R., Dey, N. and Hassanien, A.E. (eds) *Security Issues and Privacy Threats in Smart Ubiquitous Computing: Studies in Systems, Decision and Control*, Vol. 341. Springer, Singapore. Available at: https://doi.org/10.1007/978-981-33-4996-4_16 (accessed 21 May 2024).

Bloomberg (2023a) Russia's key economic sectors shrug off sanctions. *Bloomberg News*. Available at: https://www.bloomberg.com/news/articles/2023-11-15/russia-s-war-economy-sees-key-sectors-shrugging-off-sanctions (accessed 15 November 2023).

Bloomberg (2023b) China's record-high youth unemployment rate likely to worsen. *Bloomberg News*. Available at: https://www.bloomberg.com/news/articles/2023-05-16/china-s-youth-jobless-rate-jumps-to-record-20-4-in-danger-sign (accessed 17 May 2023).

Buzdalov, I.N. (2015) Agricultural sector. *Great Russian Encyclopaedia* (in Russian), Vol. 29. Scientific Publishing House 'Great Russian Encyclopaedia' pp. 708–709.

Douglas, L. (2024) Number of US farms falls and size increases, census shows. *Reuters*. Available at: https://www.reuters.com/world/us/number-us-farms-falls-size-increases-census-shows-2024-02-13/ (accessed February 2024).

Eurostat (2022) Farms and farmland in the European Union: Statistics. Available at: https://ec.europa.eu/eurostat/statistics-explained/index.php?title=Farms_and_farmland_in_the_European_Union_-_statistics (accessed 21 May 2024).

Eurostat (2023) Trade in agricultural goods down 3.2% in 2023. Eurostat. Available at: https://www.fao.org/sustainable-development-goals-data-portal/data/indicators/212-prevalence-of-moderate-or-severe-food-insecurity-in-the-population-based-on-the-food-insecurity-experience-scale/en (accessed 24 August 2024).

Food and Agriculture Organization (FAO) (n.d.) SDG Indicators Portal: Indicator 2.1.2 Prevalence of moderate or severe food insecurity in the population, based on the Food Insecurity Experience Scale. Available at: https://www.fao.org/sustainable-development-goals-data-portal/data/indicators/212-prevalence-of-moderate-or-severe-food-insecurity-in-the-population-based-on-the-food-insecurity-experience-scale/en

Food and Agriculture Organization (FAO) (2014) *The State of Food and Agriculture 2014. Innovation in Family Farming*. Food and Agriculture Organization of the United Nations, Rome.

Food and Agriculture Organization (FAO) (2023) *The State of Food Security and Nutrition in the World*. Food and Agriculture Organization, Rome.

Glauber, J.W. (2023) Ukraine one year later: Impacts on global food security. International Food Policy Research Institute (IFPRI). Available at: https://ebrary.ifpri.org/digital/collection/p15738coll2/id/136745 (accessed 21 May 2024).

Government of India (2016) State of Indian Agriculture 2015-2016. Government of India. Ministry of Agriculture & Farmers Welfare Department of Agriculture, Cooperation & Farmers Welfare Directorate of Economics & Statistics, New Delhi.

GRFC (2024) FSIN and Global Network Against Food Crisis (GRFC), 2024, Rome. Available at: https://www.fsinplatform.org/report/global-report-food-crisis-2024

Hasell, J. and Roser, M. (2013) Famines. Our World in Data. Available at: https://ourworldindata.org/famines (accessed 21 May 2024).

Hoshino, M. (2019) Preferential policies for China's ethnic minorities at a crossroads. *Journal of Contemporary East Asia Studies* 8(1), 1–13. Available at: https://doi.org/10.1080/24761028.2019.1625178

Isaev, N. (2018) Russia lost one fifth of its territory (россия потеряла пятую часть территории). Available at: www.svpressa.ru (accessed 24 August 2024).

Jadhav, R., Bhardwaj, M. and Patel, S. (2023) India imposes major rice export ban, triggering inflation fears. *Reuters* New Delhi. Available at: https://www.reuters.com>markets>commodities>global-rice-supplies (accessed 20 July 2023).

Lowder, S.K., Sánchez, M.V. and Bertini, R. (2021) Which farms feed the world and has farmland become more concentrated? *World Development* 12, 105455. Available at: https://doi.org/10.1016/j.worlddev.2021.105455 (accessed 24 August 2024).

Lustgarten, A. (2020) How Russia wins the climate crisis. *The New York Times Magazine and ProPublica*. Available at: https://www.nytimes.com/interactive/2020/12/16/magazine/russia-climate-migration-crisis.html (accessed 22 May 2024).

Rippon-Butler, H. (2023) Opinion: Working together to solve the US land access crisis. *Agri-Pulse*. Available at: https://www.agri-pulse.com/articles/19797-opinion-working-together-to-solve-the-us-land-access-crisis (accessed 8 August 2023).

Sachs, J.D., Lafortune, G. and Fuller, G. (2024) The SDGs and the UN Summit of the Future. *Sustainable Development Report*. Dublin University Press. DOI: 10.25546/108572.

Schooh, M., Baah, S.K.T., Lakner, C. and Friedman, J. (2022) Half the population lives on less than US$6.85 per person per day. *World Bank Blogs*. Available at: https://blogs.worldbank.org/en/ (accessed 8 December 2022).

Serova, E. (ed.) (2020) Special issue: challenges for Russia's agriculture. *Russian Journal of Economics* 6(1).

Shahbandeh, M. (2023) Agriculture in the US: statistics and facts. *Statista*. Available at: https://www .statista.com/topics/1126/us-agriculture/#editorsPicks (accessed 21 May 2024).

Statista (2022) Production of meat worldwide from 2016 to 2023. Statista Insights. Statista Inc, New York.

Statista (2023) Value of United States agricultural exports in 2022 and 2023. Statista Insights. Statista Inc, New York.

United Nations (n.d) Sustainable Development Goals: 2 Zero Hunger. Available at: https://www.un.org /sustainabledevelopment/hunger/ (accessed 21 May 2024).

United Nations (2023a) India poised to become the world's most populous nation. *UN News*. Available at: https://news.un.org/en/story/2023/04/1135967 (accessed 7 September 2024).

United Nations (2023b) 2 Zero Hunger. Available at: https://unstats.un.org/sdgs/report/2023/goal-02/ (accessed 21 May 2024).

USDA (2007) US Farms: Numbers, Size, and Ownership. *USDA ERS Structure and Finance of US Farms: Family Farm Report 2007*. Edition/EIB-24 Economic Research Service /USDA, Washington DC. Available at: https://www.ers.usda.gov/webdocs/publications/44153/11865_eib24b_1_.pdf?v=8529.3

USDA (2024) Farms and Land in Farms. USDA Agricultural Statistics Service, US Department of Agriculture. *February* 16, 2024. Available at: https://www.nass.usda.gov/Publications/Methodology_ and_Data_Quality/Farm_and_Land_In_Farms/02_2024/farmqm24.pdf

World Commission on Environment and Development (WCED) (1987) *Our Common Future* (Brundtland Report). Report of the World Commission on Environment and Development (WCED), United Nations, New York.

World Food Programme (2023) WFP World Annual Review 2022. World Food Programme, Rome. Available at: https://www.wfp.org/publications/wfp-annual-review-2022 (accessed 24 August 2024).

Zipser, D. and Po, F. (eds) (2021) China Consumer Report 2021, Special edition, November 2020. Understanding Chinese Consumers: Growth Engine of the World. McKinsey & Company.

2 Agriculture Turning International

Abstract

Initially, humans picked wild vegetable food, fruit and berries. Later, they used sticks to cultivate the soil and planted crop seeds on a small plot of land, often cleared using fire. This slash-and-burn agriculture is still used by some 50 million farmers in the remote and harsh environments of the tropics. Beans and cowpeas were early domesticated crops. In regions with a cooler climate, farmers began to keep animals in small barns. The manure and urine could be spread on the crop fields. From about 3300 **BCE**, agricultural production was intensified. Higher yields could be support larger communities in several parts of the world. This led to trading of agricultural produce since not all community members were now farmers. This food trade then developed over a wider area, between the Sumerians in Mesopotamia and people of the Indus valley. Later, the Arab caravans traversing the Silk Road, not only carried wares, but also food and spices from distant places. Agriculture had now turned truly international.

For their livelihood, the first humans fed on wild vegetables and grains, and the meat of wild animals, birds, fish and insects. In Africa, people have exclusively been foragers for more than 99% of their existence (Philipson, 2005). About 12,000 years ago, humans switched from nomadic, hunter–gatherer lifestyles to farming; the Neolithic Revolution. People began altering flora and fauna communities through means such as fire-stick farming and forest gardening. Gradually, crops, and later animals, were integrated into farming systems, leading to the creation of villages and smaller communities. Some 9000 years ago, an estimated global population of around 5 million started to grow faster.

Fire-stick farming was carried out with simple tools. An area was cleared of vegetation, often after fire. A portable wooden stick was used for planting. The small plot was cultivated for a few years when nutrients from the ash of the burnt vegetation gave some yield, albeit low. Then, that area was abandoned for a new area and not used again until the fertility had been naturally restored. Other terms for this system are swidden agriculture or slash-and-burn agriculture. This system is still operated by some 50 million farmers in the remote and harsh environments of the tropics.

Shifting cultivation systems are ecologically viable when there is enough land for restorative fallows, at least 10–20 years. The system ceases to be sustainable when the number of people grows to a point where demand for land in a specific location reduces the fallow period below 10 years. Instead, it is becoming detrimental to the environment. Gradually the simple wooden stick was replaced by improved digging tools, for example, the chakitaqlla in the Inca Empire or the soil knife in mountainous regions in Japan.

In regions with winter season, farmers began to keep their animals in small barns. Then, their manure and urine could later be spread on the fields with crops. This provided nutrients for soil fertility. The use of animal manure required a special tool for turning the soil to cover the manure. In addition, some land was kept as fallow for some years, or as grassland. Thus, the first plough was invented and later designed to be pulled by animals. This development took place without any research but from learning by experience. Later, simple machinery for harvesting and threshing was developed to be drawn by animals – long before the era of fossil energy.

Farming started in various parts of the world for several reasons. For example, it is believed that increased pressure on natural food resources may have forced people in East Asia to settle and begin farming. In contrast, climatic changes at the end of the last Ice Age brought long dry seasons in the Middle East. Together with abundant water, these conditions favoured annual plants, that died off in the dry periods, leaving a tuber or a dormant seed. This enabled hunter–gatherers to use storable grains and pulses as a basis for starting small, settled villages and domesticating animals.

Pulses were easy to store, had a long shelf life and could thus prevent famines. Among the first major pulse crops grown were broad beans, chickpeas, peas and lentils. Beans and cowpeas were domesticated in ancient times, and were grown in Central and South America and Africa. Through farmers´ selective breeding, domesticated pulses have features not found in wild varieties: thinner peel, increased seed size, no pod shattering and seed dormancy mechanisms (De Ron , 2015).

Cereals were grown in Syria some 9000 years ago, being part of the eight Neolithic founder crops being cultivated in the Levant (Zeder, 2011). They were emmer wheat, einkorn wheat, hulled barley, peas, lentils, bitter vetch, chickpeas and flax (Weiss and Zohary, 2011). Wheat was the first crop to be grown and harvested on a larger scale.

Starting during the Bronze Age, from about 3300 BCE, agricultural production was intensified and large civilizations developed in several parts of the world. Initial agricultural developments gave rise to higher yields, which could support larger communities. That led to a need for the trade of agricultural produce since not all members of a community were farmers. During ancient Greek culture, small urban centres developed which encouraged the Greeks to engage in trading wheat between these centres. Gradually, this trade expanded and included other products as well as food, from more distant places and from neighbouring cultures. Agriculture had turned international.

Internationalization began with trade links between the Sumerians and the Indus Valley civilization in the third millennium BCE (Frank, 1998). The trade on the Silk Road opened long-distance political and economic interactions between states. It connected Asia to Europe and Africa and had been established by the emperors of the Han dynasty between the 2nd century BCE and the 1st century AD. It was a network of several routes connecting with different provinces of China.

The Arabs dominated the trade through camel caravans with their extensive experience of the desert and of harsh conditions. Arab merchants at many trading points exchanged goods, such as silk, cotton, sugar, salt, spices, Chinese porcelain, ivory and precious stones. The caravans included not only traders, but also craftsmen, missionaries, robbers, refugees and artists. Exchanges also took place in religions, philosophies, diseases, diplomacy and culture. Some scholars even claim this to be a form of very early globalization.

There were great movements of trade across Asia and the Indian Ocean from the 15th century onward. Trade links increased with the rise of several European empires. The Portuguese Empire was first, followed by the Spanish Empire, and later came the Dutch and British Empires. In the 17th century, world trade developed further when chartered companies were founded (Chapter 4). This development was based on expansionism and the trading of new commodities, such as sugar, but in particular slaves. Moreover, there was an important transfer of animal stock, plants, crops and epidemic diseases. During this period European, Muslim, Indian, South-east Asian and Chinese merchants were involved in both trade and communications.

There was another dramatic change during this era. From the time of Aristotle (4th century BCE), until the colonial empires of the 16th and 17th centuries, it was believed that self-restraint separated humans from animals. A life of moderation was a sign of civilization and maturity. Greed was considered a sin and interest was immoral. Such a view had not benefitted economic growth but now there was a notable change. Humans wished to treat nature as an infinite resource. It would be mastered, not adapted to as had been the case for many centuries. Such a view was supported by Francis Bacon, who has been called the father of empiricism. An infinite resource could be exploited by humans, and gradually be exploited more and more effectively because the desires of humans were infinite. This required ever-increasing economic growth, a view later supported by Adam

Smith. It was believed that through innovations and human creativity, people would be able to cope with nature and at the same time stimulate growth. This concept neglected, however, that innovations had already been developed long before that time, created by humans and completely adapted to nature from many centuries before (Chapter 8).

Some early opposition to the idea of growth was expressed by, among others, Jean-Jacques Rousseau. His view of country life was that it was the path to happiness and harmony for humans, society and nature, which should be respected. This idea failed to influence politicians or the economic system but his ideas on the social contract theory and natural education had an impact on political thought, for example, in the French Revolution. The period between 1600 and 1800 was characterized by advances in transportation and communications technology. Hopkins (2002) identified this as a 'proto globalization' period.

Karl Marx predicted the universal character of the modern world society. He embraced ideas of scarcity where human desires, rather than nature, must be mastered to achieve the social good, influencing how humans might interact with both nature and the economy. His theories declared the development of human societies via class conflict. When production is in the capitalist mode, this is manifested in the conflict between those that control the means of production (the ruling classes, often referred to as the bourgeoisie) and those that enable the means of production by selling their labour power for wages (the working classes, often referred to as the proletariat). Marx's prediction was that, just like earlier socio-economic systems, capitalism would produce internal tensions. These tensions would lead to the self-destruction of capitalism and it would be replaced by the socialist mode of production. In his Manifesto of the Communist Party from 1848 he had stated that 'the bourgeoisie has, through its exploitation of the world market, given a cosmopolitan character to production and consumption in every country'. Marx had concluded that 'all old-established national industries have been destroyed or are daily being destroyed' (Marx and Engels, 1848).

There have been various internationalization theories that are especially relevant to trade.

Long ago, Adam Smith claimed that a country should specialize in, and export, commodities in which it had an absolute advantage (Smith, 1776). Thus, it was able to produce a commodity with less cost per unit produced than its trading partner. In turn, it should import commodities in which it had an absolute disadvantage. But it is not necessary to have an absolute advantage to gain from trade, only a comparative advantage.

The gravity model of trade was a theory, predicting bilateral trade flows based on the economic sizes of measurements of the size of gross domestic products (GDPs) and distance between two units. The Heckscher–Ohlin model (the factors proportions development) essentially implied that countries will export products that utilize their abundant and cheap factor(s) of production and import products that utilize the countries' scarce factor(s). Hymer (1960) and Kindleberger (1969) were early in suggesting that multinational corporations owed their existence to market imperfections. Those market imperfections were, however, structural imperfections in markets for final products (Pitelis and Sugden, 2000).

New Trade theorists challenged the assumption of diminishing returns to scale. They stressed the importance of firms rather than sectors in understanding the challenges and the opportunities countries face in globalization. The British economist, Alfred Marshall, was one of the founders of the neoclassical school in economics with his work on the theory of value and the theory of the firm. Within any industry, some firms thrive among international competition while others are unable to cope. This results in much more pronounced intra-industry reallocations of market shares and productive resources. Other New Trade theorists argued that the use of protectionist measures to build up a huge industrial base in certain industries will allow those sectors to dominate the world market.

Later Paul Samuelson, called the 'Father of Modern Economics', together with others, formed a basis for the present hegemonic politics of growth. When awarded the Nobel Memorial Prize in Economic Sciences, the Swedish Royal Academies stated that 'he has done more than any other contemporary economist to raise the level of scientific analysis in economic theory' (Frost, 2009). Since the emergence of the neoclassical economic doctrine, the idea of growth has been completely dominant.

Years ago, there was no agreed definition of internationalization. It can be described as a strategy for companies that seek horizontal integration globally. Business entrepreneurs interested in internationalization must have the ability to think globally, with a good understanding of international cultures, beliefs, values and behaviours, to succeed in internationalization with different companies within other countries. They must search for top quality innovations and be committed to corporate social responsibility. Thus, internationalization turned globalization into a key word, including not only individual types of agricultural produce and the food industry, but also agriculture as a sector.

References

De Ron, A.M. (ed.) (2015) *Grain Legumes (Handbook of Plant Breeding, 10)*. Springer Science & Business Media, New York.

Frank, A.G. (1998) *ReOrient: Global Economy in the Asian Age*. University of California Press, Berkeley, CA.

Frost, G. (2009) Nobel-winning economist Paul A. Samuelson dies at age 94. *MIT News*. Available at: https://news.mit.edu/2009/obit-samuelson-1213 (accessed 31 May 2024).

Hopkins, A.G. (ed.) (2002) *Globalization in world history*. Norton, New York.

Hymer, S.H. (1960) The international operations of national firms: A study of direct foreign investment. PhD dissertation, Published posthumously. The MIT Press, 1976, Cambridge, MA.

Kindleberger, C.P. (1969) *American Business Abroad: Six Lectures on Direct Investment*. Yale University Press, New Haven, CN.

Marx, K. and Engels, F. (1848) Manifesto of the Communist Party. Marxists Internet Archive. Available at: https://www.marxists.org/archive/marx/works/1848/communist-manifesto/ch01.htm (accessed 31 May 2024).

Philipson, D. (2005) *African Archaeology*, 3rd ed. Cambridge University Press, Cambridge, UK.

Pitelis, C. and Sugden, R. (eds) (2000) *The Nature of the Transnational Firm*, 2nd ed. Routledge, London and New York.

Smith, A. (1776) *An Inquiry into the Nature and Causes of the Wealth of Nations*. W. Strahan and T. Cadell, London.

Weiss, E. and Zohary, D. (2011) The Neolithic Southwest Asian founder crops: Their biology and archaeobotany. *Current Anthropology* 52(S4). Available at: https://doi.org/10.1086/658367 (accessed 24 August 2024).

Zeder, M.A. (2011) The origins of agriculture in the Near East. *Current Anthropology* 52(S4), S221–S235. Available at: https://doi.org/10.1086/659307 (accessed 3 June 2024).

3 Globalization

Abstract

There have been various definitions of globalization, but it can briefly be described as the increasing connectedness and interdependence of world cultures and economies. Primarily, it is an economic process of interaction and integration associated with social and cultural aspects. Politically, it has been described as the emergence of a transnational elite and a phasing out of the nation-state. Power is moved from local people to giant multinational companies. It is also a process of interaction and integration among people, companies and governments worldwide. Other aspects of globalization are higher education, research, culture and communications. Artificial intelligence (AI) is of growing interest, reinforcing individualization rather than strengthening society for the good of all. Free globalized trade has contributed to economic development but it might not be fair. Transnationalism is a related concept, but more limited in scope.

A Term of Old Origin

Large-scale globalization of the current type began in the 19th century (O'Rourke and Williamson, 2002), apart from the chartered companies which were founded during the colonial period beginning in the 17th century. The industrialization process led to economies of scale, mass and standardized production of various household items to meet an increasing demand for commodities. The invention of shipping containers advanced global trade. Also, globalization was influenced by imperialism in both Africa and Asia. Globalization developed further in the late 19th century and early 20th century. It created more economic growth, gradually leading to increased consumption.

In the 1980s, globalization spread rapidly through the expansion of both capitalism and neoliberal ideology. More focus was given to the concept which has influenced both the scientific and political debate on increasing global trade by large corporations operating worldwide. The implementation of neoliberal policies enabled the privatization of public industry, and the deregulation of laws or policies that interfered with the free flow of the market, as well as cuts to governmental social services (Klein, 2008).

When the neoliberal policies were introduced to low-income countries, structural adjustment programmes (SAPs) were implemented by the World Bank and the International Monetary Fund (IMF). They served as global financial market regulators, promoting neoliberalism and free markets. Countries receiving monetary aid through the SAPs had to open their markets to capitalism, free trade and the operation of the transnational corporations (TNCs). They also had to allow the privatization of publicly owned industry, frequently combined with a reduction of social services, such as health care and education (Beneria *et al.*, 2015). A serious drawback was that the IMF and the World Bank neglected environmental and climate issues in their requirements.

Low-income governments gradually came under the influence of the TNCs. By using their investment funds, the TNCs could demand higher profits to participate in national companies and influence different countries to allow competition for jobs. They could force them to reduce taxes or promise subsidies in the form of free land, buildings and transport systems. Capital could pass freely across all national borders. National politicians lost power over the market, whereas the TNCs gained more control. In general, this development was approved by most politicians in many countries as a positive feature of globalization.

Already, in the late 1990s, some effects of globalization had been emphasized (Martin and Schumann, 1996). The power elite of the world estimated that one-fifth of the working-age population would be enough to keep the world

© Bo Malte Ingvar Bengtsson 2025. *Agricultural Innovation for Societal Change: Towards Sustainability* (B.M.I. Bengtsson)
DOI: 10.1079/9781800627802.0003

economy running. The remaining 80% would have no permanent employment. Low wages in China, India and other countries in Southeast Asia or Eastern Europe were attractive, drawing job opportunities from the West (Gunter and Wilcher, 2020).

Liberalization meant a strengthening of global interconnectedness which grew quickly in the world's economies and cultures. Advances and developments in communications technology, such as the Internet and smartphones, became major factors in the globalization process. In turn, they generated further interdependence of economic and cultural activities around the globe. This increase of global interactions led to growth not only in international trade but also in the exchange of ideas, beliefs and culture.

Migration increased between low-income countries and the most-developed countries. Workers moved to areas with higher wages, exemplified by migration from Africa to Europe and from Mexico to the southern border areas of the United States. In the United States, the controversial exception rule, Title 42, was lifted in May 2023 which was assumed to have increase the number of migrants since. Since the start of the Title 42 policy, Border Patrol has expelled migrants more than 2.4 million times according to the US Customs and Border Protection (Villagran, 2022).

Some Definitions of Globalization

The term globalization can briefly be described as the increasing connectedness and interdependence of world cultures and economies. Primarily, it is an economic process of interaction and integration associated with social and cultural aspects. In business, globalization is the process by which commercial companies or other organizations develop international influence or operate on an international scale. It is also a process of interaction and integration between people, companies and governments worldwide.

There have been various early definitions of globalization. Albrow and King (1990) defined globalization as 'all those processes by which the people of the world are incorporated into a single world society'. As a sociologist, Robertson (1992) described globalization as 'the compression of the world and the intensification of the consciousness of the world as a whole'. Sassen (1991) popularized the term global city.

Held *et al.* (1999) elaborated a model of globalization that was assumed to have superseded existing models. It assessed how the processes of globalization had operated in different historical periods in respect to political organization, military globalization, trade, finance, corporate productivity, migration, culture and the environment. The model concentrated on some advanced capitalist societies (United States, United Kingdom, Sweden, France, Germany and Japan). For comparative purposes, reference was given to low-income economics. The new concept of globalization was to include four specific elements: intensity, extensity, impact and velocity. This simplicity may seem attractive but requires further elaboration to include local, regional and national features. Moreover, the political fatalism in discussions of globalization was confronted by elaborating the possibilities for democratizing and civilizing a global transformation.

Political globalization has been described as the emergence of a transnational elite and a phasing out of the nation-state. James (2005) defined globalization as 'the extension of social relations across world-space'. Steger (2009) identified four main empirical dimensions of globalization: economic, political, cultural and ecological, with an ideological dimension, cutting across the other four. The fifth dimension was filled with a range of norms, claims, beliefs and narratives about the phenomenon itself. Later, James and Steger (2014) stated that the concept of globalization 'emerged from the intersection of four interrelated sets of communities of practices: academics, journalists, publishers/editors, and librarians'.

Despite a range of definitions and models, discussions about globalization seldom relate to future sustainable societies. Moreover, trade is of the highest relevance to industrialized countries with commodities for export. In low-income countries, agriculture is still a large component of the gross domestic product (GDP) with food production for domestic use and plantation crops for export. In these countries, future agricultural sustainability will mean more national food production, cost reductions on trading and access to food for all. Although there are positive effects from international trade in low-income countries, there are also indications that such effects may be rare in many of them.

Globalization and transnationalism are related concepts, but transnationalism is more limited in scope than globalization. Transnational processes are anchored in and transcend one or more nation-states. The term transnationalism refers to the exchange of people, ideas, technology and money between nations. The term became popular during the 1990s to explain the migrant diaspora, culturally mixed communities and the complicated economic relations that increasingly characterized a changing world. This is in addition to past migration over several decades to large capitals, mainly in low-income countries, resulting in millions of slum dwellers. They live in poor housing with limited access to water, electricity and other services. The Habitat for Humanity and the United Nations (UN) has estimated that there are some 1.6 billion people lacking access to adequate housing and expect the number to grow to 2 billion by 2030 (Balla, 2023). According to the UN and Habitat about half of this growth in slum population will take place in eight countries: Philippines, Pakistan, India, Ethiopia, Tanzania, Egypt, Nigeria and the Democratic Republic of Congo. But the lack of housing is also a problem in rich countries in Europe according to the European homelessness agency, FEANTSA, and the Abbé Pierre Foundation (FEANTSA, 2022).

Slum dwellers are largely overlooked by the authorities, living according to their own social traditions, religions and customs, and practising traditional laws, often resulting in criminality. Also, in Sweden, there were 59 vulnerable areas (The Police Authorities, 2023). These areas geographically were defined places with low socioeconomic status, and high levels of security and criminality.

The phenomenon of different ethnic groups wanting to live in their own areas is not uncommon. This has given rise to migrant communities, based on their own cultural and ethnic references such as religion or their culture of origin. It is different from the historical approach when migrant groups moved from one nation to another and were expected to assimilate and adopt the prescribed moral and political values of their country of residence. This originally happened in the United States, where migrants were assimilated after a generation.

Today, transnationalism, in contrast to nationalism, would force policy makers to look beyond the interests of their nation-state when creating new policies and regulations. Some countries liberalize laws to allow dual citizenship or rights and privileges for non-citizen groups who wish to reside permanently within their borders. Others may formulate more exclusionary immigration policies.

Free Trade of Food and Sustainability

In low-income countries, there are commentators who view the globalization of the food trade as the single most important reason for the growing hunger and poverty across the world (Dembitzer, 2021). The dominance of a small number of companies in the food industry is problematic. Globalization, in this context, means the removal of power from local people and placing it in the hands of giant food commodity and chemical seed companies, and even aid donors. A major reason for starvation is lack of access to food, pointed out explicitly decades ago by Sen (1982). Instead, food sovereignty should be the primary political focus in all countries aiming for sustainability. Without access to food, there can be no long-term survival for all. This raises the fundamental question of whether current views on globalization are beneficial to all people, particularly in low-income societies.

The food trade is centuries old. From the 16th century to the 19th century, agriculture focused on the products from colonial plantations sold at a cheap price in the global market. Europe's sovereign class made huge profits, and that continent prospered. The price was the exploitation of the natural resources of the colonies, using slave labour.

Nowadays, certain countries are better off, and others are not. People in low-income countries can eat new food, such as white bread, which is not as nutritious as earlier consumption and not sustainably produced. Likewise, food imports to industrialized countries offer a range of new options for interested consumers. Those living in high-income countries benefit from large wage differences between the rich countries and the low-income countries. So far, few seem to want to renounce such privileges, so world trade of the current nature is highly celebrated in both the market and in technology.

Free trade has contributed to economic development, but the question is whether it is fair. In 2017, the world's richest benefitted from much of the economic growth according to Oxfam International. Three years later, the world's more than 2100 billionaires had more wealth than 4.6 billion of the world's people (Silvio, 2022). Two-thirds of all wealth generated since 2020 has gone to the richest 1% (Oxfam, 2023). The number of billionaires grew by 13.4% in 2020 according to the research firm Wealth-X (2021). At the same time, the poorer half of humanity's accumulated wealth did not increase at all. In 2023, Oxfam called for 'billionaire-busting' policies to halve the wealth and number of billionaires between now and 2030 through taxation and other moves to achieve a 'fairer, more rational distribution of the world's wealth'.

Free trade in agricultural products is different from trade with industrial products. Agriculture is based on biology and produced during varying climatic conditions. One feature of future sustainable food production will be reduced transportation costs, not only within countries but also between continents. Moreover, the informal food sector makes up a large part of trade in agricultural products, often staple crops, in low-income countries. That sector contributes to the bulk of female employment, not least in African countries, where women often deal with trade issues at the local level. In sub-Saharan Africa, about half of the population used to derive their income from informal border trade, in contrast to one-tenth deriving income from the informal food sector in high-income countries. Now, the size of the informal sector varies widely across regions and countries. It still accounts for about one-third of low- and middle-income countries' economic activity as compared to some 15% in advanced economies (Deléchat and Medina, 2020).

David Livingstone, the British explorer and missionary, had already called for trade to open up Africa, arguing for Commerce, Christianity, and Civilization. In the 1880s, most of Africa was ruled by Africans but there was to be a quick change. By 1902, 5 European powers and one extraordinary individual had set up 30 new colonies and protectorates with some 110 million inhabitants (Pakenham, 1991). Trade expanded, although globalization was later pitted against nationalism prior to the First World War.

At about the same time, Japan had to allow free trade through unfavourable agreements with Western industrialized countries. They were able to sell their goods freely in Japan's open ports. This led to peasant protests and to bloody rice riots in 1918 when the price of rice had doubled.

Over time, various international trade negotiations have taken place. The General Agreement on Tariffs and Trade (GATT) led to certain agreements to remove trade barriers. Its successor, the World Trade Organization (WTO), was to provide a framework for negotiating and formalizing trade agreements and a dispute resolution process. It focused on market efficiency and economics, but little attention was given to environmental aspects and almost none to poverty reduction.

According to the Global Policy Forum (2003), exports nearly doubled from 8.5% of total gross world product in 1970 to 16.2% in 2001. But the failure of the Doha Development Round of trade negotiation stopped the use of global agreements to advance trade. Instead, many countries have turned to bilateral or multilateral trade agreements between a few countries.

Although the WTO has led to a doubling of exports, representatives of low-income countries have considered the WTO as more of a brake than an asset for trade in agriculture. India, for example, has claimed little benefit from the WTO. India and China have proposed that agricultural subsidies of about US$160 billion within the United States, European Union (EU), and other wealthy nations must be stopped prior to other reforms in global trade (Suneja, 2017). They found it strange that the EU spends up to 40% of its budget on agricultural purposes, while global financial institutions urge low-income countries to avoid subsidies. In the joint proposal to the WHO on 17 July, 2017, India and China argued that subsidies for many items provided by the developed world were over 50% of the value of production (Business Standard, 2017).

In 2017, the EU and Brazil had agreed on a proposal to the WTO to minimize agricultural subsidies to create an improved system of international trade. When the United States introduced tariffs on steel and aluminium, Germany proposed the need for a new trade agreement on industry between the EU and the United States. The extensive and postponed Transatlantic Trade and Investment Partnership (TTIP) agreement was found unrealistic.

A conspicuous feature of recent world trade is the rise of trade in intermediate goods and services. In an Organisation for Economic Co-operation and Development (OECD) study it was found that intermediate inputs represented 56% of goods trade and 73% of services trade (Miroudot *et al.*, 2009). This was a result of the fragmentation of the production and the increasing importance of outsourcing due to a rapid decrease in the costs of transportation, transaction and tariffs. Such a continuing trend may be questionable.

When the World Bank and the IMF introduced the SAPs, they almost forced African governments to limit state interference in agriculture. There should be no subsidies to agriculture, no guarantees for purchases of agricultural products. No price regulation was allowed. Depending on the mode of production and local wage level, the price of a locally produced chicken would be higher locally than after transportation with modern refrigeration and transportation systems worldwide. Sovereign national governments lost most of their conventional regulation of food provision, whereas large multinational companies were able to put limits on governmental interventions. Simultaneously, other actors, such as food retailers and non-governmental organizations (NGOs) took up roles in governing food provision.

In 2008, 12 South American countries founded the Union of South American Nations (UNASUR) on the model of the EU, though several counties withdrew their membership in 2018. Two free trade systems emerged: the Mercosur and Andean Community Customs Union. There are plans for a single currency, a parliament and a military defence council. Mercosur is the fifth largest world economy with some 295 million people in the full membership countries (MERCOSUR, 2024). Argentina, Brazil, Paraguay and Uruguay are active members, and Venezuela is a full member but has been suspended since 2016. In addition, there are seven South American states who are associated members. Bolivia was given full membership in 2023 (CFR, 2023). More than 90% of tariffs will be abolished for the four countries. After 20 years of negotiations, an agreement in principle was signed in summer 2019 between the EU and Mercosur countries (Blenkinsop and Kihara, 2019). The deal was planned at the 2019 G20 Osaka summit but the text was criticized by nongovernmental organizations (NGOs), scientists, unions, farmers and Indigenous people. The final texts have not been finalized, signed or ratified. The draft trade deal is part of a wider Association Agreement between the two blocs.

The EU had hoped for a 'milestone' in trade with the Mercosur countries at its summit in Brussels in 2023, despite protests from environmentalists who considered it harmful to the environment. Negotiations were still underway as of May 2024 after meetings in the European Parliament in October 2023 between its Committee on International Trade and the European Commission. These meetings covered the ongoing negotiations with MERCOSUR and the Committee on Agriculture and Rural Development in November 2023 (European Parliament, 2024). Protectionist-minded EU countries are concerned about competition from the large countries, such as Brazil and Argentina. Russia has been actively working on the admission of Brazil and Mercosur to the competing Eurasian Economic Union (EAEU).

European farmers' organizations and Swedish farmers are worried that the MERCOSUR agreement is selling out farmers. They must pay for European car manufacturers, pesticide industries, manufacturers and pharmaceutical companies to be able to sell on better terms in the four Mercosur countries. There will be increased competition for agriculture. Imports of beef and chicken will be restricted in a quota system, that also includes ethanol and sugar. The agreement is questionable from an environmental point of view because the Mercosur countries use more pesticides and antibiotics in animal production compared to a country like Sweden.

In 2019, the countries of the African Union (AU) signed an agreement, creating the world's largest free trade area. It should be a single continental market for goods and services, with free movement of businesspersons and investments. The agreement for the African Continental Free Trade Area (AfCFTA) became operational in 2020 (East Africa Community, 2021). Each individual country must ratify the agreement.

The Regional Comprehensive Economic Partnership (RCEP), one of the world's largest, entered into force in 2022 after ten years of negotiations. China, Japan, South Korea, Australia and New Zealand, as well as the ten Association of Southeast Asian Nations (ASEAN) countries, agreed to reduce or remove tariffs in a wide range of trade sectors to reach 2.2 billion potential consumers. The aim is to increase consumption and economic growth, which may not lead to any sustainability. India withdrew in 2019, concerned that tariff reductions would harm local producers (Vu and Nguyen, 2020).

The United States did not participate in the free trade deal. It was a direct competitor to the Trans-Pacific Partnership (TPP), a highly contested proposed trade agreement of 2015 between 12 Pacific Rim economies. It aimed to lower non-tariff and tariff barriers to trade, covering 40% of the global economy. The final agreement was judged by US authorities to have led to net positive economic outcomes for signatories but the US trade deficit with RCEP countries had expanded across all sectors. Thus, the United States withdrew formally when President Donald Trump signed a presidential executive memorandum in early 2017 (Diamond and Bash, 2017). Another development was when the North American Free Trade Agreement (NAFTA) practically ceased to work in 2020, replaced by United States–Mexico–Canada Agreement (USMCA) (Dialogue, The, 2023).

The COVID-19 pandemic had a devastating impact on trade patterns because trade in services accounted for two-thirds of global GDP. In 2021, the WTO had drawn some conclusions on the inequities in the production and distribution of vaccines, and intellectual property rights to enable a more inclusive global economic recovery (WTO, 2022). Still, most Western politicians argue for more free trade, especially with the large African market. It accounts only for 3% of global trade but has 17% of the world population.

Globalization, Higher Education and Research

Academic literature divides globalization into three areas; economics, political and cultural. Economically, it comprises both the globalization of production of goods and services around the globe and the globalization of markets. The phenomenon of economic globalization can be viewed as either positive or negative but does not include the exploration of natural resources required for sustainable food production.

The growth of the worldwide political system in both size and complexity is referred to as political globalization. This includes national governments, and governmental, international non-governmental and social movement organizations. In consequence, nation-states decline in importance with other actors appearing on the political landscape. Humanitarian initiatives are led by philanthropic organizations whose missions are global in nature; for example, the Bill and Melinda Gates Foundation, Accion International and the Acumen Fund (now Acumen). Philanthropy and the business model are now combined in these charities, and in turn, supporting organizations, such as the Global Philanthropy Forum and the Global Philanthropy Group, are formed.

An increase in international education and the development of global cross-cultural competence in the workforce is associated with globalization. A greater number of students now go to foreign countries for their higher education, a phenomenon also found in agriculture. This can not only lead to improved language skills and the broadening of social horizons but also an improved acceptance and understanding of other cultures. Between 1963 and 2006, the number of students studying in a foreign country increased ninefold (Varghese, 2008).

In higher education, internationalization may bridge the gap between diverse cultures and countries (Adel *et al.*, 2018). The term was earlier used to describe the global life of the mind and in international relations (James, 2005) to describe the extension of the European Common Market. James (2005) has argued that the oldest dominant form of globalization is the movement of people.

Cultural Globalization

The transmission of ideas, knowledge, meanings, information and values around the world to extend and intensify social relations is known as cultural globalization. Among the earliest cultural phenomena to globalize were religions. They spread through missionaries, evangelists and imperialists, by force and by migration, in addition to traders. As of today, this process is diffused by the Internet, popular culture media, satellite television, and international travel, ideas, styles, technologies, languages, etc. People's individual and group cultural identities are associated with the shared knowledge and norms which are formed through cultural globalization. Globalization has also much influenced sport.

Globalization has not only increased cross-cultural contact between different cultures and populations but also decreased the uniqueness of once isolated communities. Cultural diffusion can create a homogenizing force, where globalization is seen as synonymous with that force through the interconnectedness of trade, culture and politics.

Long ago, cultural goods came from East Asia but in recent times, Western countries have been the main exporters of cultural goods. Eastern Asia is gradually becoming more culturally influential again. Globalization critics argue that when the culture of a dominating country is introduced into a recipient country it can be threatening to local cultural diversity. The disappearance of a local culture is a negative factor over a long-term perspective. Traditional music can be lost, although, against this trend, both jazz and reggae began locally but have become international.

Global Communications

Until the mid-19th century, communications were reliant on courier services. Long-distance communications relied on mail, with ships taking weeks to carry it across the ocean. In 1895, the first wireless telegraphy transmitters were developed. Today, the Internet and artificial intelligence (AI) are vital in global journalism and are connecting people across geographical boundaries.

According to the International Telecommunication Union (ITU) some 67% of the world's population was on online in 2023; a growth of 4.7% since 2022 (Table 3.1). There is a marked difference in the use of the Internet in 2023 between low-income countries (27%) and high-income countries (93%).

The digital network of public infrastructure may be a tool allowing rural people to open bank accounts and receive wages faster and more easily. It enables entrepreneurs to rapidly reach customers more widely or those who are far away. One problem is that about 850 million people worldwide have no official form of identification (Zholudev, 2023). Without an ID, health care, education and meaningful employment are often out of reach as the Bill & Melinda Gates Foundation has observed.

Initially, AI felt like science fiction but nowadays ChatGPT, DALL-E and other models have rapidly become part of everyday life. AI is predicted to occur in every human context, a most important and rapid technological advancement, by automating certain tasks and streamlining companies for increased growth, which may hardly lead to sustainability. Probably, functions linked to administrative work, and finance, legal and business management are among the

Table 3.1. Internet users by region in 2005–2023 (International Telecommunication Union (ITU), 2024).

Region	2005 (%)	2010 (%)	2017 (%)	2021 (%)	2023 (%)
Africa	2	10	21.8	39.7	37
The Americas	36	49	65.9	83.2	87
Arab States	8	26	43.7	70.3	69
Asia and the Pacific	9	23	43.9	64.3	66
Commonwealth of Independent States	10	34	67.7	83.7	89
Europe	46	67	79.6	89.5	91

most vulnerable to being replaced by AI. It can be helpful to local innovators, researchers and entrepreneurs.

There are advances in AI-driven drug discovery which potentially may give hope for new treatments of cancer, HIV/AIDS, tuberculosis and malaria. There is an increasing understanding of the gut microbiome and there may be a breakthrough with a probiotic supplement, transforming the lives of millions. Some research is also focused on AI tools to combat antibiotic resistance.

AI is also claimed to be an innovation for the care of older people and can increase the capacity of health care workers. In addition, it may help in levelling up reading abilities for students. One noticeable effect is that it has become increasingly common for Swedish students at certain colleges and universities to cheat with the help of AI; a total of 90 students were warned in 2023 compared to none in 2022.

The use of AI may lead to less learning in schools. According to a recent survey from the Swedish Youth Barometer, over half of young Swedish people who have used AI in schoolwork state that they have done it in a way that they believe has not been allowed (Swedish Youth Barometer, 2023). There is a need for a regulatory framework on how to use AI, both in schools and for students. More tests are required on site in classrooms with more focus on books rather than digital tools.

AI may, however, have dramatic effects on human life and even pose a threat to humanity. Already a decade ago, the British theoretical physicist, Stephen Hawking, told the BBC that the development of full AI could spell 'the end of humans as a species' (Lewis, 2014). AI creates its own world and can assist closed societies in monitoring their citizens but threatens open democratic societies. Since social media is person-obsessed, AI will exacerbate digital manipulation in the form of deep fakes, filter bubbles and polarization. In consequence, AI reinforces individualization rather than strengthening society for the good of all. This becomes a global dilemma beyond society's control which is the responsibility of politicians (Sträumke, 2023). The problem, however, is the concentration of power in a few technology giants, which are not governed by states, who are thus unable to influence the development

of AI. In low-income countries unemployment is already a problem and the new AI technology will be labour-saving. Such development is questionable when unemployment is a problem, even in developing countries.

The police and intelligence services of several EU countries have started using AI technology or are planning to do so. The UK government has proposed that police officers should be able to compare AI footage with the national database of driving licence photos to identify suspected criminals (Boffey, 2023).

In late 2023, the EU initiated regulation. AI technology is neither good nor bad. It depends on how it is used, which became evident when the EU formulated its European Artificial Intelligence Act in 2024. The riskier the technology, the stronger the obligations of its developers. They must label all AI-generated content, so that it is evident it is not real, by designing a system that can detect whether content has been generated or manipulated by AI.

Measurements of Globalization

Measurements of economic globalization focus on variables such as trade, foreign direct investment (FDI), GDP, portfolio investment and income. Other attempts to measure globalization include variables related to political, social, cultural and even environmental aspects of globalization. The KOF Index of Globalization is an index of the degree of globalization of countries, using three main criteria: economic, political and social. Unlike the Maastricht Globalization Index, it does not consider environmental factors. Other examples are the A.T. Kearney/ Foreign Policy Globalization Index and the DHL Global Connectedness Index which measures globalization based on international trade.

There are some dimensions of globalization that are only occasionally discussed by scholars. One example is environmental globalization in the form of international treaties on environmental protection. Another is military globalization, both in growth at the global scale and in the scope of security. A third example may be global diseases. Instead of comparing countries novel approaches are being developed to manage issues of globalization.

References

Adel, H.M., Zeinhom, G.A. and Mahrous, A.A. (2018) Effective management of an internationalization strategy: A case study on Egyptian–British universities' partnerships. *International Journal of Technology Management & Sustainable Development* 17(2), 183–202.

Albrow, M. and King, E. (1990) *Globalization, Knowledge and Society*. SAGE, London.

Balla, N. (2023) Slum populations are set to surge as the housing crisis bites. World Economic Forum in cooperation with *Reuters*. Available at: https://www.reuters.com/article/business/feature-worlds -slum-populations-set-to-surge-as-housing-crisis-bites-idUSL4N3800P6/ (accessed 25 August 2024).

Beneria, L., Berik, G. and Floro, M.S. (2015) *Gender, Development and Globalization. Economics As If All People Mattered*. Routledge, New York.

Blenkinsop, P. and Kihara, L. (2019) EU, Mercosur strike trade pact, defying protectionist wave. *Reuters*. Available at: https://www.reuters.com/article/business/eu-mercosur-strike-trade-pact-defying-prote ctionist-wave-idUSKCN1TT2KC/ (accessed 25 August 2024).

Boffey, D. (2023) Surveillance technology is advancing at pace – with what consequences? Plans for facial recognition searches across UK driving licence records could threaten idea of policing by consent. *The Guardian*. Available at: https://www.theguardian.com/uk-news/2023/dec/20/surveillance-techn ology-is-advancing-at-pace-with-what-consequences (accessed 25 August 2024).

Business Standard (2017) India, China call for eliminating agri-subsidiesin developed countries. *Press of Trust India*, New Delhi. Available at: https://www.business-standard.com/article/current-affairs/india -china-call-for-eliminating-agri-subsidies-in-developed-countries-117083100516_1.html (accessed 25 August 2024).

Council on Foreign Relations (CFR) (2023) Mercosur: South America's Fractious Trade Bloc. CFR. Available at: https://www.cfr.org/backgrounder/mercosur-south-americas-fractious-trade-bloc (accessed 25 August 2024).

Deléchat, C. and Medina, L. (2020) What is the informal economy? Having fewer workers outside the formal economy can support sustainable development. *Finance & Development Magazine* 54–55. Available at: https://www.imf.org/en/Publications/fandd/issues/2020/12/what-is-the-informal-econ omy-basics (accessed 25 August 2024).

Dembitzer, B. (2021, 3 December) Poor countries mustn't open economies until they are strong. *The Guardian*.

Dialogue, The (2023) How well is the USMCA working after three years? *Latin American Advisor*. Available at: https://www.thedialogue.org/wp-content/uploads/2023/06/LAA230623.pdf (accessed 25 August 2024).

Diamond, J. and Bash, D. (2017) Trump signs order withdrawing from TPP, reinstate "Mexico City policy" on abortion. *CNN*. Available at: https://edition.cnn.com/2017/01/23/politics/transpacific-partnersh ip-trade-deal-withdrawal-trump-first-executive-action-monday-sources-say (accessed 24 January 2017).

East Africa Community (2021) African Continental Free Trade Area (AfCFTA) Agreement. Available at: https://www.eac.int/trade/international-trade/trade-agreements/african-continental-free-trade-area -afcfta-agreement (accessed 10 June 2024).

European Parliament (2024) EU-Mercosur Association Agreement. In 'A Stronger Europe in the World', EuropeanParliament, Strasbourg, France. Available at: https://www.europarl.europa.eu/legislative -train/theme-a-stronger-europe-in-the-world/file-eu-mercosur-association-agreement (accessed 25 August 2024).

FEANTSA (2022) Seventh Overview of Housing Exclusion in Europe 2022. The European Federation of National Organisations Working with the Homeless (FEANTSA), Brussels. Available at: https://www .feantsa.org/public/user/Resources/reports/2022/Rapport_Europe_GB_2022_V3_Planches_Correct ed.pdf (accessed 25 August 2024).

Global Policy Forum (2003) Comparison of Growth of World Exports, Gross World Product and World Production 1950-2001. Global Policy Forum, New York. Available at: https://archive.globalpolicy.org /component/content/article/104-tables-and-charts/47379-comparison-of-growth-of-world-exports -gross-world-product-and-world-production.html (accessed 25 August 2024).

Gunter, B.G. and Wilcher, B. (2020) Three decades of globalisation: which countries won, which lost. *The World Economy* 23(4), 1076–1102. Available at: https://doi.org/10.1111/twec.12915 (accessed 25 August 2024).

Held, D., McGrew, A.G., Goldblatt, D. and Perraton, J. (1999) *Global Transformations: Politics, Economics and Culture*. Stanford University Press, Redwood City, CA.

International Telecommunication Union (ITU) (2024) Measuring digital development: Facts and figures 2023. Telecommunication Development Bureau, ITU, Geneva, Switzerland.

James, P. (2005) Arguing globalizations: Propositions towards an investigation of global formation. *Globalizations* 2(2), 193–209. Available at: https://doi.org/10.1080/14747730500202206 (accessed 12 June 2024).

James, P. and Steger, M.B. (2014) A genealogy of 'globalization': the career of a concept. *Globalizations* 11(4), 417–434. Available at: https://doi.org/10.1080/14747731.2014.951186 (accessed 12 June 2024).

Klein, N. (2008) *The Shock Doctrine: The Rise of Disaster Capitalism*. Canada, Vintage.

Lewis, T. (2014) Stephen Hawking: Artificial intelligence could end human race. *LiveScience*. Available at: https://www.livescience.com/48972-stephen-hawking-artificial-intelligence-threat.html (accessed 11 June 2024).

Martin, H.-P. and Schumann, H. (1996) *Die Globalisierungsfalle. Der Angriff auf Demokratie und Wohlstand*. Rowolt Verlag GmbH, Hamburg, Germany.

MERCOSUR (2024) About us. Available at: https://www.mercosur.int/en/ (accessed 25 August 2024).

Miroudot, S., Lanz, R. and Ragoussis, A. (2009) Trade in Intermediate Goods and Services, OECD Trade Policy Working Papers, No. 93, OECD Publishing. Available at: https://doi.org/10.1787/5kmlcxtdlk8r-en (accessed 12 June 2024).

O'Rourke, K.H. and Williamson, J.G. (2002) When did globalization begin? *European Review of Economic History* 6(1), 23–50. Available at: https://doi.org/10.1017/S1361491602000023 (accessed 12 June 2024).

Oxfam (2023) Richest 1% bag nearly twice as much wealth as the rest of the world put together over the past two years. Oxfam, Canada. Available at: https://www.oxfam.ca/news/richest-1-bag-nearly-twice-as-much-wealth-as-the-rest-of-the-world-put-together-over-the-past-two-years/ (accessed 10 June 2024).

Pakenham, T. (1991) *The Scramble for Africa*. Wiedenfeld and Nicolson, London.

Robertson, R. (1992) *Globalization: Social Theory and Global Culture* (reprint). SAGE, London.

Sassen, S. (1991) *The Global City*. Princeton University Press, Princeton, NJ.

Sen, A. (1982) *Poverty and Famines: An Essay on Entitlement and Deprivation*. Clarendon Press, Oxford; Oxford University Press, Oxford.

Silvio, D.H. (2022) Free trade and global inequality: Why the World Trade Organization and the International Monetary Fund continue to advocate for free and unfettered trade. *Intergovernmental Research and Policy Journal*. Available at: https://irpj.euclid.int/articles/free-trade-and-global-inequality-why-the-world-trade-organization-and-the-international-monetary-fund-continue-to-advocate-for-free-and-unfettered-trade/ (accessed 10 June 2024).

Steger, M.B. (2009) *Globalization: A Very Short Introduction*. Oxford University Press, New York.

Sträumke, I. (2023) *Machines That Think: The Secret of Algorithms and the Path to Artificial Intelligence* (in Swedish). Polaris.

Suneja, K. (2017) India, China question US & EU's 'trade distorting' farm subsidies. *The Economic Times*. Available at: https://economictimes.indiatimes.com/news/economy/foreign-trade/india-china-question-us-eus-trade-distorting-farm-subsidies/articleshow/59689979.cms?from=mdr#google_vignette (accessed 25 August 2024).

Swedish Youth Barometer (2023) High School Report. Swedish Youth Barometer Company AB, Stockholm.

The Police Authorities (2023) The situation in vulnerable areas. Government task 2023, PMY Report, The Police Authorities, Dnr: A658.585/2023, Saknr: 492. Stockholm (in Swedish).

Varghese, N.V. (2008) *Globalization of Higher Education and Cross-border Student Mobility*. International Institute for Educational Planning, UNESCO, Paris.

Villagran, L. (2022) What is Title 42, when does it end, how does it impact US-Mexico border? Here's what to know. *USA Today*.

Vu, K. and Nguyen, P. (2020) Asia form's world's biggest trade bloc, a China-backed group excluding U.S. Reuters. Available at: https://www.reuters.com/article/idUSKBN27V03J/ (accessed 10 June 2024).

Wealth-X (2021) Billionaire Census 2021. Wealth-X, Altrata Company. Available at: https://go.wealthx .com/2021-billionaire-census (accessed 25 August 2024).

World Trade Organization (WTO) (2022) Annual report. WTO, Geneva, Switzerland.

Zholudev, V. (2023) Digital IDs and the global fight for identity. Biometric Update. Available at: https://www .biometricupdate.com/202305/digital-ids-and-the-global-fight-for-identity (accessed 11 June 2024).

4 Transnational Corporations

Abstract

Transnational corporations (TNCs) are among the world's biggest economic institutions, but early TNCs can be traced to the colonizing activities of Western Europe. These early TNCs started in the 16th century and lasted for several hundred years with transport of tropical goods. The early Industrial Revolution, beginning in the 18th century, led to a search for natural resources such as minerals, petroleum, foodstuffs and household goods. This led to more capital-intensive manufacturing processes, and better storage techniques with increasing internationalization of capital. Companies merged and made acquisitions to concentrate operations and certain features of a TNC emerged. In the last few decades, the number of TNCs has grown exponentially. The agribusiness TNCs are influencing global agriculture, especially export-oriented crops, sometimes displacing local food crop production. They are active in the production of seed and agrichemicals. Some of the trade-offs involving TNCs are discussed and also their influence, at the national level, on politics, international negotiations and controversies are highlighted.

Transnational corporations (TNCs) are among the world's biggest economic institutions. A rough estimate in 2000 indicated that the 300 largest TNCs owned or controlled at least one-quarter of the entire world's productive assets (Coolgeography.co.uk, n.d.). But the figures are very approximate and the TNCs have expanded significantly since then. The TNCs tend to dominate in industries where output and markets are oligopolistic or concentrated in a smaller number of firms, dominating the global market. Examples are car and truck manufacturers, the oil industry, the chemical sector and the food sector.

When the world's top 100 non-financial TNCs from developing and transition economies, ranked by foreign assets, were listed by UNCTAD (2020) it was revealed that 44 out of the 100 firms were Chinese, including those in Hong Kong. Taiwan was represented by seven firms.

The TNCs' total annual sales are comparable to or greater than the yearly gross domestic product (GDP) of many countries. Years ago, the Itochu Corporation's sales exceeded the GDP of Austria, while those of Royal Dutch/Shell equalled Iran's GDP (Greer and Singh, 2000).

In 2023, the world's largest food and beverages TNC is Nestlé with total reported sales of CHF93.0 billion, a decrease of 1.5% compared to 2022 (Nestlé, 2023a). In the same year, the world's largest retailer, Walmart, had net sales of US$420.6 billion (Walmart, 2023).

Early Transnational Companies

Early TNCs can be traced to the colonizing and imperialist activities of Western Europe. They began in the 16th century and lasted until the mid-20th century. Portugal and Spain were pioneers in establishing overseas empires. At the end of the 16th century, the English and the Dutch began to challenge the Portuguese monopoly of trade with Asia.

The British East India Company (EIC) and the Dutch East India Company (VOC) were chartered in 1600 and 1602, respectively. Their primary aim was to enter the spice trade. The VOC focused on the spice trade of the East Indies archipelago and the EIC on India. Soon, the textiles industry of India overtook spices in terms of profitability (Ferguson, 2004).

British chartered companies

The EIC was originally an English, and subsequently British, joint-stock company that began trading in the region around the Indian Ocean and later with East Asia. The company controlled much of the Indian subcontinent and colonized Hong Kong and parts of Southeast Asia. The EIC had its own armed forces (Erikson, 2014).

The EIC established its first Indian factories at Surat (1615) and its second at Masulipatnam (1616) on the Bay of Bengal (Tracy, 2015). By 1647, the company had 23 factories and settlements in India (Woodruff, 1954). The Mughal emperor invited English traders to trade in Bengal. This was important since the GDP of India during the Mughal era was estimated at 22% of the world economy in 1600. It was the second largest in the world, only behind China during the Ming era (Maddison, 2001).

In the 1620s, the EIC had begun using and transporting slaves in Asia and the Atlantic (Allen, 2015). Early reports about the EIC's high profits led the Crown to grant subsidiary licences to other English trading companies although the charter for the EIC was renewed for an indefinite period. Other tradesmen could establish private trading firms in India after the Deregulating Act was passed in 1694. A new company was also formed (the English Company Trading to the East Indies), but the stockholders of the EIC subscribed a generous sum of money to it and could dominate the new company.

In the early 1700s, customs duties were waived for the English in Bengal. India was producing 24.5% of the world's manufacturing output up until 1750 (Williamson and Clingingsmith, 2004). This period has been described as a form of proto-industrialization. The EIC had also been engaged in an increasingly profitable opium export trade to Qing China since the 1730s. It was illegal, outlawed by China in 1729, but it helped reverse the trade imbalances from British imports of tea (Martin, 2007).

The EIC accounted for 50% of world trade during the mid-1700s and the early 1800s (Farrington, 2002). In the late 1770s, the EIC had become the protectors of the Mughal dynasty in that it accounted for most of their

trade and had large armed forces. The trade included basic commodities, such as cotton, silk, indigo dye, sugar, salt, spices, saltpetre, tea and opium. The corporation was ruling the British Empire in India in the 19th century (Tracy, 2015).

The British West Indies provided England's most important and lucrative colonies. Early settlements were established in St. Kitts in 1624 and large sugarcane plantations were established on Barbados in 1627. Initially, white indentured labour was used but rising costs caused English traders to use imported African slaves instead. This sugar production generated great wealth and Barbados not only developed into the most successful colony but also one of the most densely populated places in the world. Sugar cultivation expanded across the Caribbean. England annexed Jamaica from the Spanish and colonized the Bahamas. The Caribbean expansion financed the development of the English colonies in North America. It also accelerated the growth of the slave trade across the Atlantic Ocean.

The Government of India Act 1858 enabled the British Crown to take direct control of all territories held by the EIC in India. Later, several crop failures caused widespread famines on the Indian subcontinent, causing the death of some 15 million people. The corporation had neglected to have any policy to deal with famines. It faced financial problems and was dissolved in 1874. The government of the British Raj assumed the Company's governmental functions in India and absorbed its armies. The British Empire had become the world's dominant colonial power.

Dutch chartered companies

The Dutch began to take a leading role at sea in the late 16th century. After defeating the Portuguese in 1640–1641, the Dutch had expanded their spice trade monopoly in the Strait of Malacca. Then, a period of intense competition between the EIC and the VOC resulted in the Anglo-Dutch Wars. By the second half of the 17th century, the Dutch dominated the culture and economy of the sea; the Dutch Golden Age. It was built on corporate colonialism by Dutch corporations. The Dutch empire comprised territories under Dutch control from the 17th to the 20th century.

The VOC fiercely protected its trade in South-east Asia. Factories of the corporation had a monopoly on the production of nutmeg, cinnamon, clove and mace for export to Europe (Parthesius, 2010). The local merchants were controlled through a tax and had to sell at a set price without any profit. The VOC had a monopoly on the export of tin from Malaysia. The market for sugar from Indian Ocean, in particular around the large Indonesian archipelago, began to decline in the early 1700s because Brazil could produce it cheaper. This caused financial difficulties for the VOC. It was defunct in 1799 and the territories it controlled became the Dutch East Indies (TANAP, 2006).

Some traders in Amsterdam had tried to circumvent the VOC monopoly but failed. Instead, they had formed, together with foreign investors, another chartered company in 1621. These merchants were primarily interested in taking over the Spanish colonies in the Caribbean and South America. The new company, the Dutch West India Company (WIC), received a trade monopoly in the Dutch West Indies in the Republic of the Seven United Netherlands according to Zeeuws Archief (Emmer, 2019). In addition, it was given jurisdiction over Dutch participation in the Atlantic slave trade, Brazil, the Caribbean and North America. The WIC made Curaçao a centre of the Atlantic slave trade. In contrast to the VOC, the WIC had no right to deploy military troops.

When in debt in 1674, the WIC was dissolved but the slave trade between West Africa and the Dutch colonies in the Americas required a new chartered company. Thus, the New West India Company was formed (Law, 1994). From the 1730s, this new WIC focused on the slave trade and the last slave ship entered Willemstad in 1775 (Emmer, 2011). The new company was dissolved in 1791.

colonialism and a large slave trade. They expanded into new territories, sometimes using force. Ultimately, all the early chartered companies faced financial difficulties. When finally defunct, the company activities were nationalized by their respective governments.

The corporations made great profits from trade. In the early decades of the 17th century, the VOC was the wealthiest commercial operation in the world. Williams (1964) claimed that the profits that Britain received from its sugar colonies, or from the slave trade between Africa and the Caribbean, contributed to the financing of Britain's Industrial Revolution and port cities such as Bristol, Liverpool and London. After legal threats from the British state and attacks by the Royal Navy, the EIC ended its slave trade in 1834. Two decades later, the British government reduced protectionism for sugar from the British West Indian colonies. Instead, it promoted free trade in cheaper sugar from Cuba and Brazil.

There were limited developmental benefits to African countries, apart from those involved in the slave business (Baten, 2016). One specific effect of the trade was that more efficient currencies were adopted by the West African merchants. During the transport, the slaves experienced harsh and unhygienic conditions. The Portuguese used the fruits of the oil palm to feed the slaves at sea to keep them alive. The average mortality rate during the Middle Passage from Africa to the Americas was one in seven (Marshall, 1998). When the slave trade was suppressed, palm oil replaced the traffic in slaves (Purseglove, 1972).

Trade gave rise to the production of new crops in Europe. Columbus brought the sweet potato to Spain and the potato arrived in Europe around 1570. The Atlantic trade also brought new crops to Africa, for example, the pineapple. Introduced by returning explorers, it later reached Madagascar (1542), Java (1599), West Africa (1602) and South Africa (1660), according to Purseglove (1972).

How the early chartered companies affected trade

The early chartered companies dominated trade during a period characterized by expansionism,

The Early Industrial Revolution Period

With an approaching Industrial Revolution, crops and goods produced by slavery became less important to the British economy. Instead, there was a search for natural resources such as minerals, petroleum, foodstuffs and household goods. Many new companies were formed, and not only in Britain. The formation of the British company Unilever can serve as an illustration. It was established in 1929 by a merger of the operations of the Dutch manufacturer Margarine Unie and the British soap maker Lever Brothers. The latter was a British manufacturing company founded in 1885 that had entered the United States market a decade later. Margarine Unie had been formed in 1927 by the merger of four Dutch margarine companies. One of them started in 1867, beginning large-scale production of margarine after having acquired the patents and rights to Hippolyte Mège-Mouriès' invention in 1871 (Schiff, 2016).

Gradually this development led to larger and more capital-intensive manufacturing processes and better storage techniques. Increasing internationalization of investment and trade meant that companies could expand their markets, also internationally. There were mergers and acquisitions to concentrate operations by transnationals in sectors, such as petrochemicals and food.

In Japan, this period witnessed the growth of the *zaibatsu* (financial clique), including the Mitsui and Mitsubishi corporations in the late 1800s. These corporations worked in alliance with the Japanese state and had oligopolistic control of the country's industrial, financial and trade sectors. As of today, the Mitsubishi corporation is a global integrated business enterprise with subsidiaries worldwide and a global network of around 1,700 group companies.

Some large companies started to make investments in low-income countries, such as the US agribusiness giant, United Fruit Company. It controlled 90% of US banana imports at its start in 1899 (Dunning, 1993). The company was formed by a merger of the banana trading company, Tropical Trading and Transport Company and its rival, the Boston Fruit Company. United Fruit primarily traded bananas cultivated on Latin American plantations. In the 1970s, there were some 400 investigations of bribery of foreign officials of American companies by the US Securities and Exchange Commission and discussions in the House Committee on Interstate and Foreign Commerce (Staggers, 1977). One example was United Fruit which had been accused of bribing government officials in exchange for preferential treatment, trying to avoid taxes to governments in countries where the company operated, and aiming for a monopoly. The company disclaimed such reports and merged with the AMK Corporation in 1970 to form the United Brands Company. In 1984, that company was transformed into Chiquita Brands International, and following further acquisitions and changes it became today's Swiss-domiciled American producer and distributor of bananas, Chiquita Brands International Sàrl.

Features of a Transnational Corporation

Technical definitions of TNCs vary. In this context, the term 'transnational corporation' means a for-profit enterprise. It engages in many business activities – including sales, distribution, extraction, manufacturing, and research and development – outside the country of origin. Thus, it is dependent financially on operations in two or more countries and its management makes decisions based on global or regional alternatives.

A TNC can be a public corporation, trading its shares of stock at stock exchanges or brokerage houses. Its shareholders can include both individuals and institutions, such as banks, insurance companies and pension funds. Private TNCs have no public trading of shares and are frequently family controlled, for instance, Cargill Inc. In 1865, it started with a grain warehouse in Iowa, USA. Five years later, it established a headquarters in Albert Lea, Minnesota. Its business rapidly expanded to include coal, flour, feed, lumber and seeds, as well as investments in railroads, land, water irrigation and farms. In 1940, the company operated in four countries and began to expand globally in the 1950s. It is currently operating in 70 countries with 160,000 employers and selling to 125 markets (Cargill, 2023).

A parent company, located in the TNC country of origin, exercises an authoritative, controlling influence over a subsidiary in another country, either directly, if it is private, or by owning some or all the shares if it is public (Greer and Singh, 2000). Control can be exercised in different ways. Parent corporations can exert a controlling power even with small shareholdings in subsidiaries. Subsidiaries can have a different name to the parent company and can also operate in the same country as the parent.

A few examples may illustrate developments over time. The American H.J. Heinz Company was founded in 1869, developing thousands of food products. In 2013, Heinz agreed to be purchased by Berkshire Hathaway and the Brazilian investment firm 3G Capital. Two years later, they facilitated for Kraft Foods Group Ltd to merge with Heinz. This resulted in the Kraft Heinz Company; the fifth-largest food and beverage company in the world with net sales of some US$26 billion in 2022 (Kraft Heinz, 2022).

In 1907, Shell Transport and Trading Company was merged with the Royal Dutch Petroleum Company. In 2022, Shell, formerly Royal Dutch Shell, operated in more than 70 countries with 93,000 employees (Shell, 2022). In 1988, the Swedish ASEA (founded 1883) merged with the Swiss company BBC (founded 1891) to create ABB Ltd. It has facilities in 140 countries with some 105,000 employees (ABB, 2023).

The British company, Imperial Chemical Industries (ICI), was formed in 1926 by the merger of four leading British chemical companies. In the early 1990s, the company was de-merged due to increasing competition. In 2008, ICI was taken over by the Dutch company, AkzoNobel, now operating in 150 countries. During 2021–2023 it acquired Grupo Orbis in Latin America and the Huarun decorative paints business in China (AkzoNobel, 2023).

Around the year 2000, most of the 296 largest TNCs, excluding those from the financial industry, had a presence in Africa. However, they had less than 10% of their global foreign assets there (UNCTAD, 1999). Some TNCs have operated in Africa for a long time. For example, General Electric, Volkswagen and Standard Bank Group. More recently, Vodafone and Jumia (the e-platform) have expanded on the African continent.

In China, many multinational corporations are operating, such as AT&T, General Electric, Apple, Intel, Walmart, Target, KFC, McDonald's and Starbucks. Also, non-Western multinationals are active there, for example, Toyota, Mitsubishi, Subaru, Samsung, Hyundai, Lucky Goldstar and Kia.

In 1999, the Chinese government decided on a Go Out Policy to promote Chinese investments abroad. Chinese direct foreign investments rose from US$3 billion in 1991 to a peak of more than US$196 billion in foreign investments in 2016. Large Chinese multinational companies include DJI Innovations, Haier Group Corporation, Huawei, Datang Telecom, NIO and Lenovo (Investopedia, 2022). According to Euromonitor International, Haier, founded in 1984, had been the number one global major appliance brand in volume sales in the world for the last 12 consecutive years (Euromonitor International, 2021).

Formed in 1987, Huawei is a private Chinese company with 207,000 employees, providing information and communications technology infrastructure to over 170 countries (Huawei, 2022). Otherwise, many of China's largest companies are state-owned enterprises under the influence of the Communist Party. In 2022, the Fortune's Global 500 list of the world's largest corporations included 145 Chinese companies in total (Fortune, 2022). Forbes (2022) reported that three of the world's ten largest public companies were Chinese.

The Increasing Number of Transnational Corporations

The number of TNCs, including parent companies and subsidiaries, has grown exponentially, especially corporations as subsidiaries (Table 4.1). This is a recent change and has led to the rapid emergence of complex multi-tiered

Table 4.1. Estimated number and growth of transnational corporations and their affiliates during 1970–2007 (Greer and Singh, 2000; UNCTAD, 1994, 2008, 2020).

Period/Year	Transnational corporations	Affiliates
1970	7,000	Not available
Early 1990s	37,000	170,000
2007	79,000	790,000

corporate structures. Affiliated entities can collectively conduct the business of the enterprise. A subsidiary can be owned by multiple parent corporations. These developments have changed the operation of business in the TNCs. They may result in the potential for human rights violations on a global scale (Skinner *et al.*, 2020).

Since the companies' structures have become more complex, it is more difficult to determine the ownership of TNCs' various affiliated companies or to compile complete financial data on TNCs. The lack of transparency on ownership and the existence of various TNC affiliates exacerbates this problem. Since the early 1990s, the subsidiaries' global sales have surpassed worldwide trade exports.

The current trends mean that the TNCs' role in global commerce is of a much more significant scale than before. Their substantial number can be misleading because the wealth of transnationals is concentrated. In 1906, there were two or three leading firms with assets of US$500 million. In 1971, there were 333 such large corporations, one-third of which had assets of US$1 billion or more. In 1992, the top 100 firms had US$3.4 trillion in global assets, of which approximately one-third was held outside their home countries (Deng *et al.*, 2009).

Recent estimates for 2023 indicate there might be a small reduction of the numbers of both the TNCs and their foreign affiliates compared to 2007 (Fortune, 2023).

The most recent list comprises the world's top 50 largest companies by consolidated revenue, according to the Fortune Global 500 rankings for 2023 (Table 4.2). All listed companies have annual revenues exceeding US$130 billion for the most recent fiscal year. The list is, however, not complete, as not all companies disclose information. Sixteen are state-owned; China (13) and one each in Russia, Saudi Arabia and Germany. The American retail corporation Walmart has been the world's largest company by revenue since 2014 (Fortune, 2024). There are no agribusiness TNCs on the list, although the US company, Cargill Inc., is listed as a conglomerate.

Agribusiness Transnational Corporations

TNCs are influencing global agriculture. They have controlled most of the land cultivated worldwide for export-oriented crops, sometimes displacing local food crop production.

Table 4.2. The world's top 50 largest companies by consolidated revenue for the most recent fiscal year (2022/2023) (Fortune, 2023).

Headquarters	Oil and gas	Automotive	Health care	Electronic and Elec	Steel and Construction	Retail and Financials	Conglomerates
USA	5	1	6	2	–	3+1	3
China	3	–	–	1	4	+ 4	1
Germany	–	2	–	1	–	1	
Switzerland	–	–	–	–	–	–	2
United Kingdom	2	–	–	–	–	–	–
Japan	–	1	–	–	–	–	1
Russia	1	–	–	–	–	–	–
Saudi Arabia	1	–	–	–	–	–	–
France	1	–	–	–	–	–	–
Netherlands	–	1	–	–	–	–	–
Singapore	–	–	–	–	–	–	1
Taiwan	–	–	–	1	–	–	–
TOTAL	13	5	6	5	4	4+5	8

Global seed and chemical companies have long been aware of the future food situation and invested heavily in research and development. These TNCs have access to some of the world's genetic seed stocks. They finance much of worldwide biotechnology research and nanotechnology, and control the patenting of various life forms of microbes and biomolecules. This includes research on genetically modified crops, such as maize, soybeans and cotton with resistance to Roundup (glyphosate) to protect against weeds. About 20 major TNCs account for most sales of pesticides, such as polychlorinated biphenyls (PCBs), DDT, dioxins, furans and chlorinated solvents. In the late 1980s, it was not unusual for chemical companies, including TNCs, to export products to low-income countries that were banned, unregistered, cancelled or withdrawn in the US and other industrialized countries.

Firms from developed countries still dominate among TNCs in agribusiness. A decade ago, 12 of the top 25 agriculturally based TNCs were headquartered in low-income countries and 13 in developed countries (UNCTAD, 2010). The top position was occupied by Sime Darby Berhad of Malaysia, followed by Dole plc or in Ireland.

Several global agrifood companies have a long history, as previously exemplified by Cargill Inc. Another is the German company, Bayer AG, founded in 1863. Today, it is one of the world leaders in the chemical–technical industry. The American Monsanto Company was founded in 1901 as a chemical and pharmaceutical company with PCBs as an early profitable product. The PCBs did not easily break down or degrade, which made them attractive to industries and in consumer products manufactured from 1935. They were considered carcinogenic and were banned in the US by the Toxic Substances Control Act in 1976, as they were deemed to be stored in nature and food. PCBs were internationally banned by the Stockholm Convention on Persistent Organic Pollutants in 2001.

The Archer Daniels Midland Company (ADM), formed in 1902, has more than 270 industrial plants and an extensive global transport network for transporting cereals. A younger US food company is Tyson Foods, Inc., founded in 1935. It is the world's second-largest

producer and marketer of poultry, pork and beef meat. The Brazilian company, JBS SA, was established in 1953 and began to expand within the country when the new capital, Brasilia, gave it a new market. It bought America's third-largest meat and pork company, Swift & Company, in early 2000 and later Smithfield Foods was bought. Then, JSB USA became the second-largest company in meat and pork and JBS-owned Pilgrim the second-largest company in chicken meat.

JBS SA's acquisition of the European chicken company Moy Park in 2017 made the company as one of the world's largest livestock producers with 150 industrial plants worldwide. In the following year JBS slaughtered some 77,000 cattle, 116,000 pigs and 13.6 million poultry every day (Wasley *et al.*, 2019). Its operations have, however, been criticized. The Bureau of Investigative Journalism revealed, in partnership with *The Guardian* and *Repórter Brasil*, its illegal use of deforested land and bribery of many candidates from across the political spectrum (Wasley *et al.*, 2019). In spite of the EU restricting Brazilian beef in 2008, JBS had managed to use its Australian subsidiary to continue its exports (Blankfeld, 2011). In 2020, both JBS SA and J&F Investimentos SA were fined for extensive bribery according to the US Securities and Exchange Committee in cooperation with Ministerio Publico Federal and the Procuradoria-Geral da Republica in Brazil (US Department of Justice, 2020).

BRF SA is one of the world's largest food companies with over 30 brands. It is the result of the merger in 2013 between Sadia and Perdigão, two of Brazil's most important food companies. The company's products are chicken and pig, sold in over 150 countries on all continents from approximately 50 factories. In 2017, BRF's subsidiary, OneFoods Holdings Ltd began focusing on the halal market. Headquartered in Dubai, it is the world's largest halal-animal protein company.

Fonterra is a New Zealand multinational cooperative dairy, owned by more than 10,000 farmers and their families. Fonterra Co-operative Group Limited was established in 2001 following the merger of the country's two largest dairy producers: New Zealand Dairy Group and Kiwi Co-operative Dairies. The company accounts for a quarter of New Zealand's exports to about 140 countries, equivalent to about one-third of the world's dairy exports.

As of today, the agriculture-related TNCs include the much larger food processors, manufacturers, retailers, traders and suppliers of inputs. Most of the TNCs in food and beverages are large and the retailing and supermarket TNCs also play a role in international agricultural supply chains. The majority of the 25 largest TNCs in this industry sector are from Western countries (Table 4.3). Recently, a change took place when the China National Chemical Corporation bought the Swiss company, Syngenta. Simultaneously, Monsanto was taken over by Bayer and the American companies, DuPont Pioneer and Dow Chemicals, have merged. As a state owner, China is one of the leading nations in agribusiness since 2017, together with the German company, BASF. The concentration of power means that global trade in inputs for global agriculture and forestry will be dominated and controlled by the supply of a few giant multinationals. In turn, this leads to reduced competition, exemplified when Bayer had to sell off part of its business to BASF to obtain European Union (EU) approval.

Transnational Corporations and Contract Farming

Contract farming is an old approach for producing export crops, such as tea, palm oil and rubber, on plantations. They were grown on company-owned estates, surrounded by independent growers under contract. As an example, smallholder tea cultivation started in Kenya in the 1950s, operated through a management agreement with multinational tea companies. This was radically changed when the Kenya Tea Development Authority (KTDA), established in 1964, took over the management from the multinational tea companies. In 2009, the KTDA became a private company, owned by about 600,000 smallholder tea producers. The farmers are shareholders in 54 tea companies that own the KTDA and its subsidiary companies (Kenyan Tea Development Authority, n.d.).

Estate production has suited the operations of TNCs. Contracting firms are always relatively large processors, exporters, or supermarket chains because large, fixed costs are associated with contracting. The TNCs use the same large-scale model and commercial producers that colonists introduced when they established plantations for cash crops. Often, both crops and labourers were brought to plantations from different continents. Thus, colonial agriculture caused global displacement of many people and plants, and it is still taking place. As a consequence, Traditional farming for food production was abandoned or even prevented.

As of today, TNCs are engaged globally in contract farming activities and other non-equity forms in more than 100 countries. One example is the Nestlé company. Some years ago, it had contracts with more than 600,000 farms in over 80 developing and transition economies

Table 4.3. Top ten agribusiness companies in the world in 2018 (Sekulich, 2019).

Company	Headquarters	Revenue (US$ billion)	Number of employees
Cargill Inc.	USA	114.7	160,000
DowDuPont	USA	85.97	98,000
Archer Daniels Midland Co.	USA	64.34	32,300[a]
Bayer AG	Germany	46.7	116,998
Deere & Co.	USA	38.4	74,000
CNH Industrial NV	Netherlands	29.7	64,625
Nutrien (Agrium and PotashCorp)	Canada	19.6	20,300
Syngenta AG	Switzerland	13.5	28,704
Yara International	Norway	12.9	16,757
BASF	Germany	6.8	115,490

[a]2016

as direct suppliers of various agricultural commodities (Nestlé, 2023b).

Another example is the Singapore-based Olam. It is a major food and agri-business public company in farming, processing and distribution operations. It was formed in 1989 by the Kewalram Chanrai Group as a non-oil-based export operation out of Nigeria to secure hard currency. It expanded and moved to Singapore in 1995, and is now operating in 60 countries, supplying food and industrial raw materials to over 20,900 customers worldwide (Olam, 2021). It has a global contract farming network and is one of the world's largest suppliers of cocoa beans, coffee, rice and cotton. Its ownership includes Temasek Holdings (54%) and Mitsubishi Corporation (17%).

Contract farming arrangements by TNCs cover a broad variety of commodities, from cash crops to livestock and staple food crops. Olam deals globally with 17 agricultural commodities. Agricultural crops constitute two-thirds of Unilever's raw materials for a variety of consumer goods. Both smallholders and large farms in low-income countries are involved in Unilever's work, as well as other suppliers.

In Brazil, 75% of poultry production and 35% of soybean production are sourced through contract farming involving TNCs. Major exporters of soybeans are Bunge and Cargill (35%) and Outros (20%). Currently, Brazil is the world's largest soybean producer growing on some 37 million ha, having surpassed the United States (Statista, n.d.). According to Brazilian scientists, the growth of soybean production volume is based on increased productivity rather than an increase of the planted area. It has even been argued that this Brazilian production system is environmentally friendly (Gazzoni *et al.*, 2019).

In 1967, contract farming with tobacco was initiated in Thailand. In 2018, the Thai Government released a new law: the Contract Farming Promotion and Development Act (FAO, 2018). Also in South-east Asia, after the Vietnamese government had introduced land laws, granting farmers land-use rights, a national policy was adopted in 2002 to promote contract farming (Nguyen, 2005). The policy included state enterprises, foreign companies and oral contracts between farmers, private merchants and service cooperatives. All major agricultural products were included. As of today, some 90% of cotton and fresh milk are purchased through farming contracts in Vietnam (Phan, 2022). In Kenya, about 60% of sugarcane is produced through this mode. Contract farming is now used by one-quarter of Kenyan farmers. In recent years, it has gained popularity in avocado farming. Contract-growing avocado farmers are earning 35% more income than those selling their produce in the open market (Davis, 2023).

Foreign Direct Investment

Global foreign direct investment (FDI) in 2021 was US$1.58 trillion, up 64% from the low pre-pandemic level in 2020 according to the World Investment Report (UNCTAD, 2022). The recovery benefitted all regions. Almost three-quarters of growth was concentrated in developed economies, more than double the 2020 level. The flow of investment into low-income countries was at a record high level (30%) (Table 4.4), largely due to strong growth in Asia. According to the International Monetary Fund (IMF), the United States was the largest recipient of FDI in 2019

Table 4.4. Foreign direct investment inflows globally and in regions in selected years between 2008 and 2021 (US$) (World Investment Report; UNCTAD/FDI/MNE database).

Area	2008	2014	2020	2021
Globally (US$ trillion)	1.8	1.2	1.0	1.6
Regionally (US$ billion)				
Africa	53	54	40	83
Asia	249	465	535	619
Latin America and the Caribbean	126	159	88	134

and 2020. Major investors were Japan, Germany and the Netherlands.

Together with private bank loans, the TNC investments have grown much more than development aid or multilateral bank lending. Low commodity prices, structural adjustment and unemployment have led many governments in low-income countries to search for TNCs to assist in their development and in privatizing public-sector industries. These low-income countries have offered market expansion and lower wages to the TNCs, together with fewer health and environmental regulations than industrialized nations.

Even though the FDI flows to developing economies grew more slowly than those to developed regions they still increased by 30% according to World Investment Reports (Table 4.4). The increase was the result of sturdy growth in Asia, certain recovery in Latin America and the Caribbean, and an increase in Africa. In 2021, European investors were the largest holders of foreign assets in Africa, led by the United Kingdom and France (Oluwole, 2022).

With regard to investments in agriculture, the Chinese Go Out Policy has led Chinese companies to be more active in purchasing arable land in other countries around the globe. Chinese investments have recently increased in the Mekong region of South-east Asia on maize, sugar and rubber production, including processing, purchasing and trade. Chinese investors have also secured several large-scale land concessions in Cambodia (Grimsditch, 2017). In Laos, the number of land concessions is smaller since companies have signed contract farming or land rental agreements with local landholders. Moreover, Chinese companies, supported by the Chinese Communist Party, are buying food companies.

In Mexico, Grupo Bimbo has expanded across Latin America and the Caribbean over many years to become the world's largest baked-goods company. It started in 1943 by introducing the Super-Pan Bimbo, wrapped in cellophane. It became the largest bakery in Latin America in the late 1970s and spread to South America from 2000 (International Directory of Company Histories, 1998). Thereafter, it has expanded into the East Asian market and US coastal areas. The company expanded into Brazil after the Chocolates La Corona and El Globo bakeries

were integrated into the Group (Grupo Bimbo, 2022). According to the consulting agency Kantar, Bimbo occupied first place among the preferred brands in Central and South America in the food category. Also, it ranked first in the category of Fast Moving Consumer Goods in a review of 23,500 brands from 54 countries (Kantar, 2021).

Zambeef is a young company, starting in a butcher shop in Lusaka in 1991 and noted on the Zambia Stock Exchange in 2003. It is involved in the production, processing, distribution and retailing of beef, chicken, milk, dairy products, stock feed and flour. It also grows maize, soybeans and wheat on some 21,000 ha as fodder for its meat production. The company, with a staff of around 7000 Zambians, has recently expanded to operate two butcheries in Nigeria and Ghana. Current plans indicate expansion and the doubling of Mponge Farm's cropping capacity according to Zambeef's Annual Report (2022).

Some Trade-offs Involving Transnational Corporations

Overall effects

Too often, TNCs are viewed either as the heroes or the villains of the globalized economy. Critics consider them as threats to national identities and wealth. They have been accused of neglecting national laws and of exploiting cheap labour. Nonetheless, national governments compete for FDI from such companies.

In one analysis, the overall conclusion was that multinational enterprises are a force for the promotion of prosperity in the world economy (Navaretti and Venables, 2004). This is most likely accurate but confined to economics and with less concern for the social and environmental aspects of sustainability. TNCs operate to grow with the objective of making the highest profit possible, and this economic prosperity is seldom shared equally. Low-income countries have been left behind since 2000.

In recent decades, global economic growth has been achieved, but with negative consequences for the environment and hardly in a sustainable manner. The TNCs have played an influential role in greenhouse gas emissions from the industrial sectors they are involved in, which is seldom observed but has been noted for some time (Rowlands, 2021). It was claimed that the largest 500 TNCs generate more than half of the greenhouse gas emissions produced annually (Elliott, 1997). This was actively denied by the companies (Oreskes and Conway, 2010). Currently, 13 out of the world's top 50 largest companies are oil and gas producers. In 2021, global CO_2 emissions from energy combustion and industrial processes reached their highest ever annual level; a 6% increase from 2020 (IEA, 2022).

The influence of transnational corporations at the national level

The operations of TNCs may offer a certain political stability in a low-income country. They can contribute to the industrial sector, giving rise to new employment opportunities for national citizens. This may contribute to improving the education and the skills of the employees. Consumers will have access to a choice of more products from an international market. If they are cheap, they may benefit the local consumers, but locally produced goods can be hampered, and even the local culture is affected.

TNCs can also contribute to the improvement of infrastructure for transportation and airports. They can assist in the exploitation of national resources, especially where TNCs may prefer to follow local requirements which are less strict than international rules. This applies to the TNCs' involvement in agriculture in low-income counties, where lower wages are attractive. TNCs can operate from various countries, which makes it difficult for partner governments to control any negative effects. If problems arise, a TNC can suddenly switch operations to another country.

Low-income countries are open to TNC involvement in agricultural production but there are differences between individual countries. So far, TNCs have been less inclined to invest in the production of staple foods, improve national food security and provide food sovereignty. Since a corporation is usually dominating the market, it may prevent local small-scale businesses from competing. So far, there is no clear trend discernible as to whether there are overall positive impacts of TNC involvement in agriculture in low-income countries. When TNCs promote modernization of the current type of agriculture, its long-term impact is likely to accelerate a reduction in farm employment.

Current agricultural innovations of large TNCs are not geared towards staple food producers in low-income countries and not towards sustainability. One particular concern is the asymmetry in the relationship between numerous small farmers and a restricted number of large buyers, raising serious issues of competition.

TNC involvement also raises social and political issues, for example, when companies own, or control large tracts of agricultural land. One example is one of the world's largest palm oil producers, the Malaysian company, Sime Darby. It signed a 63-year lease agreement with the Liberian government in 2009 to invest in an oil palm plantation of some 311,000 ha to be planted on the farmland and forests used by local communities. The company planned to provide employment for around 30,000 Liberians and cultivate 220,000 ha within twenty years (Friends of the Earth, 2013). But local farmers and communities had not been consulted before their lands were taken which led to protests, reaffirming their land ownership. The agreement for land concessions had also violated international guidelines for multinational corporations to which Liberia has been a signatory (OECD, 2011: Lomax et al., 2012). They include the International Covenant on Civil and Political Rights (ICCPR), the International Covenant on Economic, Social and Cultural Rights (ICESCR), and the International Convention on the Elimination of all Forms of Racial Discrimination (ICERD). Only 10,300 hectares were planted by 2020, when Sime Darby Plantation (Liberia) Inc. was sold to Liberian Mano Palm Oil Industries Limited. The deal was completed for a cash consideration of US$1, plus an earn-out payment to be determined by the average future crude palm oil price and future crude palm oil production of Sime Darby Plantation (Liberia) in 2022 (The Star, 2020).

The influence of transnational corporations on politics

Long ago, TNCs could seek direct assistance from governments in industrialized countries to protect their interests in low-income nations. Early examples were when the US invaded Guatemala in 1954 to prevent the government taking the unused land of the United Fruit Company for redistribution to peasants (Swamy, 1980), or the failed attempts by International Telephone and Telegraph (ITT) to finance the US Central Intelligence Agency to fund a campaign against Salvador Allende in the Chilean national elections (Barnet and Müller, 1974). These events led some low-income countries of the United Nations to initiate the drafting of a Code of Conduct to establish guidelines for corporate behaviour (Deng *et al.*, 2009).

A second initiative was the request by the UN Secretary-General at the World Economic Forum in 1999. The objective was to create international principles aimed at companies on human rights, labour standards, the environment and corruption. A UN Global Compact was launched with nine principles: a set of core values. Only at the first Global Compact Leaders' Summit in 2004, was corruption added in accordance with the United Nations Convention Against Corruption, adopted in 2003, a legally binding instrument against universal anti-corruption. Most UN member states are parties to this Convention. The Global Compact has given top priority to the UN Sustainable Development Goals, which have been accepted by several TNCs, such as Cargill Inc.

In 2011, the Organisation for Economic Co-operation and Development (OECD) adopted similar guidelines, providing non-binding principles and standards for responsible business conduct in a global context in consistency with applicable laws and internationally recognized standards. The guidelines 'reflect expectations from governments to businesses on how to act responsibly' (OECD, 2011). They also include human and labour rights, the environment, bribery, consumer interests, science and technology, competition and taxation.

Today, TNCs can exert leverage in many ways in individual countries. This can be done by employing government officials, participating in national economic policy making committees, offering financial contributions to political parties and even by bribery. Evidence has suggested that a Monsanto employee might have been involved in the writing of Malawi's new seed policy of 2018, requiring high-quality seed from commercial companies, as noted by the biotech industry (ISAAA, 2018). But some 80% of the seed for crops grown in Malawi comes from local seed, which the Food and Agriculture Organization (FAO) had underlined. The importance of local crops and varieties for food and nutrition security was neglected and the legislation promoted the commercial seed sector (Andersen *et al.*, 2022). At the same time, improved seeds are vital for increased agricultural production.

The influence of transnational corporations on international negotiations

Since the 1980s, TNCs have become actively involved at international political negotiations. They were lobbying early on for reduced barriers to trade and the investment of capital flows to get free markets, such as the EU, North American Free Trade Agreement (NAFTA) and the General Agreement on Trade and Tariffs (GATT). TNCs gave great importance to the Agreement on Trade-Related Aspects of Intellectual Property Rights (TRIPS), negotiated during the Uruguay Round of the GATT. This agreement facilitated the corporations in privatizing and patenting the plant and other genetic resources of low-income countries, rich in biodiversity.

Another instrument was the Agreement on Trade-Related Investment Measures (TRIMS). Its rules restrict preference being given to domestic firms and consequently enable international firms to operate with less difficulty in foreign markets. Policies and trade balancing rules that have been traditionally used to both favour the interests of domestic industries and fight against restrictive business practices were banned. They came under the authority of the World Trade Organization (WTO), which hardly benefitted the low-income countries. On the other hand, the WTO assisted in the creation of the current globalized trade structures according to the

OECD. Until recently, trade rows between nations have been prevented from escalating into large global trade wars.

TNCs have claimed that deregulated trade and investment will produce enough growth to end poverty and generate resources for environmental protection, including in low-income countries (Greer and Singh, 2000). This has been strongly advocated by many politicians in wealthy countries, although the relative power of TNCs has increased whereas the policy-making strength of nations has weakened. The profits of private companies are expected to be reinvested in increased production to increase common prosperity, but the political leadership in low-income countries seems to forget that TNCs and their shareholders have a constant need for more profits.

Today's capitalism is based on the present at the expense of the future. It assumes that global resources are in inexhaustible. Unrestricted free trade and investment-based growth has led to overexploitation of land and natural resources in many countries. This trend is alarming but nature itself can react indirectly in 'revolt' against the ecological effects caused by rapacious human activity, such as global warming, and it may be too late for humans to reverse the reaction, or people can react directly by revolting against investments for more growth causing pollution as high price for short-term economic progress. This was highlighted long ago by World Bank staff members Daly and Goodland (1993) in a working document: 'the dream that growth will raise world wages to the current rich country level, and that all can consume resources at the US per capita rate, is in total conflict with ecological limits that are already stressed beyond sustainability'. That view was, however, not repeated in official World Bank documents. Nonetheless, a similar view was expressed by the American-British economist Kenneth E. Boulding when he concluded that: 'anyone who believes in unlimited growth on a finite planet is either an idiot or an economist'.

Still, modern economics claim that human beings are innately possessed of infinite wishes. Thus, ruling politicians must facilitate endless growth and consumption irrespective of the limits of the natural resources and nature's capability. This is a drastic change from earlier times when generations saved and circulated resources to be able to live within nature's constraints. World leaders and politicians still believe in modern economic growth; an approach that has characterized human beings only during 0.0007% of human history (Jonsson and Wennerlind, 2023). They have suggested a search for alternative ways to get a flourishing Earth, although they have not provided any concrete proposals.

Also, history gives some perspective on free trade when low-income countries were not yet industrialized. In the 18th century, Britian was the first nation to industrialize but – for quite some time – strongly controlled trade, prices and investments. After that period, the country was safe to open for international competition and trade and could avoid exploitation. This may serve as a lesson.

TNC lobbying has become common at UN meetings, for example, at the 1992 United Nations Conference on Environment and Development (UNCED) in Rio de Janeiro. Some sections of the summit's key documents were undermined and TNC pressure led to the removal of some proposals to regulate the practices of global corporations (Greenpeace, 1992). Another example was the UN climate meeting in Glasgow (COP 26) in 2021. Over 100 fossil-fuel companies were represented by 30 trade associations and membership organizations (Corporate Accountability, 2021). The fossil-fuel lobby alone had around 500 delegates, arriving in individual private jet planes. This figure was based upon an analysis of the UN's provisional list of named attendees at COP 26 by Corporate Accountability, Corporate Europe Observatory (CEO), Glasgow Calls Out Polluters and Global Witness. The fossil-fuel companies were also present at the 28th meeting on climate issues (COP28) in Dubai among 80,000 other delegates and participants.

Transnational corporations and controversies

Over the years, TNCs have been associated with many controversies, facing criticism and boycotts. Some examples can be given to illustrate this. One was the marketing of baby formula as an alternative to breastfeeding in low-income countries. Nestlé began aggressively marketing infant formula as superior to breast milk (Hicks, 1981). The company was accused of unethical marketing practices (Dobbing, 1988) and causing infant deaths among mothers without clean water access. It was criticized since water is usually scarce and such marketing conflicted with Nestlé's promotion of bottled water. Another criticism related to Nestlé's reliance on child labour in the early 2000s in cocoa production on plantations in the Ivory Coast and Ghana (Wolfe, 2005). Later, the NGO Mighty Earth found that cocoa used in chocolate produced by Nestlé and other companies was grown illegally in national parks and protected areas in those two countries (Wessel and Foluke Quist-Wessel, 2015).

Monsanto, which is now owned by Germany's Bayer Group, has been accused of selling PCBs, causing health problems in US schools. PCBs were commonly used in caulking, light fixtures and other parts of buildings from the 1950s to the 1970s, according to the Bureau of Climate and Environmental Health in Massachusetts. In 1979, the US government banned PCBs due to their toxicity and after discovering links to cancer. PCBs are one of around 200 substances that are carcinogenic, toxic and difficult to break down. Over time, this has caused Monsanto serious problems, even though the company has claimed that there is no evidence or test results to prove that any individuals have been exposed to harmful levels of PCBs.

In one school, in the state of Washington on the west coast of the United States, former students and parent volunteers have claimed that exposure to PCBs has caused health problems, including brain damage and autoimmune disorders. In December 2023, a jury in the United States ordered Monsanto to pay US$73 million compensation to seven people who claimed to have been exposed to PCB substances and had become ill from them and US$784 million in damages (Picchi, 2023). The verdict concluded that the chemicals were not safe and lacked warnings. The Monsanto company had been negligent and liable for selling the PCBs used in the Sky Valley Education Centre in Washington state.

The PCB case is not the only costly legal battle for Monsanto. It was also accused over its advocacy campaign to get glyphosate approved in the EU, and the problem of glyphosate has continued (see Chapter 9, Pesticides). In 2018, the US Environmental Protection Agency (EPA) had announced that glyphosate was unlikely to be either carcinogenic or have other effects on human health if used as directed. The EU Chemicals Unit has also concluded that glyphosate cannot be classified as carcinogenic. Despite this, Monsanto was sentenced in August 2018 by a jury in the Superior Court of California to pay US$289 million to a dying man with blood cancer. The jury awarded the man US$250 million in punitive damages and about US$39 million in compensatory damages. But Monsanto asked the San Francisco Superior Court Judge to overrule the jury's entire award for punitive damages. Ultimately, the judge reduced the punitive damages award to about US$39 million which was accepted. Some similar cases have been settled but there are over 3000 other Roundup cases awaiting trial (Massey and Llamas, 2024). As a result of litigation over the harmful effects of the herbicide Roundup, Bayer has set aside US$16 billion to cover damages.

Monsanto has also been criticized for its aggressive marketing of genetically modified organisms (GMOs). Bayer and Monsanto, along with donors and banks, have funded a non-profit organization, the International Service for the Acquisition of Agri-biotech Applications (ISAAA) to promote biotechnology (Powerbase, 2010). In 2016, a so-called 'opinion tribunal' was conducted in The Hague against Monsanto. It reported three dangerous substances marketed by Monsanto and criticized the agro-industrial model (International Monsanto Tribunal, 2017).

In 2011, Unilever was fined by the European Commission for establishing a price-fixing cartel for washing powder in Europe, along with Procter & Gamble and Henkel (BBC, 2011). Five years later, Unilever and Procter & Gamble were fined by Autorité de la concurrence in France in for price-fixing on personal hygiene products (The Connexion, 2016).

Syngenta was sentenced by a US court to pay over 7000 Kansas farmers nearly US$218 million because the company had commercialized the Agrisure Viptera GM maize variety before China had approved it. Thus, the American farmers could not export it in 2013 because China refused to accept it (Cronin Fisk and Bross, 2017).

References

ABB (2023) Integrated Report 2023. Zurich, Switzerland. Available at: https://new.abb.com/news/detail/112910/abb-publishes-2023-integrated-report (accessed 25 August 2024).

AkzoNobel (2023) Report 2023. Amsterdam. Available at: https://www.akzonobel.com/content/dam/akzonobel-corporate/global/en/investor-relations-images/result-center/archive-annual-reports/2029-2020/akzonobel-annual-report-2023.pdf (accessed 25 August 2024).

Allen, R.B. (2015) *European Slave Trading in the Indian Ocean, 1500–1850*. Ohio University Press, Athens, OH.

Andersen, A., Vásquez, V.M. and Wynberg, R. (2022) *Improving Seed and Food Security in Malawi: The Role of Community Seed Banks*. FNI Policy Brief 1/2022. Fridtjof Nansen Institute, Lysaker, Norway.

Barnet, R.J. and Müller, R.E. (1974) *Global Reach: The Power of the Multinational Corporations*. Simon and Schuster, New York.

Baten, J. (ed.) (2016) *A History of the Global Economy: 1500 to the Present*. Cambridge University Press, Cambridge, UK.

BBC (2011) Unilever and Procter & Gamble in price fixing fine. Available at: https://www.bbc.com/news/business-13064928 (accessed 18 June 2024).

Blankfeld, K. (2011) JBS: the story behind the world's biggest meatproducer. *Forbes*. Available at: https://www.forbes.com/sites/kerenblankfeld/2011/04/21/jbs-the-story-behind-the-worlds-biggest-meat-producer/ (accessed 25 August 2024).

Cargill (2023) Reimagine what's possible. *2023 Cargill Annual Report*. Cargill Inc. Available at: https://www.cargill.com/about/doc/1432242761261/2023-cargill-annual-report.pdf

Coolgeography.co.uk (n.d.) Transnational corporations. Available at: https://www.coolgeography.co.uk/A-level/AQA/Year%2013/Development%20%26%20Globalisation/TNCs/Transnational_Corporations.htm (accessed 15 June 2024).

Corporate Accountability (2021) Fossil fuel lobbyists outnumber any countrydelegation to COP26. London & Glasgow. Available at: https://corporateaccountability.org/media/release-fossil-fuel-lobbyists-outnumber-any-country-delegation-to-cop26/ (accessed 31 August 2024).

Cronin Fisk, M. and Bross, T. (2017) Syngenta loses $218 million verdict in first GMO trial test. *Bloomberg Media*. Available at: https://www.bloomberg.com/news/articles/2017-06-23/syngenta-ordered-by-jury-to-pay-218-million-to-kansas-farmers (accessed 31 August 2024).

Daly, H. and Goodland, R. (1993) *An Ecological-Economic Assessment of Deregulation of International Commerce under GATT*. World Bank Environment Department, Spring draft, Washington, DC.

Davis, K. (2023) Contract Farming in Kenya: Opportunities,Advantages and Disadvantage – Green Lifo. Available at: https://greenlifo.com/contract-farming-in-kenya/#google_vignette (accessed 25 August 2024).

Deng, H., Higgs, L. and Chan, V. (2009) Redefining transnational corporations. *Transnational Corporations Review* 1(2), 69–80. Available at: https://doi.org/10.1080/19186444.2009.11658195 (accessed 17 June 2024).

Dobbing, J. (1988) Infant Feeding: Anatomy of a Controversy, 1973–1984. Springer, Cham, Switzerland.

Dunning, J. (1993) *Multinational Enterprises and the Global Economy*. Addison-Wesley Publishing Company, Reading, MA.

Elliott, L. (1997) *The Global Politics of the Environment*. Red Globe Press, London.

Emmer, P. (2019) West India Company, Dutch. Encyclopedia.com. Available at: https://www.encyclopedia.com/history/news-wires-white-papers-and-books/west-india-company-dutch (accessed 25 August 2024).

Emmer, P.C. (2011) *De Nederlandse slavenhandel 1500–1850*. Singel Uitgeverijen, Amsterdam.

Erikson, E. (2014) *Between Monopoly and Free Trade: The English East India Company, 1600–1757*. Princeton University Press, Princeton, NJ.

Euromonitor International (2021) Haier Group in Consumer Appliances. Euromonitor International, London.

Farrington, A. (2002) *Trading Places: The East India Company and Asia 1600–1834*. British Library.

Ferguson, N. (2004) *Empire: The Rise and Demise of the British World Order and the Lessons for Global Power*. Basic Books, New York.

Food and Agriculture Organization (FAO) (2018) New contract farming law released in Thailand. Available at: https://www.fao.org/in-action/contract-farming/news-cf/news-detail/en/c/1145200/ (accessed 17 June 2024).

Forbes (2022) Forbes Global 2000 list 2022: The top 200. Forbes Media, New York. Forbes, 2022. Available at: https://www.forbes.com/sites/forbesstaff/2022/05/12/forbes-global-2000-list-2022-the-top-200/ (accessed 25 August 2024).

Fortune (2022) Global 500 Ranking 2022. Available at: https://fortune.com/ranking/global500/2022/ (accessed 17 June 2024).

Fortune (2023) Global 500 Ranking 2023. Available at: https://fortune.com/ranking/global500/2023/ (accessed 17 June 2024).

Fortune (2024) Walmart company profile. Available at: https://fortune.com/company/walmart/ (accessed 17 June 2024).

Friends of the Earth (2013) Sime Darby and land grabs in Liberia. Available at: http://www.foeeurope.org/sites/default/files/press_releases/foee_sime_darby_and_its_eu_financers_240613.pdf (accessed 18 June 2024).

Gazzoni, D.L., Cattelan, A.J. and Nogueira, M.A. (2019) *Does the Brazilian Soybean Production Increase Pose a Threat on the Amazon Rainforest?* Embrapa Soja, Londrina, Brazil.

Greenpeace (1992) *The Greenpeace Book on Greenwash*. Greenpeace International, Amsterdam.

Greer, J. and Singh, K. (2000) *A Brief History of Transnational Corporations*. Global Policy Forum, New York. Available at: https://archive.globalpolicy.org/component/content/article/221-transnational-corporations/47068-a-brief-history-of-transnational-corporations.html (accessed 25 August 2024).

Grimsditch, M. (2017) *Chinese Agriculture in Southeast Asia: Investment, Aid and Trade in Cambodia, Laos, and Myanmar*. Heinrich-Böll-Stiftung Southeast Asia.

Grupo Bimbo (2022) Our Financial Results 2022. Mexico City. Available at: https://bimbo.xdesign.mx/informe-anual/2022/en/financial-results.html (accessed 31 August 2024).

Hicks, G.M. (1981) The infant formula controversy. *Backgrounder* 142, 1–9.

Huawei (2022) Company Facts. Huawei Technologies Co. Ltd., Longgang, Shenzhen. Available at: https://www.huawei.com/en/media-center/company-facts (accessed 25 August 2024).

International Directory of Company Histories (1998) Grupo Industrial Bimbo. International Directory of Company Histories. Available at: https://www.company-histories.com/Grupo-Industrial-Bimbo-Company-History.html (accessed 31 August 2024).

International Energy Agency (IEA) (2022) *Global Energy Review: CO2 Emissions in 2021*. International Energy Agency (IEA), Paris. Available at: https://www.iea.org/reports/global-energy-review-co2-emissions-in-2021-2 (accessed 20 June 2024).

International Monsanto Tribunal (2017) Advisory Legal Opinion. International MonsantoTribunal, The Hague, Netherlands. Available at: https://en.monsantotribunal.org/Conclusions (accessed 31 August 2024).

International Service for the Acquisition of Agri-biotech Applications (ISAAA) (2018) Malawi Releases New Seed Policy. Biotech Updates, International Service for the Acquisition of Agri-Biotech Applications (ISAAA). Available at: https://www.isaaa.org/kc/cropbiotechupdate/article/default.asp?ID=16514 (accessed 20 June 2024).

Investopedia (2022) How Many Multinational Corporations Operate in China? Available at: https://www.investopedia.com/ask/answers/021015/how-many-multinational-corporations-operate-china.asp (accessed 20 June 2024).

Jonsson, F.A. and Wennerlind, C. (2023) *Scarcity. A History from the Origins of Capitalism to the Climate Crisis*. Harvard University Press, Cambridge, MA.

Kantar (2021) Brand Footprint 2021. Kantar Group and Affiliates. London.

Kenyan Tea Development Authority (n.d.) A global leader in quality teas. Available at: https://ktdateas.com/about-us/ (accessed 17 June 2024).

Kraft Heinz (2022) Kraft Heinz Annual Report 2022. Chicago.

Law, R. (1994) The slave trade in seventeenth century Allada: A revision. *African Economic History* 22, 59–92. Available at: https://doi.org/10.2307/3601668 (accessed 20 June 2024).

Lomax, T., Kenrick, J. and Brownell, A. (2012) Sime Darby oil palm and rubber plantation in Grand Cape Mount county, Liberia. Available at: https://www.forestpeoples.org/sites/fpp/files/publication/2012/11/liberiasimedarbyfpic_0.pdf (accessed 31 August 2024).

Maddison, A. (2001) *The World Economy: A Millennial Perspective*. Organisation for Economic Co-operation and Development(OECD), Paris.

Marshall, P.J. (1998) *The Eighteenth Century. The Oxford History of the British Empire*. Oxford University Press, Oxford, UK.

Martin, L.C. (2007) Tea: The Drink that Changed the World. Tuttle Publishing, North Clarendon, VT.

Massey, K. and Llamas, M. (2024) Roundup lawsuit. *Drugwatch*. Available at: https://www.drugwatch .com/legal/roundup-lawsuit/ (accessed 1 November 2024).

Navaretti, G.B. and Venables, A.J. (2004) *Multinational Firms in the World Economy*. Princeton University Press, Princeton, NJ.

Nestlé (2023a) Annual Review 2023. Vevey, Switzerland. Available at: https://www.nestle.com/sites /default/files/2024-02/2023-annual-review-en.pdf (accessed 25 August 2024).

Nestlé (2023b) Creating Shared Value and Sustainability Report 2023. Advancing regenerative food systems at scale. Vevey, Switzerland. Available at: https://www.nestle.com/sites/default/files/2024 -02/creating-shared-value-sustainability-report-2023-en.pdf (accessed 25 August 2024).

Nguyen, H.H. (2005) *Contract Farming in Vietnam*. National Institute of Agricultural Planning and Projection, Ministry of Agriculture and Development, Vietnam. Available at: https://www.fao.org/fileadmin /user_upload/contract_farming/presentations/Contract_farming_in_VietNam.pdf (accessed 25 August 2024).

Olam (2021) *Transforming to Serve a Changing World: Annual Report 20*. Olam, Singapore.

Oluwole, V. (2022) Investment flows to Africa reached a record $83 billion in 2021. *Business Insider Africa*. Available at: https://africa.businessinsider.com/local/markets/investment-flows-to-africa-reached -a-record-dollar83-billion-in-2021/lkxdgw1 (accessed 18 June 2024).

Oreskes, N. and Conway, E.M. (2010) *Merchants of Doubt: How a Handful of Scientists Obscured the Truth on Issues from Tobacco Smoke to Global Warming*. Bloomsbury Publishing, New York.

Organisation for Economic Co-operation and Development (OECD) (2011) *OECD Guidelines for Multinational Enterprises*. OECD Publishing, Paris.

Parthesius, R. (2010) *Dutch Ships in Tropical Waters: The Development of the Dutch East India Company (VOC) Shipping Network in Asia 1595-1660*. Amsterdam University Press, Amsterdam.

Phan, T.T. (2022) Agricultural land conversion and land rights in Vietnam: A case study of farmers' resistance in the peri-urban areas of Hanoi. In: Tran, T.A.-D. (ed.) *Rethinking Asian Capitalism: The Achievements and Challenges of Vietnam Under Doi Moi*. Palgrave Macmillan, Cham, Switzerland.

Picchi, A. (2023) Monsanto ordered to pay US$857m to Washington school students and parentvolunteers over toxic PCBs. *CBS News*. Available at: https://www.cbsnews.com/news/monsanto-verdict -pcb-857-million-washington-school/ (accessed 31 August 2024).

Powerbase (2010) International Service for the Acquisition of Agri-biotech Applications. Available at: https://powerbase.info/index.php/International_Service_for_the_Acquisition_of_Agri-Biotech _Applications (accessed 18 June 2024).

Purseglove, J.W. (1972) *Tropical Crops. Monocotyledons and Dicotyledons*. Longman Group Ltd, London.

Rowlands, M. (2021) World on Fire. Humans, Animals, and the Future of the Planet. Oxford University Press Inc, New York.

Schiff, E. (2016) *Industrialization Without National Patents: The Netherlands, 1869–1912; Switzerland, 1850–1907*. Princeton Legacy Library, Princeton University Press, Princeton, NJ.

Sekulich, T. (2019) Top ten agribusiness companies in the world. *Tharawat Magazine*.

Shell (2022) Annual Report and Accounts 2022. Our People. London. Available at: https://reports.shell .com/annual-report/2022/strategic-report/powering-lives/our-people.html (accessed 25 August 2024).

Skinner, G., Chambers, R. and McGrath, S. (2020) *Transnational Corporations and Human Rights: Overcoming Barriers to Judicial Remedy*. Cambridge University Press, Cambridge, UK.

Staggers, H.O. (1977) Unlawful corporate payments Act of 1977. Report Together with Minority Views To accompany H.R.3815, Nr 95-640, September 28,1977, House Committee on Interstate and Foreign Commerce House of Representatives, Washington D.C. Available at: https://www .justice.gov/sites/default/files/criminal-fraud/legacy/2010/04/11/houseprt-95-640.pdf (accessed 25 August 2024).

Statista (n.d.) Leading soybean producing countries worldwide from 2013/14 to 2023/24. Available at: https://www.statista.com/statistics/263926/soybean-production-in-selected-countries-since-1980/ (accessed 17 June 2024).

Swamy, D.S. (1980) *Multinational Corporations and the World Economy*. Alps Publishers, New Delhi.

TANAP (2006) The End of VOC. Towards a New Age of Partnership (TANAP). Database of VOC Documents. Available at: https://globalise.huygens.knaw.nl/source-corpus

The Connexion (2016) Huge price fixing fine is upheld. Available at: https://archive.ph/20170209165431/ http://www.connexionfrance.com/court-appeal-price-fixing-fine-upheld-consumer-companies -gillette-loreal-procter-gamble-sc-johnson-18567-view-article.html (accessed 18 June 2024).

The Star (2020) Sime Darby Plantation completes sale of Liberia operations. *The Star*. Petaling Jaya. Malaysia. Available at: https://www.farmlandgrab.org/post/29430-sime-darby-plantation-completes -sale-of-liberia-operations (accessed 31 August 2024).

Tracy, J.D. (2015) Dutch and English trade to the East. In: Bentley, J., Subrahmanyam, S. and Wiesner-Banks, M.E. (eds) *The Cambridge World History. Volume VI: The Construction of a Global World, 1400–1800CE, Part 2: Patterns of Change*. Cambridge University Press, Cambridge, UK, pp. 240–262.

United Nations Conference on Trade and Development (UNCTAD) (1994) *The World Investment Report. Transnational Corporations, Employment, and the Workplace*. UNCTAD Division on Transnational Corporations and Investment, Geneva, Switzerland.

United Nations Conference on Trade and Development (UNCTAD) (1999) *World Investment Report*. UNCTAD, Geneva, Switzerland.

United Nations Conference on Trade and Development (UNCTAD) (2008) *World Investment Report 2009*. Transnational Corporations, Agricultural Productionand Development, Geneva, Switzerland.

United Nations Conference on Trade and Development (UNCTAD) (2010) *World Investment Report 2009*. Transnational Corporations, Agricultural Production and Development, Geneva, Switzerland.

United Nations Conference on Trade and Development (UNCTAD) (2020) *World Investment Report 2009*. Transnational Corporations, Agricultural Production and Development, Geneva, Switzerland.

United Nations Conference on Trade and Development (UNCTAD) (2022) *World Investment Report. Transnational Corporations, Employment, and the Workplace*. UNCTAD Division on Transnational Corporations and Investment, Geneva, Switzerland.

US Department of Justice (2020) J&F Investimentos S.A. Pleads Guilty and Agrees to Pay Over $256 Million to Resolve Criminal Foreign Bribery Case. Office of Public Affairs, press release. Available at: https://www.justice.gov/opa/pr/jf-investimentos-sa-pleads-guilty-and-agrees-pay-over-256-million -resolve-criminal-foreign (accessed 25 August 2024).

Walmart (2023) 2023 Annual Report. Walmart Inc. Bentonville, AR. Available at: https://corporate.walma rt.com/content/dam/corporate/documents/esgreport/reporting-data/tcfd/walmart-inc-2023-annual -report.pdf (accessed 25 August 2024).

Wasley, A., Heal, A., Michaels, L., Phillips, D., Campos, A. *et al.* (2019) JBS: The Brazilian butchers who took over the world. *The Bureau of Investigative Journalism, in partnership with Repórter Brasil*. Available at: https://www.thebureauinvestigates.com/stories/2019-07-02/jbs-brazilian-butchers-took -over-the-world/ (accessed 25 August 2024).

Wessel, M. and Foluke Quist-Wessel, P.M. (2015) Cocoa production in West Africa: A review and analysis of recent developments. *NJAS: Wageningen Journal of Life Sciences* 1–7. Available at: https://doi .org/10.1016/j.njas.2015.09.001 (accessed 20 June 2024).

Williams, E.E. (1964) *Capitalism and Slavery*. Andre Deutsch, London.

Williamson, J.G. and Clingingsmith, D. (2004) *India's Deindustrialization in the 18th and 19th Centuries*. Global Economic History Network, London School of Economics.

Wolfe, D. (2005) *Naked Chocolate: The Astonishing Truth about the World's Greatest Food*. Simon & Schuster, New York.

Woodruff, P. (1954) *The Men Who Ruled India: The Founders of Modern India*, Vol.1. St. Martin's Press, New York.

Zambeef (2022) Annual report 2022. Lusaka, Zambia. Available at: https://zambeefplc.com/annual -reports/ (accessed 18 June 2024).

5 National Political Action Instead of Rhetoric

Abstract

After several UN conferences regarding food and agriculture in the 1970s, the world's governments have regularly reiterated within the UN the need to reduce hunger, malnutrition and poverty. Experience shows that political declarations are inadequate if not implemented within an effective national agricultural policy. In 2000, the world's leaders formulated a political vision to fight poverty, translated into eight Millennium Development Goals (MDGs). Fifteen years later, extreme income poverty had been halved, although China accounted for most of the reduction. However, millions of people were still hungry. In 2015, the international community adopted the UN Agenda 2030 with 17 Sustainable Development Goals (SDGs). Poverty and hunger should be eradicated by 2030. The SDGs aim at economic, social and environmentally sustainable development in all countries. In 2021, the UN convened a Food Systems Summit, launching actions for progress on the SDGs, but the World Bank projected that 8.6% of the population lived in extreme poverty. The insecurity of these recent times thus justify a need for a Global Board for Food Security.

Food for All in 2050

Agricultural development is one of the powerful tools that can be used to end extreme poverty and it is crucial to economic growth. According to earlier estimates by the Food and Agriculture Organization (FAO) in 2009, the world will need to produce 60% more food to feed a world population of 9.3 billion in 2050. This was a first step in preparing for the World Summit on Food Security with world leaders in late 2009. This challenge is amplified by the fact that the middle-class population is growing. This implies increased demand for meat over grain, legumes and wheat. In addition, general consumption will increase. At the beginning of the 1980s, the average world citizen had 0.5 ha at their disposal for food production. The corresponding figure for 2020 was estimated at 0.09 ha, an ordinary Swedish villa plot! That number will be significantly lower when planet Earth houses another 2 billion more people by 2050.

The challenge is huge since the current food system accounts for 26% of total global greenhouse gas emissions (Ritchie *et al.*, n.d.). Agriculture, forestry and other types of land use make up 18% whereas the rest comes from packaging, refrigeration and transport. Water is of great global importance because 90% of all food comes from cultivated plants on the ground and only 10% from the oceans and inland waters (Crippa *et al.*, 2021). New policies, together with agricultural technologies, must give more emphasis to promoting dietary diversity and reducing environmental externalities (Qaim, 2017).

A business-as-usual approach would take too heavy a toll on global natural resources. Future agricultural techniques must be more in tune with agricultural ecosystems, minimizing the use of chemical inputs and assisting farmers in coping with adaptation to a changing climate and in enhancing their resilience. Food sovereignty must be the goal, providing access to nutritious food for all. This kind of farming must be practical and accessible to many small-scale farmers by being adapted to the varying conditions they face, and also harnessing traditional knowledge. In future, industrial-scale farmers must be encouraged to show much greater environmental awareness and given incentives to adopt sustainable practices or be penalized for using unsustainable ones.

Addressing food loss and waste can improve food and nutrition security, since one-third of food is lost or wasted. Food insecurity can

DOI: 10.1079/9781800627802.0005

worsen diet quality and lead to undernutrition. The UNICEF report *The State of Food Security and Nutrition in the World 2020* urged the transformation of food systems to reduce the cost of nutritious foods and increase the affordability of healthy diets. This must be implemented with specific solutions for individual countries, and even within them. Governments were called upon to mainstream nutrition in their approach to agriculture. This means cutting cost-escalating factors in the production, storage, transport, distribution and marketing of food, and includes a reduction of food loss and waste. Local small-scale producers should receive greater support to grow and sell more nutritious foods and to secure better access to markets (UNICEF, 2020).

UN Attempts at Action

For decades, studies by researchers and reports from influential actors, such as the World Bank and UN agencies, have theoretically described how poverty should be measured and defined. Since the 1940s, there have been specialized UN agencies dealing with food production and security.

The main purpose of the Food and Agriculture Organization (FAO) is to make sure people have regular access to enough high-quality food for healthy lives. Its goals are eradication of hunger, food insecurity and malnutrition, the elimination of poverty, the driving force of economic and social progress for all, and the sustainable management and utilization of natural resources for the benefit of present and future generations. The FAO also issues a food price index.

Investment in agriculture and rural development to boost food production and nutrition is a priority for the World Bank Group. It works with partners to improve food security and build a food system to feed all. Activities include encouraging climate-smart farming techniques and restoring degraded farmland, breeding more resilient and nutritious crops, and improving storage and supply chains for reducing food losses (UN, 2023). The World Bank has made up to US$30 billion available as part of the global response to the food crisis (World Bank, 2022a).

The International Fund for Agricultural Development (IFAD) focuses on rural poverty reduction in rural populations in developing countries. IFAD has supported about 518 million people in rural areas since 1978 (IFAD, 2021a; McCarthy, 2021). The World Food Programme (WFP) aims to bring food assistance to more than 80 million people in some 80 countries. It is continually responding to emergencies (UN, 2023). The WFP raised a record of US$14.2 billion in 2022.

Early major food conferences

After the Second World Food Congress in 1970, the UN Conference on the Human Environment in 1972 and the First World Food Conference in 1974, the world's governments have regularly reiterated, within the UN, the need for concrete measures to reduce hunger, malnutrition and poverty. Experience shows that such political declarations are inadequate if they are not supplemented by actions within a framework of an effective national agricultural policy. Countries have different preconditions for agriculture. The US is one of the largest exporters of food, while others import their food, for example, Saudi Arabia (80%) and Sweden (50%).

Since 1974, there has been a special UN committee on food security. In hindsight, its influence seems to have been marginal. The Committee on World Food Security (CFS) was reformed in 2009. The CFS develops and endorses policy recommendations and guidance on a wide range of food security and nutrition topics (IFAD, 2021b). They are founded upon scientific and evidence-based reports from a High-Level Panel of Experts and/or supported technically by the FAO, the IFAD, the WFP, and representatives of the CFS Advisory Group (Food and Agriculture Organization (FAO), 2023).

Millennium Development Goals (MDGs)

In 2000, the world leaders at the UN formulated a political vision to fight poverty, translated into 8 MDGs with 18 specific targets (UN, 2000). According to the MDG Report 2015 (UN, 2017) the 8 goals of the Millennium Declaration in 2000 was a positive development. Extreme income poverty was halved compared to the 1990 level, although only a few countries, such as China, accounted for the larger share of the reduction. However, millions of people were still hungry.

Highlights (UN, 2017)

a. The number of people living in extreme poverty declined from 1.9 billion in 1990 to 836 million in 2015.
b. The number of people living on more than US$4 a day nearly tripled between 1991 and 2015.
c. The proportion of undernourished people in developing regions dropped by almost half since 1990.
d. The number of out-of-school children of primary school age worldwide fell from 100 million in 2000 to an estimated 57 million in 2015.
e. Gender parity in primary schools has been achieved in most countries.
f. The mortality rate of children under 5 years was cut by more than half since 1990.
g. Maternal mortality has fallen by 45% worldwide since 1990.
h. Over 6.2 million malaria deaths have been averted between 2000 and 2015.
i. New HIV infections fell by approximately 40% between 2000 and 2013. By June 2014, 13.6 million people living with HIV were receiving antiretroviral therapy: an increase from around 800,000 in 2003.
j. Tuberculosis prevention, diagnosis and treatment interventions saved an estimated 37 million lives between 2000 and 2013.
k. Worldwide, some 2.1 billion people gained access to improved sanitation.
l. Globally, 147 countries met the MDG drinking water target, 95 countries met the MDG sanitation target and 77 countries met both.
m. Official development assistance increased 66% in real terms from 2000 and 2014, reaching US$135.2 billion.

Shortfalls

In preparing for Agenda 30 and the Sustainable Development Goals (SDGs) there were detailed discussions within the UN of the progress made by the MDGs, but several shortfalls remained as illustrated in the MDG Goals Report 2015 (UN, 2017). There was, however, no mention of the specifics of prominent shortfalls. One target of the MDGs was to 'halve, between 1990 and 2015, the proportion of people who suffer from hunger'. Yet the Global Hunger Index (International Food Policy Research Institute (IFPRI), 2015) reported there were 795 million hungry people in the world at that time. There was no progress on the target of controlling population growth. Other targets that were not met included 'to halve, between 1990 and 2015, the proportion of people who suffer from hunger' and to 'have achieved by 2020 a significant improvement in the lives of at least 100 million slum dwellers'. Finally, one of the key targets was 'to implement the concept of sustainability and integrate the principles of sustainable development into country policies and reverse the loss of environmental resources'. The concept of sustainability was the one that world leaders had agreed to at the Earth Summit (UN, 1993).

The Global Agriculture and Food Security Program (GAFSP)

Agriculture and food security have recently gained more political importance on the international policy agenda. One major development was the establishment of the UN High-Level Task Force (HLTF) on the Global Food Security Crisis in 2008. Partners included 22 different organizations, funds, programmes and other entities within the UN family, the Bretton Woods institutions, the World Trade Organization (WTO), and the Organisation for Economic Co-operation and Development (OECD).

The aim of the HLTF was to create a prioritized plan of action to address the global food crisis and coordinate its implementation. The HLTF developed the Comprehensive Framework for Action (CFA). It was to address the threats and opportunities resulting from rising food prices, create policy changes to avoid future food crises, and contribute to country, regional and global food and nutritional security (FAO, 2008). The GAFSP was designed 'to build resilient and sustainable agriculture and food systems in low-income countries'. The fund included over US$1.5 billion in grant financing to countries, US$476 million to support private sector development, and US$47 million for farmers' or producer organizations. More than 16 million people, including 7 million women, have benefitted from the public sector support. In early 2023, the GAFSP announced US$220 million in new agricultural investment grants for 15 national governments in low-income countries (World Bank, 2023).

Zero hunger challenge

In 2012, the UN Secretary-General launched the Zero Hunger Challenge during the Rio + 20 World Conference on Sustainable Development. It was to inspire a global movement towards a world free from hunger within a generation. It calls for zero stunted children under the age of two; full access to adequate food all year round; all food systems to be sustainable with a 100% increase in smallholder productivity and income; and no loss or waste of food (UN, 2023).

Sustainable Development Goals (SDGs)

In 2015, the international community adopted the SDGs (UN, 2015); a continuation of the eight MDGs with one significant difference. This UN Agenda 2030 aims at economic, social and environmentally sustainable development in both high-income and low-income countries. All forms of poverty and hunger should be eradicated everywhere by 2030. Moreover, the SDGs enlist 169 specified subgoals focusing on the environment, sustainability, gender equality, growing gaps and issues such as governance and basic human rights. Some of the subgoals are extremely ambitious (Annex II). This will require a profound change of not only the global food and agriculture system but also the whole of society.

Certain observers find the SDGs ambitious, too detailed and question their full implementation in the current political reality. This is reflected in the fact that the world is making little progress on the SDGs. None of the SDG targets relate to reduced consumption and how it may be achieved in wealthy societies. This implies a naïve political hope for sustainable growth without any profound consequences for the present generation. The high number of subgoals seems far from a practical approach and more of a political illusion, considering the difficulties for several low-income countries in managing such a statistical challenge.

In 2020, lockdowns imposed to limit the spread of COVID-19 pushed 166 countries into recession (World Bank, 2020). This affected the number of hungry people as shown by the World Bank´s Poverty and Inequality Platform (PIP). It is an interactive computational tool, providing quick access to the World Bank's estimates of poverty, inequality and shared prosperity for most economies worldwide. The April 2022 update to the PIP involved several changes to the data underlying the global poverty estimates (Aguilar et al., 2022). In consequence, the global poverty estimate for 2018 was revised, showing a rate of 8.6%, down from the previously reported 9.1% (World Bank, 2022b). However, this increasing global poverty trend, continued to 8.9% in 2019 and 9.7% in 2020, due to the COVID-19 pandemic; the first increase in global poverty in decades (Aguilar et al., 2024). In 2023, 691 million people were projected to live in extreme poverty (below US$2.15/day) according to the World Bank (Yonzan et al., 2023). This meant little progress in reducing poverty since 2019.

In 2021, an estimated 44% of people below the poverty line lived in countries on the World Bank's fragile and conflict-affected situations list. Two-thirds of the poor live in middle-income countries on less than US$6.85 per day (World Bank, 2024). To help turbocharge implementation of the SDGs in these contexts, the UN required flexible and pooled funding platforms that can incentivize and facilitate integrated programming across sectors. Then, long-term funding is necessary but multilateral institutions of the UN System, have seen a decline in the share of assessed and core funding (UN, 2022). This undermines the UN's multilateral character and hampers its ability to rapidly address critical development needs. Also, it weakens the UN's leadership role.

Food Systems Summit

In 2021, the UN Secretary-General convened a Food Systems Summit to launch actions to deliver progress on all 17 SDGs. Participants included key players from the worlds of science, business, policy, health care and academia, as well as farmers, Indigenous people, youth organizations, consumer groups and environmental activists (UN, 2023).

At the political level, one means of strengthening the resilience of the global food system would be to elevate this issue systematically to the heads of state. This means that future food security should be placed at the same level as global macroeconomic stability. The latter is handled by the Financial Stability Board, established in 2009. It gives impartial advice to national governments on how to ensure the stability and resilience of the financial system (Davey, 2022). Today, there is still no equivalent board for food security. In these critical times, it thus seems highly justified that there be one global mechanism for guidance instead of a number of different UN agencies, each one with a specific mandate.

For those who have followed and even participated in these international events and conferences it is sad to note that most of the same major development issues remain on the international agenda. When new brains enter the arenas and meeting tables, calls for new conferences are inevitable. Instead, there is a need to reconsider past conclusions and implement those that are still relevant as practical action.

Agrarian policy takes a long time to implement. The time perspective is often neglected by politicians, in contrast to farmers. They are aware of the ecological nature of practical farming operations. There is no need for more rhetoric but there is a need for realistic action on urgent priorities, instead of ad hoc, quick solutions at times of crisis.

References

Aguilar, R. A. C., Aron, D. V., Diaz-Bonilla, C., Betran, M.G.F., Foster, E.M. *et al.* (2022) April 2022 update to the Poverty and Inequality Platform (PIP). Global Poverty Monitoring Technical Note. Development Data Group, World Bank Group, Washington DC. Available at: https://documents1.worldbank.org /curated/en/099422404082231105/pdf/IDU02070690808ee7044720a1e6010de398e6a75.pdf (accessed 31 August 2024).

Aguilar, R.A.C., Diaz-Bonilla, C., Fujs, T.H., Jolliffe, D., Lakner, C. *et al.* (2024) March 2024 global poverty update from the World Bank: First estimates of global poverty until 2022 from survey data. World BankData Blogs. World Bank Group, Washington DC. Available at: https://blogs.worldbank.org/en/ opendata/march-2024-global-poverty-update-from-the-world-bank--first-esti (accessed 31 August 2024).

Crippa, M., Solazzo, E., Guizzardi, D., Monforti-Ferrario, F., Tubiello, F.N. *et al.* (2021) Food systems are responsible for a third of global anthropogenic GHG emissions. *Nature Food* 2, 198–209. Available at: https://doi.org/10.1038/s43016-021-00225-9 (accessed 31 August 2024).

Davey, E. (2022) Globalisation of food production has left millions hungry. *The Guardian*. Available at: https://www.theguardian.com/world/2022/may/23/globalisation-of-food-production-has-left-millions -hungry (accessed 20 June 2024).

Food and Agriculture Organization (FAO) (2008) *High-Level Task Force on the Global Food Crisis: Comprehensive Framework for Action, July 2008*. FAO, Rome. Available at: https://www.fao.org /fileadmin/templates/worldfood/Reports_and_docs/FINAL_20CFA_20July_202008.pdf (accessed 20 June 2024).

Food and Agriculture Organization (FAO) (2009) *How to Feed the World in 2050*. FAO, Rome. Available at: https://www.fao.org/fileadmin/templates/wsfs/docs/expert_paper/How_to_Feed_the_World_in _2050.pdf (accessed 31 August 2024).

Food and Agriculture Organization (FAO) (2023) Committeeon World Food Security Communication Strategy 2020-2023. FAO, Rome. Available at: https://www.fao.org/fileadmin/templates/cfs/Docs1718 /Evaluation/Updated_CFS_Communication_Strategy_2018_Final_Draft.pdf (accessed 20 June 2024).

International Food Policy Research Institute (IFPRI) (2015) *2015 Global Hunger Index: Armed Conflict and the Challenge of Hunger*. Published jointly by IFPRI, Washington DC;Concern Worldwide, Dublin; Welthungerhilfe, Bonn, Germany; and the World Peace Foundation/Tufts University. Available at: https://www.globalhungerindex.org/pdf/en/2015.pdf (accessed 31 August 2024).

International Fund for Agricultural Development (IFAD) (2021a) *IFAD Annual Report 2020*. IFAD, Rome. Available at: https://www.ifad.org/en/ar2020/ (accessed 31 August 2024).

International Fund for Agricultural Development (IFAD) (2021b) Committee on world food security (CFS). IFAD, Rome. Available at: https://www.ifad.org/en/web/latest/-/committee-on-world-food-security-49 (accessed 20 June 2024).

McCarthy, J. (2021) Countries worldwide are stepping up to support rural farmers living in poverty. *Global Citizen*. Available at: https://www.globalcitizen.org/en/content/farmers-rural-poverty-ifad -replenishment/ (accessed 31 August 2024).

Qaim, M. (2017) Globalization of agrifood systems and sustainable nutrition. *Proceedings of the Nutrition Society* 76(1), 12–21. Available at: https://doi.org/10.1017/S0029665116000598 (accessed 20 June 2024).

Ritchie, H., Rosado, P. and Roser, M. (n.d.) Environmental impacts of food production. *Our World in Data*. Available at: https://ourworldindata.org/environmental-impacts-of-food (accessed 21 June 2024).

United Nation International Children's Emergency Fund (UNICEF) (2020) *The State of Food Security and Nutrition in the World 2020*. UNICEF, New York. Available at: https://www.unicef.org/reports/state -of-food-security-and-nutrition-2020 (accessed 20 June 2024).

United Nations (UN) (1993) Report of the United Nations Conference on Environment and Development. UN, New York. Available at: https://documents.un.org/doc/undoc/gen/n92/836/55/pdf/n9283655 .pdf (accessed 31 August 2024).

United Nations (UN) (2000) *Development Goals for a New Millennium*. UN, New York. Available at: https://www.un.org/en/conferences/environment/newyork2000 (accessed 21 June 2024).

United Nations (UN) (2015) *Transforming our World: The 2030 Agenda for Sustainable Development*. New York, UN. Available at: https://sdgs.un.org/2030agenda (accessed 31 August 2024).

United Nations (UN) (2017) The Millennium Development Goals Report 2015. UN, New York. Available at: https://www.un.org/millenniumgoals/2015_MDG_Report/pdf/MDG%202015%20rev%20(July%201) .pdf (accessed 21 June 2024).

United Nations (UN) (2022) Financing the UN Development System: Joint Responsibilities in a World of Disarray. New York, UN. Available at: https://financingun.report/annual-report/financing-un-develop ment-system-joint-responsibilities-world-disarray-2022 (accessed 20 June 2024).

United Nations (UN) (2023) *Global Issues: Food*. New York, UN. Available at: https://www.un.org/en /global-issues/food (accessed 20 June 2024).

World Bank (2020) Global Economic Prospects: June 2020. World Bank Group, Washington DC. Available at: https://documents.worldbank.org/en/publication/documents-reports/documentdetail/50299159 1631723294/global-economic-prospects-june-2020 (accessed 31 August 2024).

World Bank (2022a) Annual Report. Helping Countries to Adapt to a Changing World. Washington, D.C.

World Bank (2022b) *Poverty and Inequality Platform*. Washington, D.C. Available at: https://pip.worldbank .org/home (accessed 20 June 2024).

World Bank (2023) Global agriculture and food security program provides new funding for countries to strengthen the resilience of smallholder farmers. ReliefWeb. Available at: https://reliefweb.int/report /world/global-agriculture-and-food-security-program-provides-new-funding-countries-strengthen -resilience-smallholder-farmers (accessed 20 June 2024).

World Bank (2024) *Poverty, Prosperity and Planet Report 2024: Pathways Out of the Polycrisis*. Washington, D.C. DOI: 10.1596/978-1-4648-2123-3.A.

World Food Programme (WFP) (2022) WFP Annual Review 2022. WFP Nordic Office, UN City Communications, Copenhagen. Available at: https://www.wfp.org/publications/wfp-annual-review -2022#:~:text=As%20a%20global%20food%20crisis,economic%20aftershocks%20of%20COVID %2D19 (accessed 31 August 2024).

Yonzan, N., Mahler, D.G. and Lakner, C. (2023) Poverty is back to pre-COVID levels globally, but not for low-income countries. *World Bank Blogs*. Available at: https://blogs.worldbank.org/en/opendata /poverty-back-pre-covid-levels-globally-not-low-income-countries#:~:text=Using%20this%20appr oach%2C%20we%20find,the%20start%20of%20the%20pandemic (accessed 20 June 2024).

6 Agricultural Innovations and Development Changes among Resource-Poor Farmers from the Mid-1960s to the 2020s

Abstract

Resource-poor farmers were met and interviewed together during field investigations in the mid-1960s in Trinidad and Ethiopia. In 1980, most of them were revisited and new ones were consulted for updates on agricultural innovations, together with meetings and data collection for some 100 farmers in a parish in southern Sweden between 1925 and 1980. This formed a platform for a comparative analysis of the adoption of agricultural innovations. In the early 2000s, attempts were made to revisit all these farmers and to contact an equal number of new farmers for agricultural field data. Altogether, some 200 farmers were included from one tropical island, one African highland country and one highly industrialized country. Twenty years later, relevant facts and developments from these countries form the background for a discussion on agricultural innovations and their adoption by small farmers over a long time period and whether these developments have led to societal change with long-term sustainability.

Background

Studies and discussions of the future of agriculture deal with international developments or detailed studies of social change. It is, however, important to include technical and social changes at the farm level across a longer-term perspective, through an international perspective and with a view towards future sustainability. This discussion at the microlevel will be an integral part of the investigation of steps towards agricultural sustainability. The data is based upon information from 222 resource-poor farmers in Trinidad and Tobago (61), Ethiopia (92) and Sweden (69), the majority being small tenant farmers. They have been consulted using field investigations and interviews with a starting point in the mid-1960s and revisited in 1980 and 2002–2003. In addition to meeting revisited farmers, an equal number of new farmers were consulted. The period since 2002–2003 will be discussed using official data, academic publications and other documentation.

Most farmers visited were smallholders, a term that often has been used without precision, such as treating family farms and small farms interchangeably. In general, farm size increases with average income levels, which is relevant since most farms are small in low-income countries. Precision matters in relation to how best to develop and design public policies and investment for family farming. About 99% of farms in high-income countries are larger than 5 ha compared to only 28% in low-income countries (Rural21, 2021). It was an early goal of the Millennium Development Goals (MDGs) to increase the productivity of smallholders and improve rural livelihoods. A similar focus was endorsed by the United Nations Decade of Family Farming 2019–2028 (Food and Agriculture Organization (FAO), 2021).

Visits and Consultations with Farmers

The first consultation in the 1960s

In 1965, resource-poor aroid farmers were visited at three locations in Trinidad: a tropical

DOI: 10.1079/9781800627802.0006

West Indian island. Some field observations were also made. These farmers grew various crops on a larger scale than just for household production. Ten farmers in each location were interviewed about their production of dasheen and eddoes (*Colocasia* spp.), and tannia (*Xanthosomas sagattifolium*) respectively. Except for eddoes, the other aroids were grown as crop mixtures, sometimes including cocoa, at one remote location. The average aroid crop comprised about 1.6 ha for dasheen and 0.5 ha for tannia (Bengtsson, 1966). No information was gathered on land ownership, but squatting was common in two locations.

In Ethiopia, farmers were approached in the early phase of a development project supported by Swedish aid. The Chilalo Agricultural Development Project (CADU) started in 1966 in the highlands around Asella, at an altitude of 2400 m above sea level 200 km south of Addis Ababa. The goal was to develop all aspects of the rural community in Chilalo Awraja. Most farmers visited in 1967 were tenants under landlords with crop-sharing. Major crops were wheat, barley, broad beans, maize, field peas, the native teff (*Eragrostis tef*) and grassland. Farmers were interviewed and continuously monitored through field observations of their ploughing, seeding and weeding, and crop sampling of yield levels (Bengtsson, 1968).

Farmers used the ard for ploughing. Soil burning (*guie*) was common and no farmer used fertilizer, but manure was applied to some crops. Farmers used their own seed and hand weeding was common. Farmland was calculated in 'timads': the area a pair of oxen would plough in 1 day, making it difficult to estimate the average farm size.

The first revisit to farmers in Trinidad and Ethiopia in 1980 and consultations with Swedish farmers active during 1950–1980

In 1980, farmers were revisited in Trinidad (17%). Some squatting remained although one-third of the farmers rented their land and ten farmers had purchased their farms (Bengtsson, 1983).

In the same year, Ethiopian farmers were revisited (58%). Field investigations were repeated and so was crop sampling of the yields. Farmers were no longer tenants and Swedish aid to CADU had ceased after the imperial government had been overthrown in the Ethiopian revolution in 1974. In 1980, the average farm size of visited farmers was estimated at 2.9 ha of cropland and 1 ha of grassland. The revisited farmers had 3.7 ha compared to 2.2 ha for newly met farmers. This was a result of the land reform process (Bengtsson, 1983). Revisited farmers had been able to exert power in the peasant organizations, having been earlier selected as so-called 'model farmers' by the CADU project. In 1980, the average family size was 5.7 persons.

To compare the rate of innovation adoption, it was decided to investigate how this process had worked in a high-income country. In the late 1970s, all former and active farmers were interviewed in the former parish of Gråmanstorp. It is situated in the north-west of the province of Scania in southern Sweden, where the author had grown up. Data collection and discussions covered the period between 1925 and 1980.

In 1925, there were 112 farms with a tenancy rate of 90%, comprised of many small farms (<10 ha), a few medium-sized and large farms (< 50 ha) and one large estate which also had 1300 ha of forested land. The terms of lease turned gradually into cash instead of day-labour obligations. Between 1925 and 1975, the cost of the terms of lease per hectare increased by almost 1200%, higher for small farms. In contrast, the prices of wheat and barley purchased by the farmers increased only by about 200% and that of nitrate of soda by 150%.

In 1980, there were 19 active farmers with a tenancy rate of 58%. Eight farms had more than 50 ha. The average household was composed of 2.5 persons in contrast to 4.6 persons in 1925. Common crops were winter and spring wheat, barley, rye, oats, sugar beet, field peas, flax, potatoes, rapeseed, hay from clover and grassland. Fallows were common up to the 1960s.

Adoption of agricultural innovations 1965–1980

Facts from the field studies in the three countries indicated that it took about ten years for most farmers to adopt agricultural innovations. Mineral fertilizers and improved crop cultivars were most easily adopted. In 1980, Swedish farmers could easily adopt new crop varieties from the breeding institutes at Svalöf and Weibullsholm. Then, a trend had already begun, where farmers adopted new crop cultivars of wheat 2 or 3 years after their acceptance in the national list of certifications. Further details about all countries during this period are given by Bengtsson (1983).

The development project in Ethiopia started early to introduce improved seed for wheat and barley. The use of improved seed increased to 78% for wheat and 33% of barley farmers between 1967 and 1980. The farmers found it expensive.

In Trinidad, there was no introduction of new, improved varieties of aroids and no breeding work was carried out on these crops. No herbicide had been specifically designed for aroids. Dicotyledonous weeds were more common in Sweden (92%) and Ethiopia (60%) compared to Trinidad (20%). The time for acceptance of insecticides was much longer than for fertilizers and even herbicides, due to the increased risks in handling. Since 1965, fertilizers had been applied to eddoes only, which were cultivated close to the capital, Port of Spain (Table 6.1). No cow manure was used on dasheen or tannia.

No mechanical innovation was developed for aroids. The development project in Ethiopia tried to introduce an improved plough and a

harrow to replace the ard. It was rejected by almost all farmers (96%). It was too heavy for their oxen. No harrow was in use in 1980.

In Sweden, there were a range of mechanical innovations, most of them being adopted within ten years (Bjersgård hurdle, tractor on rubber tyres, fertilizer spreader and forage harvester). It took a little more than ten years to adopt the milking machine and the Artturi Ilmuri Virtanen (AIV) silage method. Smaller farmers were quicker to adopt the milking machine than the larger ones, but the latter bought combine harvesters more rapidly. The towed combine harvester took more than ten years for most farmers to adopt. Farm size was often more important for quick adoption than ownership of land.

Although resource-poor farmers and tenants constituted most farmers up to the 1980s, their problems were given little attention by the research establishment, except genetic improvements of new crop varieties. For a long time, the research focused on fertilizer experiments. Agricultural extension agents visited few farmers in the parish (10%) up to the mid-1960s, in contrast to annual visits by the agricultural sales representatives of private companies.

The second revisit to farmers

In 2003, aroid farmers in Trinidad were revisited (54%). Production had declined, mostly for dasheen (by 50%) and less for eddoes and by 20% for tannia. The use of mineral fertilizer on eddoes was common but less so on the other root crops. All farmers in both Ethiopia and Sweden used fertilizer. A comprehensive discussion with details for the period between 1980 and the early 2000s, for all farmers, is contained in Bengtsson (2007). Some highlights are given below.

Herbicide use in Trinidad had increased from one aroid grower in 1980 to some 50% of growers but the use of insecticides on eddoes had declined since 1980. Erosion had increased and all farmers claimed that less priority was given to agriculture than in the past.

In 2004, Ethiopian farmers were revisited (52%). One-third of the farmers visited in 1967 were found. Some farmers had adopted cross-bred cattle. Insecticides were being used (20%),

Table 6.1. Percentage of resource-poor farmers adopting agricultural chemicals in Trinidad, Ethiopia and Sweden in the mid-1960s and 1980 (Bengtsson, 1983).

Chemical	Trinidad (%)	Ethiopia (%)	Sweden (%)
Fertilizers			
1965–1967	30	no use	100
1980	44	92	100
Herbicides			
1965–1967	no use	no use	72
1980	33	12	100
Insecticides			
1965–1967	no use	no use	10
1980	28	no use	22

together with herbicides (96%). No farmer used the improved plough or harrow. Major problems were the increasing costs of fertilizers, which were often unavailable in time, and the land tax.

Three Swedish farmers remained active in agriculture on large farms at Gråmanstorp. Two of them had formed a limited company and the third leased his farm. They cultivated most of the arable land in the parish. High-tech agricultural machinery was in common use.

Agricultural Developments Since the Millennium

Trinidad and Tobago

For many years, the petroleum industry dominated the economy in Trinidad and Tobago together with tourism and manufacturing. The share of value added by the agriculture, forestry and fishing sector to the gross domestic product (GDP) is around 0.5%. In the early 2020s, some 3% of the population worked in the agricultural sector. Farmers are ageing. By 2003, three-quarters of the revisited aroid farmers had already found agriculture unprofitable and predicted doom in the future. They could not find suitable land and there was a shortage of qualified farm labour. The arable land area has decreased to about 25,000 ha compared to 40,000 ha in the mid-1990s, according to the most recent Food and Agriculture Organization (FAO) electronic statistics ().

Yield levels of aroids have gradually declined as farmers had already noted at the first revisit in the early 1980s. They also noted that aroids produced smaller corm sizes due to declining soil fertility. A consecutive harvest of this root crop normally led to a yield reduction of some 200 kg per harvest. Aroid production in Trinidad has continued to decline due to squatting encroachment, land abandonment by farmers and the development of infrastructure (T. Gopaul, personal communication, Port of Spain, 2023). Aroids are produced at the household level unless irrigation is applied in areas close to the capital.

Current annual production is estimated at about 2000t of eddoes and 3000t of dasheen but there are no longer official yield levels (T.

Gopaul, personal communication, Port of Spain, 2023). In Tobago, the cultivation continues of both dasheen and tannia and is officially recorded. Eddoes are, however, not preferred on this island. Pesticides are required to protect crops against infestations of giant African snails, locusts and coconut mites but are not used on aroids.

Traditionally important agricultural export commodities have included not only sugarcane, cocoa (of high quality), coffee, fruit and vegetables but also poultry. The production of cocoa has fallen over many years, although the country is one of only a few to be certified with extra fine cocoa quality for the Criollo, Forastero and Trinitario varieties and with access to one of the two cocoa gene banks. The government established the Cocoa Development Company of Trinidad and Tobago in the mid-2010s to increase production.

Between 85–100% of the food consumed in the country comes from crops that are not native to the region (Crop Trust, n.d.). The country is highly dependent upon food imports, about 40% coming from the US. The country is the second-largest market in the English-speaking Caribbean for US agricultural exports. Nowadays, even sugar is imported from Guyana. A consumer shift (10%) towards certified organic products can also be noted.

A new crop appeared in Trinidad and Tobago when the new cannabis bill was enacted by Parliament in 2022. The Cannabis Control Act provides regulatory control of the handling of cannabis for certain purposes, and authorizes the establishment of the Trinidad and Tobago Cannabis Licensing Authority and other connected matters (Republic of Trinidad and Tobago, 2022). Applicants must be 18 years or older. They will be allowed to cultivate four plants at home. Also, companies, firms and cooperative societies can make applications. Medical marijuana sales are considered as a potential growth area, though illegal, for the whole Caribbean agricultural industry. Globally, the legal cannabis market size is estimated to grow by US$100.6 billion between 2023 and 2028 (Technavio, 2024). Barclays earlier estimated the US weed market would be US$28 billion and, if legalized today, would grow to US$41 billion by 2028 (Sheetz, 2019). Jamaica has invested significantly in the medical cannabis industry, and is funding research into its medicinal properties, as part of its diversification drive. This global trend is demonstrated by Germany, the third European Union (EU) country to legalize cannabis för personal use in early 2024 (Le Monde, 2024). Three plants are allowed to be cultivated in one´s living quarters; hardly a major step towards sustainability on a global context.

Although there is a National Food Production Action Plan (NFPAP) and Vision 2030 as part of the UN approach towards future sustainability, there has been little action on strengthening food security. One example is a plea for community-based agriculture (Flemming *et al.*, 2015). Land reform and greater food production remain the key issues according to the Oxford Business Group (2023). After the closing down of the sugar industry in Trinidad, irrigation is a low priority for the government, a problem for some farmers but not for aroid producers. Many farmers complain that crops, machinery and materials are being stolen. Drainage and irrigation at Caroni remain a concern, as already noted by farmers in the mid-1960s.

So far, private industry has taken some initiatives towards sustainable farming but there are few examples of the Trinidad and Tobago government doing so. There has been an increase in small projects using, for example, hydroponic farming techniques. This would require replication on a larger scale to stimulate farmers and these techniques require water and irrigation. Both severe and mild droughts have led to crop losses during the 2010s. Rising sea levels could lead to a loss of habitats, property and livelihoods, and agricultural land (Oxford Business Group, 2023).

Ethiopia

After their introduction, fertilizers have been in use in Chilalo Awraja but their use stopped during the Derg period of 1974–1987 and for some farmers during times of poor economics and internal conflict, particularly since 2020 (M. Bekele, personal communication, Addis Ababa, 2023). They are also considered too expensive in contrast to herbicides and pesticides, which are commonly used, and easily available at government and private shops.

Since the late 2010s, most farmers have also been using tractors instead of oxen. Neighbouring farmers on the plains have formed a group and bought one or two tractors using a bank loan. The well-off farmers continue to use Massey Ferguson tractors as they did in the past (G. Tedla, personal communication, Addis Ababa, 2023). In 1992, the Adama Agricultural Machinery Industry (AAMI) company was established in the town of Adama. It manufactures small and more generous-sized tractors, as well as motorized threshers.

Small farmers persist with the ard, the old practice for seed covering. Farmers at higher altitudes in Chilalo are still using a pair of oxen and weed by hand (X. Ogatao, personal communication, Chilalo, 2023). These are still important production practices for many small farms in parts of the country.

Yields of Ethiopian crops have increased since the mid-1960s, although they have fluctuated over time but less than the results achieved in field experiments (Table 6.2).

The Ethiopian government formulated a national seed policy in 1992 and created the Ethiopian Seed Enterprise to encourage seed production and marketing by the private sector. It is now part of the Ethiopian Agricultural Business Corporation (EABC), a recently established public corporation. In 2004, revisited farmers in Chilalo complained about the shortage of improved seed of their choice, in particular wheat. The Ethiopia Seed Association, established in 2006, carries out seed quality inspections and certification (Access to Seeds, 2019).

Maize yields in Ethiopia have improved by some 50% since 2008. In 2013, four new varieties for high-potential and drought-prone maize growing areas were released by the Ethiopian Institute of Agricultural Research (EIAR) in collaboration with International Maize and Wheat Improvement Center (CIMMYT). Since Ethiopia must import about 20% of its wheat, increased domestic production is a national priority for the second-largest wheat producer in sub-Saharan Africa (Hodson *et al.*, 2020). In 2018 and 2019, about 20 advanced genotypes of wheat were evaluated by the National Wheat Research Program. In 2021, it led to the registration of

Table 6.2. Estimates of yields and crop samplings of selected crops for farmers under study during the 1960s to 2003–2004 and the average national yield level in 2022 (kg/ha) (Bengtsson, 1968; Bengtsson, 1983; Bengtsson, 2007; CSA, 2018; Hailegebrial and Adane, 2018; Mihretie *et al.*, 2022; Shikur, 2022; Tefera *et al.*, 2022; Swedish Board of Agriculture, 2023).

Crop	1960s (kg/ha)	1980 (kg/ha)	2003–2004 (kg/ha)	Average national yield in 2020–2022 (kg/ha)
Aroids (1st harvest)				
Dasheen	8000	6600	8000	no official data
Eddoes	5500	11,000[a]	10,000	no official data
Tannia	n.a.	2600	n.a.	no official data
Wheat				
Ethiopia[b]	980	1230	2340	2900[c]
Sweden (Scania) (winterwheat)	3200[d]	4800	6250	8890
Barley				
Ethiopiab	1130	1520	2100	2400
Sweden (Scania)	2400[d]	4000[e]	5000	7350
Other crops in Ethiopia[b]				
Sorghum	1050	2070	n.a.	2100
Teff	970	450	n.a.	1000
Broad beans	1500	2120	2300	2100[f]

[a]Wet season or under irrigation.
[b]Yields adjusted to dry matter content for 1960s and 1980.
[c]2019.
[d]1950.
[e]Late 1970s.
[f]2018.

Abay, a new bread wheat variety (Solomon, 2022).

Yield levels also depend on cultivation practices by farmers, exemplified by teff (Mihretie *et al.*, 2022). Teff yields varied significantly across sites although the average reached some 1000 kg/ha. This was low compared to the national average grain yield for the Quncho teff variety, which has been reported to be 2800 kg/ha (Kebebew, 2011). Quncho tef has been widely grown by some 6 million farmers, who have planted teff on more than 3 million ha. In 2015, researchers identified three promising high-yielding lines within the 'Teff Improvement Project', leading to the release of the Tesfa variety in 2017.

Beans have potential in Ethiopia ranging between 2.3–3.9 t/ha using recommended practices according to the Ministry of Agriculture and Rural Development. But scarcity of bean varieties is a problem, although several varieties are high-yielding in experiments – about 3.5 t/ha (Ketema, 2022).

Nationally, agriculture has expanded in certain regions but been hampered in others by internal war. About 38.5 million ha of agricultural land was used for crop and animal production in 2020; more than one-third of the total land area (Statista, 2021). There were 15.6 million agricultural households with an average farm size of 0.95 ha (FAO, 2016). This is a profound change from some 16.6 million ha under cultivation in the early 2000s. It has led to increased land fragmentation and a shrinking cultivated area per household. The system of mixed agriculture in the Ethiopian Highlands includes livestock and sustains all family farms, which average five persons according to the World Bank (2016). They keep around two livestock units per family farm from cattle, poultry and sheep. Pastoralism is common in the lowlands. The average farm size is likely to decline further with an increasing population (Table 6.3).

The Ethiopian government has agriculture at the top of the priority sectors in an ambitious economic development plan (2021–2030). The agriculture sector is projected to grow annually at 6.2% over the next ten years by developing unutilized arable land, modernizing production systems and the use of innovative technology. This includes efforts to tackle land degradation, reduce greenhouse gas emissions, increase forest protection and produce electricity from renewable sources (ITA, 2022). The government has established an Agricultural Transformation Institute (ATI) to address systemic bottlenecks in the agriculture sector and enhance the capability of the Ministry of Agriculture. Formerly the Agricultural Transformation Agency, it focuses on improving the livelihoods of the smallholder farmers on income, resilience, and sustainability.

The top priorities in agriculture are increasing the productivity of crops and livestock, and improving agricultural production methods with consideration of natural resource management. Other critical areas are small- and large-scale irrigation developments, financing agricultural inputs and a research-based food

security system. Cash crops for export, such as coffee, oilseeds, pulses, fruit and vegetables, are of special importance. There are plans to involve the private sector in processing some of these commodities but the export market for fruit, vegetables and animal production is limited (Wendimu, 2021).

Sweden

A great transformation has taken place of both Swedish agriculture and society. In the early 1900s, three out of four Swedes lived in rural areas. Smallholdings accounted for 64% of approximately 350,000 farming units. Natural pasture covered 1.5 million ha compared to the current 450,000 ha. The Swedish Board of Agriculture no longer keeps statistics on meadows. About 1 million ha of arable land have become forests or waste land of no use. Swedish forests are increasingly owned by companies and urban dwellers who do not live in the forested areas.

After joining the EU in 1995, the number of farms in Sweden has further decreased with about one-third since 1990. The number of dairy farms has declined from 18,000 to less than 3000. In all, there are 58,753 farm enterprises: one-fifth are larger than 50 ha according to the Swedish Board of Agriculture (2022). Two out of three farms cultivated up to 20 ha. More than 60% of all arable land was farmed by those with more than 100 ha. The total arable area was 2.6 million ha. The average farm size had increased to 43 ha in 2021 according to the Swedish Board of Agriculture.

About 166,000 persons were active in the agricultural sector (2022). Every third own-account holder in agriculture with a sole proprietorship was aged 65 or more in 2016: an increase from 2013. About 5700 farms have an ecological focus (20% of the area) compared to 8.5% within the EU . The number of female farmers has increased slightly: currently they constitute almost 17% of all farmers (Swedish Board of Agriculture, 2023).

Grazing areas have been reduced. Milk production has fallen significantly because milk consumption has decreased by 25% since 2009. Milk has competition from the oat drinks

Table 6.3. Population in Sweden, Trinidad and Tobago and Ethiopia between 1965 and 2022 (in millions) (Bengtsson, 1968; Bengtsson, 1983; Bengtsson, 2007; UN, 2022).

Country	1965	1980	2003–2004	2022
Sweden	7.8	8.3	9.0	10.5
Trinidad and Tobago	1.0[a]	1.2	1.2	1.3
Ethiopia	23[b]	33[b]	73	123.3

[a]Estimate at independence.
[b]Estimates (no census for 1965).

made by the Oatly company. Swedish milk is the most expensive in the EU and Sweden is the only country around the Baltic Sea that imports dairy products. Pork imports have increased, as have cheese and beef imports. In recent years, animal rights activists have protested against farmers with pigs and dairy cows, resulting in the closure of production units.

Since 2000, the Swedish population has increased, through the immigration of about 1.5 million people; some 20% of the population (Table 6.3) (Organisation of Economic Cooperation and Development (OECD) iLibrary, 2022). The agrarian sector has diminished, prices have declined and many rural areas have lost much of their normal social services. People in the countryside leave for better paid, urban jobs. For many years, Sweden's ruling politicians have focused on industrial exports and a search for the cheapest possible food, regardless of production methods and transportation costs. Incomes for farmers in the 1990s were only 76% of what they were in the mid-1970s; a trend that has continued (Swedish Board of Agriculture, 2021). Another study indicated that farm households in Sweden do well from a standard-of-living perspective, but that farming is still a low-paid occupation from a return-on-skills perspective (Nordin and Höjgård, 2019).

Farmers continue to change to improved crop varieties. Yield levels of winter wheat in Scania province are now about 8000 kg/ha and for spring wheat 5300 kg/ha according to the Swedish Board of Agriculture (2023). This is a significant difference to yield levels of winter wheat of about 2000–3000 kg/ha between 1913 and 1920. But there were no major breakthroughs of new wheat varieties in Sweden during the 2010s. Recently, researchers at the Swedish University of Agricultural Sciences (SLU) have started to develop new wheat varieties adapted for an extreme and varied climate and with good baking quality. In other research, genes have been mapped for resistance to black spot disease and powdery mildew in historical Nordic winter wheat varieties and landraces. Researchers at the University of Gothenburg are reported to be attempting to develop new salt-tolerant wheat.

Field experiments have recently been carried out in Sweden to test if new crops can be grown in Sweden, such as sweet potato, quinoa and Kernza, a perennial wheatgrass (*Thinopyrum intermedium*). The non-narcotic industrial hemp (*Cannabis sativa*) is also being considered. After being banned in 1974, production was allowed in 2003. During the Middle Ages, it was a staple crop in Sweden and there are now about 200 ha in cultivation. Hemp is four or five times better than forests in binding carbon dioxide and the plant can be used for various products. It requires no action between sowing and harvesting and no plant protection products.

Swedish agriculture has received financial support from the EU since Sweden became a member. Nowadays, the Swedish Board of Agriculture pays about 13–14 billion Swedish crowns (about US$3.5 billion) to about 60,000 Swedish farmers, every year. This support aims to create low food prices and secure food production within Europe. In 2023, a new Common Agricultural Policy was adopted by all EU member countries. Each of them will develop a strategic plan with agricultural goals. In Sweden, the total budget for the strategic plan is approximately 60 billion Swedish crowns (US$5.7 billion) for the period 2023–2027 (Swedish Board of Agriculture, 2024).

In 2019, the EU directly paid European farmers €38.2 billion (about US$3.5 billion), whereas €13.8 billion (about US$1.3 billion) was allocated to rural development (European Parliament, 2021). In 2022, the Swedish Board of Agriculture paid 13.7 billion Swedish crowns (US$1.3 billion) as EU support to Swedish farmers (Swedish Board of Agriculture, 2024). Most of the subsidies to farmers have been directed to large farm enterprises. In 2019 and 2020, the 10–12 largest beneficiaries in southern Scania province received almost half of the total sum (Skånska Dagbladet, 2019). This has been a general feature of EU farm support, further exemplified by, among others, oligarchs and the power elite in Eastern Europe (Hungary, Czech Republic and Slovakia).

The administrative costs have reached almost 1 billion Swedish crowns (about US$0.09 billion) in a complex process with prohibitive costs for processing, payment, control and not least the development and maintenance of IT systems (Ringsten, 2023). Since 2008, handling costs have increased both at the Swedish Board of Agriculture (172%) and the county administrative boards (8%). This has led to a government investigation, which proposed a new organizational structure for effective and legally secure administration and a responsibility to report suspected criminality (SOU, 2024).

The EU agricultural policy gives little practical attention to how to reach agricultural sustainability. Focus is on the financial support to agriculture. For the support period 2023–2027, the EU proposed a ceiling for financial support to larger farms. This was not attractive to the Swedish government despite most Swedish farmers having considered the areal farm payment to be a security. In the EU's new budget, member countries can set their own sustainability requirements for EU support. Finally, the EU decided that a ceiling would be voluntary, but at least 10% of the financial support would have to be reallocated to smaller farms. This EU policy is an issue that Sweden intends to oppose.

Swedish agricultural policy during the added support period will include 60 billion Swedish crowns, equivalent to US$5.7 billion (June, 2024), out of which the EU will finance 75% (Swedish Board of Agriculture, 2024). It will go to 'green jobs', strengthen profitability at the farm level and as contributions to increased exports of Swedish food. The latter becomes illusory under the current economic and regulatory system for EU and Swedish agriculture. The proposed policy is 'business as usual', although one positive feature is that young farmers should receive special investment.

In 2023, a political dialogue started with all Swedish actors about the promotion of Swedish-produced food products and the strengthening of competitiveness in the food chain. Future food security is of special importance since Sweden's self-sufficiency rate is just under 50%. There is no goal for Swedish food security in times of crisis. This applies to access to fuel, fertilizers, wheat and other crops. Sweden has no emergency stockpiles and no authority with a collective responsibility for livelihoods in case of emergency. Most shelves in the stores can be empty of food after a day and require an efficient transport system, in turn adding to greenhouse gas emissions. They account for about 15% of these emissions according to the Swedish Environmental Protection Agency and the Swedish Board of Agriculture.

The use of agrochemicals and monocultures in both agriculture and forestry has caused negative effects on the environment and human health. Such costs have not been considered in food prices, including all imported food. Greenhouse gas emissions from food production in other countries are not included in the requirements for zero greenhouse gas emissions in Sweden by 2045. In consequence, future measures to meet politically agreed climate targets will lead to dietary changes and significantly higher food prices for consumers.

The current Swedish food strategy for 2030 shows few specific references to global trends and the implications for Swedish agriculture in the future. One-third of domestic production is to be organic by 2030 but sales of organic food has already started to level off since 2019. Swedish politicians have announced few practical proposals towards achieving sustainability and align with the SDGs.

The political objective to reach zero greenhouse gas emissions will require radical changes to current Swedish industrial agriculture. The effects will be more pronounced than in other countries, such as Ethiopia and Trinidad and Tobago. This has been given surprisingly little consideration. Ward (2022) noted that such a change in the UK requires a revolution in land use, farming practices and diets. This is a call for action to avoid catastrophic climate change in the future.

References

Access to Seeds (2019) Ethiopia. Available at: https://www.accesstoseeds.org/index/eastern-southern-africa/country-profile/ethiopia/ (accessed 21 June 2024).

Bengtsson, B.M.I. (1966) A survey of cultivation practices and the utilization of edible aroids in Trinidad. DTA thesis, University of the West Indies, St Augustine, Trinidad, and Tobago.

Bengtsson, B.M.I. (1968) Cultivation practices and the weed, pest, and disease situation in the chilalo awraja. CADU Publication No.1, Chilalo Agricultural Development Unit (CADU), Addis Ababa, Ethiopia.

Bengtsson, B.M.I. (1983) Rural development research and agricultural innovations: A comparative study of agricultural changes in A historical perspective and agricultural research policy for rural development. Swedish University of Agricultural Sciences, Department of Plant Husbandry, Report 115, Uppsala, Sweden.

Bengtsson, B.M.I. (2007) *Agricultural Research at the Crossroads. Revisited Resource-poor Farmers and the Millennium Development Goals.* Science Publishers, Enfield, NH.

Central Statistical Agency (CSA) (2018) Annual Agricultural Sample Survey 2017/18 (2010E.C.) Area and production of major crops. (Private peasant holdings, Meher season). The Federal Democratic Republic of Ethiopia Central Statistical Agency, Addis Ababa, Ethiopia.

Crop Trust (n.d.) Trinidad and Tobago. Available at: https://www.croptrust.org/pgrfa-hub/crops-countries -and-genebanks/countries/trinidad-tobago/ (accessed 21 June 2024).

European Parliament (2021) Agriculture in the EU in figures: financing, jobs and production (in Swedish). Strasbourgand Stockholm.

Flemming, K., Minott, A., Jack, H., Richards, K. and Morris, O. (2015) Innovative community-based agriculture: A strategy for national food production and security. Caribbean Agricultural Research and DevelopmentInstitute (CARDI), University of the West Indies, St. Augustine Campus, St. Augustine, Trinidad and Tobago. Available at: https://www.cardi.org/wp-content/uploads/downloads/2015/09 /Innovative-community-based-agriculture-strategy-for-national-food-prod-and-security-by-CARDI .pdf (accessed 31 August 2024).

Food and Agriculture Organization (FAO) (2016) AQUASTAT Country Profile – Ethiopia. FAO, Ethiopia.

Food and Agriculture Organization (FAO) (2021) Small family farmers produce a third of the world's food. *ReliefWeb.* Available at: https://reliefweb.int/report/world/small-family-farmers-produce-third-world -s-food (accessed 21 June 2024).

Hailegebrial, K. and Adane, T. (2018) *Yield performance and adoption of released sorghum varieties in Ethiopia.* Edelweiss Applied Science and Technology.

Hodson, D., Jaleta, M., Tesfaye, K., Yirga, C., Beyene, H. *et al.* (2020) Ethiopia's transforming wheat landscape: Tracking variety use through DNA fingerprinting. *Scientific Reports* 10, 18532.

International Trade Association (ITA) (2022) Ethiopia. Country Commercial Guide. Market Overview. ITA, Washington, D.C.

Kebebew, Z. (2011) *Agroforestry perspective in land use and farmers coping strategy: Experience from Southwestern Ethiopia.* Jimma University, ResearchGate.

Ketema, W. (2022) Yield performance evaluation of common bean (*Phaseolus vulgaris* L.): Varieties under rain fed in Western Ethiopia. *American Journal of Plant Biology* 7(1), 60–64. Available at: https://doi. org/10.11648/j.ajpb.20220701.19 (accessed 26 June 2024).

Le Monde (2024) Germany becomes largest EU country to legalisecannabis for recreational use. Available at: https://www.lemonde.fr/en/germany/article/2024/04/01/germany-becomes-largest -eu-country-to-legalize-cannabis-for-recreational-use_6667013_146.html (accessed 21 June 2024).

Mihretie, F.A., Tsunekawa, A., Haregeweyn, N., Adgo, E., Tsubo, M. *et al.* (2022) Exploring teff yield variability related with farm management and soil property in contrasting agro-ecologies in Ethiopia. *Agricultural Systems* 196, 103338. Available at: https://doi.org/10.1016/j.agsy.2021.103338 (accessed 26 June 2024).

Nordin, M. and Höjgård, S. (2019) Earnings and Disposable Income of Farmers in Sweden, 1997–2012. *Applied Economic Perspectives and Policy* 41, 153–173. Available at: https://doi.org/10.1093/aepp/ ppy005 (accessed 1 September 2024).

Organisation of Economic Cooperation and Development (OECD) iLibrary (2022) International Migration Outlook 2022: Sweden. Available at: https://www.oecd-ilibrary.org/social-issues-migration-health /international-migration-outlook-2022_18f59253-en (accessed 26 June 2024).

Oxford Business Group (2023) Investment in traditional crops and training bolsters agriculture in Trinidad and Tobago. Oxford Business Group Regional Office, Port of Spain, Trinidad and Tobago.

Republic of Trinidad and Tobago (2022) Cannabis Control: Act No. 10 of 2022. Parliament of Trinidad and Tobago, Port of Spain. Available at: https://www.ttparliament.org/wp-content/uploads/2020/10 /a2022-10.pdf (accessed 31 August 2024).

Ringsten, M. (2023) Swedish Board of Agriculture, 2022. Costs and measures by authorities for the administration of EU support 2023 (in Swedish). RA23:10, Swedish Board of Agriculture, Jönköping, Sweden.

Rural21 (2021) Small farms produce a third of the world's food. Available at: https://www.rural21.com /english/news/detail/article/small-farms-produce-a-third-of-the-worlds-food.html (accessed 21 June 2024).

Sheetz, M. (2019) Barclays estimates US weed market would be $28billion if legalized today, growing to $41 billion by 2028. *Yahoo! Finance*. Available at: https://finance.yahoo.com/news/barclays-estima tes-us-weed-market-133835494.html?guccounter=1&guce_referrer=aHR0cHM6Ly93d3cuZ29vZ2x lLmNvbS88&guce_referrer_sig=AQAAAGtacrEQq-j5rztOWFXdwuKXuKsz_vy-2cXQ_QUKJ_LYBprZq cFHx0bMCXdfdaDahJYpG8U3tUik6UFiNMU7pLRnYbCbdaJWelJlAowQ777F1VdoHYTS3pgdsBNq Y8XLrspojnNq9AsZdzsYffhxCVq63vAgYMj3_X8lWVMmaStZ (accessed 31 August 2024).

Shikur, H.Z. (2022) Wheat policy, wheat yield and production in Ethiopia. *Cogent Economics & Finance* 10(1), 1–20. Available at: https://doi.org/10.1080/23322039.2022.2079586 (accessed 26 June 2024).

Skånska Dagbladet (2019) 10 största mottagarna av EU-stödet i skåne (in Swedish).

Solomon, T. (2022) Selection and registration of new bread wheat variety Abay for low to mid-altitude wheat-producing areas of Ethiopia. *American Journal of Bioscience and Bioengineering* 10(2), 33–38. Available at: https://doi.org/10.11648/j.bio.20221002.13 (accessed 31 August 2024).

SOU (2024) A new organization for the administration of EU funds (in Swedish). Government Investigations, SOU 2024:22, Ministry of Finance, Stockholm.

Statista (2021) Distribution of agricultural land use in Ethiopia in 2019, by type. Published by Saifaddin Galal. Available at: https://www.statista.com/statistics/1307345/share-of-agricultural-land-use-in -ethiopia-by-type/ (accessed 1 September 2024).

Swedish Board of Agriculture (2021) Agricultural enterprises and farm entrepreneurs. Jönköping, Sweden.

Swedish Board of Agriculture (2023) Agricultural statistics. Jönköping, Sweden.

Swedish Board of Agriculture (2024) Common Agricultural Policy 2023–2027. Jönköping, Sweden.

Technavio (2024) *Legal Cannabis Market Analysis North America, Europe, APAC, South America, Middle East and Africa – US, Canada, Germany, UK, Australia – Size and Forecast 2024-2028*. Canada, Germany, UK, Australia. Available at: https://www.technavio.com/report/legal-cannabis-market -industry-analysis (accessed 31 August 2024).

Tefera, M., Zhang, W., Zhao, S. and Duan, Z. (2022) Enhancing maize yield in Ethiopia: A meta-analysis. *International Journal of Agricultural Science and Food Technology* 8(3), 193–201.

United Nations (UN) (2022) UN Population Division Data Portal, Department of Economic and Social Affairs United Nations Secretariat, New York. Available at: https://population.un.org/dataportal/ (accessed 1 September 2024).

Ward, N. (2022) *Net Zero, Food and Farming. Climate Change and the UK Agri-Food System*. Routledge, Abingdon, UK.

Wendimu, G.Y. (2021) The challenges and prospects of Ethiopian agriculture. *Cogent Food & Agriculture* 7(1), 1. Available at: https://doi.org/10.1080/23311932.2021.1923619 (accessed 26 June 2024).

World Bank (2016) *LSMS-Integrated Surveys on Agriculture Ethiopia Socioeconomic Survey (ESS)*. Washington, D.C.

7 Global Issues For Future Food Security

Abstract

The climate, one of nine global issues, encompasses rising global temperatures and greenhouse gas emissions. It is given political priority based on reports by the Intergovernmental Panel on Climate Change (IPCC) but also references to previous climate changes over time. A second issue is the increased exploitation of natural resources with expanding environmental damage and urbanization and loss of farmland. In addition, land grabbing is a growing issue. Since biodiversity has declined over the centuries, efforts must made to search for new plants and animals, relevant for changing climatic conditions. Increasing population is an issue in several countries and so is increasing migration from many low-income countries. More people require more food which will be a difficult requirement to meet when much food is wasted, and demanding when food is to be produced with sustainability. Expanding antibiotic resistance is another emerging global issue, affecting farming and livestock. Increasing consumption leads to ever-growing waste mountains but politicians view economic growth and consumerism as their main objectives, rather than sustainable lifestyles.

Rising Global Temperature and Greenhouse Gas Emissions

Some basic facts

Most geologists consider that the Earth consists of four layers; one solid core, one fluid nucleus surrounding it, a mantle of hot viscous rocks and a stony external crust of some 40–100 km thick. The distance to the Earth's surface from the middle of the planet is about 6370 km (Williams and Montaigne, 2001). The Earth's internal temperature is estimated to be between 4000–7000°C. The surface of the Earth heats the troposphere.

The atmosphere is thin, some 190 km in depth with four layers. It is composed of nitrogen (78.08%), oxygen (20.95%), argon (0.93%), carbon dioxide (0.04%), trace gases and variable amounts of water vapour. The lowest layer is the troposphere. According to the *Concise Encyclopaedia of Science & Technology* this layer contains 75% of the total mass of the planetary atmosphere and 99% of the total mass of water vapour and aerosols. Weather mostly occurs in this layer, keeping people warm and providing access to oxygen.

The troposphere is about 13 km high on average; higher in the tropics and lower in winter in the polar regions. At middle latitudes, the temperatures of the troposphere decrease from an average of 15°C at sea level to around −55°C at the tropopause. This is the atmospheric boundary layer between the troposphere and the stratosphere. In the stratosphere, air temperature increases due to the absorption and retention of ultraviolet radiation from the sun to the Earth by the ozone layer. According to the National Aeronautics and Space Administration (NASA) an 'ozone hole' forms above the South Pole every September. Active forms of chlorine and bromine in the atmosphere attach to high-altitude polar clouds each southern winter. They initiate ozone-destroying reactions at sunrise at the end of winter in Antarctica (NOAA, 2022).

Freon was invented in the 1920s and came into production in the 1930s. It is a term covering chlorofluorocarbons (CFCs), hydrochlorofluorocarbons (HCFCs) and related compounds. Initially, the transnational corporations (TNCs) had control of the production and use of these and related compounds (Makhijani, 1992). Freons became popular and were in common use until it was discovered they destroyed the ozone layer in the stratosphere. Freons are persistent for up to a century and they are more effective than carbon monoxide in influencing the greenhouse effect (Crützen and Ramanathan, 2000). Recent studies show that five CFCs increased in the atmosphere between 2010 and 2020 despite their emissions being banned by the Montreal Protocol (Western *et al.*, 2023)

For centuries, life has kept the planet stable and cool by binding carbon in the oceans, forests and vegetation. Natural processes, such as changes in the sun's energy and volcanic eruptions, have affected the Earth's climate. Some explosive volcanic eruptions can throw particles into the upper atmosphere. There, they can reflect enough sunlight back to space to cool the surface of the planet for several years (Fahey *et al.*, 2017). The Earth´s magma is still active all over the planet, to varying degrees, and there are volcanic ridges under the oceans. El Niño is part of the natural climate phenomenon called the El Niño Southern Oscillation (ENSO) together with La Niña – both of which significantly alter global weather (Poynting and Stallard, 2024).

Over the centuries, climate has varied. During the Bronze Age, around 3000–2000 BCE in Central Europe and the Near East, the climate was 3°C warmer than today. 100 BCE–100 CE was warm in Europe, followed by a colder period and a climate crisis 500 years later. It is likely that a warmer global period occurred approximately during 800–1300, followed by the Little Ice Age (c.1300–1900). Thus, one may not completely exclude that a new warmer period may be slowly approaching.

One interesting, minor effect of the increasing average annual temperature in southern Sweden was that carp gel could survive the winters after 1850. That was previously not possible because carp originate in southern Asia. Since the fish was able to reproduce in a waterlogged area, where water warmed up quickly, an entrepreneur started a very profitable carp farming business from the 1870s onwards.

Extreme climate events, such as floods or severe droughts, have not been uncommon in Europe over a longer time perspective. About 200 years of drought occurred in Europe over the period 100 BCE–100 CE in a warm climate (Ljungqvist, 2017). Some 400 years later, droughts occurred in a period of colder weather. This also happened in the early 1500s when the European climate was not very warm. Moreover, extreme weather during a year has not been uncommon in Sweden. Hot summers have occurred over the years. In Uppsala, daily temperatures have been recorded since 1722. Old records show a hot summer in 1752 (as warm as in 1997 according to Infobladet, 2018). Since the late 1800s, data on the atmosphere, lakes and seas have been collected by the Swedish Meteorological and Hydrological Institute (SMHI; https://www.smhi.si) and its predecessors. According to available data, hot summers also occurred in 1858 (in Örebro in SMHI Blog, 21 September, 2021), 1901, 1914, 1937, 1947, 1975, 1994, 2018, 2021 and 2023.

A long-term perspective is important when discussing climate change, which is an extremely complex process. The Atlantic meridional overturning circulation (AMOC) drives the Atlantic Ocean´s currents. It determines Western Europe's weather by bringing warmer, saltier tropical waters north and chillier water south, thus releasing heat into the atmosphere. But climate change is slowing this circulation since water from the melting of Greenland´s ice cap and from permafrost has made the composition less dense and salty.

The AMOC is considered to be at its weakest status in more than 1000 years. (Ditlevsen and Ditlevsen, 2023). The Community Earth System Model focuses on a physics-based, early warning signal of the AMOC tipping and indicating the climate impacts of its collapse (van Westen *et al.*, 2024). This fully coupled simulation of the Earth system consists of atmospheric, ocean, ice, land surface, carbon cycle and other components, and differs from earlier model simulations using freshwater forcing or salinity perturbations. It indicates that a tipping of the AMOC may happen sooner than previous predictions – before 2095. If this happens, the AMOC could lower temperatures by up to 10–15°C in Europe, cause ocean ecosystems to collapse and cause sea levels to rise in the eastern US. This will affect agriculture. The IPCC has, however, concluded that the AMOC would not collapse as quickly as recent studies indicate. Other scientists have serious reservations about the conclusions and there are several assumptions about how to understand the AMOC. Last time AMOC stopped and restarted was during the Ice Ages.

Since the Industrial Revolution (c.1760–1840), human activities and modern civilization have released increasing amounts of carbon dioxide and other greenhouse gases into the atmosphere (EPA, 2024a). Before industrialization, the carbon content in the atmosphere was about 0.028% (Crützen and Ramanathan, 2000). In 1958, it had increased to 0.0315% and it is now about 0.040%. It is estimated that it will reach 0.056% at the end of this century. On the other hand, the carbon cycle is assumed to correct itself with time and make the Earth stable again. Last time, it took 60,000 years according to Bryson (2003).

The Intergovernmental Panel on Climate Change (IPCC) does not believe in a long-term trend of climate change. It is convinced that the rising global temperature can be stopped by political actions. Natural processes do not explain the warming over the last century (National Academy of Sciences, 2020). The real issue is how climate can be changed by the politicians. There is a difference between long-term climate and short-term weather. Many millennia ago, people and societies were always forced to adapt to new circumstances. Now, political action far beyond the current efforts to combat climate changes are critical for the globe´s future sustainability.

Reports by the Intergovernmental Panel on Climate Change

Since 1990, the IPCC has released reports on global temperature rises, fossil-fuel emissions, climate impact and the action needed to limit global warming. Even if global greenhouse gas emissions peak before 2025 and are slashed by 43% by 2030, the IPCC finds 'it is almost inevitable' that the Paris Agreement's 1.5°C goal will temporarily be breached in the decades ahead. Yet, the IPCC has also underlined that temperatures can be brought back below the 1.5°C threshold by the end of the century. This requires rapid action by governments through deep emission cuts, together with the deployment of techniques to capture and store carbon, for example, reforesting and using direct air capture facilities.

In March 2023, the IPCC released its Synthesis Report (SYR) of the IPCC Sixth Assessment Report (AR6). The temperature has increased by 1.1°C compared to the period 1850–1900 and the current carbon dioxide level is estimated to be about 0.042% (IPCC, 2024). The report highlights widespread and

rapid changes in the atmosphere, ocean, cryosphere and biosphere. Human-caused climate change is affecting many weather and climate extremes in all regions of the globe (IPCC, 2023). Approximately 3.5 billion people live in contexts that are highly vulnerable to climate change. Climate extreme events have exposed 27 million people to acute food insecurity and reduced water security. During the last decade, human mortality from floods, droughts and storms was 15 times higher in highly vulnerable regions (IPCC, 2022). The largest adverse impacts are observed in locations in Africa, Asia, Central and South America, least-developed countries, small islands, the Arctic, Indigenous people, small-scale food producers and low-income households (UN Food Systems, 2023).

The UN Secretary-General, Antonio Guterres, recently stressed that the rich countries must stop using coal, oil and gas by 2040. Also, the world's low-income countries must change. By 2050, they should reach net zero emissions, and by 2040 they should stop using coal.

Nuclear power is another political issue. There are nuclear power plants in 32 countries, generating about one-tenth of the world's electricity (IEA, 2022). Most of them are in the US, France and China. So far, 188 reactors have been shut down, 6 of them in Sweden. By the end of 2021, there were 437 operational nuclear reactors worldwide and 10 new ones were under construction(Statista, 2022). Before nuclear power was introduced in Sweden, hydropower accounted for all of Sweden's electricity production, 95% as late as 1965. The growing need for electric power led to an increasing number of nature lovers raising objections to the continued expansion of hydropower to still untouched rivers. This led to decisions on the introduction of nuclear power, an unsustainable step in a long-term perspective.

Political priority on greenhouse gas emissions

So far, discussions on how to combat the effects of rising temperature have been narrowly focused on greenhouse gas emissions, underlined by the IPPC. They vary widely by country and sector.

Today, the top three greenhouse gas emitters – China, the US and India – contribute 42.6% of total emissions, while the bottom 100 countries only account for only 2.9% (WRI, 2023a). After 2000, emissions increased in both China (16%) and India (14%). About 15% of greenhouse gas emissions come from methane-producing microorganisms in rice fields (Qian *et al.*, 2023).

As previously stated, TNCs have played a significant role as greenhouse gas emitters for many years. Since 1988, around 100 large companies have been the source of more than 70% of the world's greenhouse gas emissions according to the Carbon Majors Report (Griffin, 2017). This report focused on fossil-fuel producers instead of the traditional approach of using greenhouse gas emissions data at a national level. It was concluded that if fossil fuels continue to be extracted at the same rate over the next 28 years, as they were between 1988 and 2017, global average temperature would be on the rise by 4°C by the end of the century (Riley, 2017).

Another study by University College London and China's Tianjin University showed that multinational companies account for one-fifth of global carbon dioxide emissions. Based on data between 2005–2016, the carbon dioxide emissions from Coca-Cola products were almost equivalent to what China emitted from its food sector. Samsung's global emissions were higher than from all electronic manufacturers in India, Thailand and Vietnam. The report concluded that direct investments by TNCs in low-income countries reduced developed countries' emissions by placing a greater emission burden on non-industrialized countries (Zhang *et al.*, 2020).

According to the World Resources Institute (WRI), China, followed by India, Brazil and the US account for 37% of greenhouse gas emissions from global agriculture. Land-use changes and forestry have decreased greenhouse gas emissions since 1990 (14%), reaching their lowest point in 2013 and since then they have been steadily increasing. Other sectors have continued to increase their emissions since 1990, such as industry (203%) and waste (19.5%) (Friedrich *et al.*, 2023).

Globally, carbon dioxide emissions reached a record high in 2022 and have increased by 4% over the last 5 years. According to the International Energy Agency (IEA), the total Organisation for Economic Co-operation and Development (OECD) production of crude oil, liquefied natural gas (LNG) and refinery feedstocks increased by 3.9% in early 2023 compared to the previous year (IEA, 2023). The production of natural gas increased by 4.9% (IEA, 2023). The demand for LNG increased sharply after the outbreak of the war in Ukraine. China signed a new gas supply contract in 2023 with Qatar with a commitment to an annual delivery of 4,000,000t of LNG to China for the next 27 years (Benny, 2023).

Biogas can be an alternative, using plant material in a recycling process. The promise by the previous Swedish government (2018–2022) to get tax exemption for biogas made Swedish farmers invest heavily in this approach. But the European Union (EU) changed its policy and decided that biogas should be taxed in the same way as oil. In consequence, this means continued dependence on fossil energy sources and is why Swedish farmers have rightly demanded a reintroduction of Swedish tax exemption on biogas.

All countries have signed the Paris Climate Agreement and some 60 countries have developed long-term plans to decarbonize their economies. Sweden is one of 18 countries that have permanently reduced the release of greenhouse gases since 2000, except for 2023 and part of 2024. Although India ranks high among the emitters, the country ranks significantly lower than the other top ten emitters, when greenhouse gas emissions per capita are considered (Friedrich *et al.*, 2023).

The IEA has warned that the world needs to back away quickly from new oil, gas and coal projects to combat the climate change. All greenhouse gas emitters must take action, including TNCs. The Clean Development Mechanism (CDM) allows emission-reduction projects in low-income countries to earn certified emission-reduction (CER) credits, each equivalent to 1t of carbon dioxide. These credits can be traded and sold, and used by industrialized countries to a meet a part of their emission-reduction targets under the Kyoto Protocol (CDM, n.d.).

COP28 in Dubai saw progress on certain issues, such as the operationalization of the Loss and Damage Fund with some initial pledges, acceleration of the phase-out of coal power and some response to the UN's Global Stocktake synthesis report. It showed that the world is still far off track from reducing emissions enough to keep temperature rises to recommended levels. There was a historic agreement to move the world away from fossil fuels, although countries will follow different pathways to shift away from them (WMO, 2023). The agreement implies that oil-producing countries can continue to export oil for a long time to come. The shift away from fossil fuels was to be fair and fast but a critical test is whether sufficient financial resources are mobilized for the low-income countries. National climate plans must include targets with deep emission cuts in alignment with the IPCC's findings that emissions must be reduced by 60% by 2035 if the 1.5°C goal is to be reached (WRI, 2023b). This will require quick action.

In addition, 22 countries agreed that nuclear power is to triple globally by 2030 without discussing its long-term safety or sustainability. A new feature at COP28 was the declaration on sustainable agriculture, resilient food systems and climate action signed by more than 150 heads of state and government. It emphasized the potential of agriculture and food systems, calling for practical action; a dimension seldom highlighted at previous climate conferences (UNFCCC, 2023).

There is a serious conflict for governments in industrialized countries dependent on fossil energy between the need to reduce greenhouse gas emissions and provide jobs and wealth to people. This applies to the oil industry in the US. Not even Russia´s invasion of Ukraine has led to new leasing activities for countries searching for new energy supplies as an alternative to Russian imports. In 2022, new exploration for oil and gas fell sharply worldwide. According to an analysis by the Norwegian energy firm Rystad, the total area of new oil and gas leases has fallen; there were only about 40 global lease sales in 2022, the lowest level since 2000. A downward trend in oil drilling owes more to actions by oil companies than by governments. There might be a small but growing long-term shift away from fossil fuels. Still, representatives of the oil companies have maintained that the world is not ready to give up fossil fuels (Nilsen, 2022).

In 2022, Norway planned to expand Arctic oil and gas drilling in a new licensing round to reduce dependence on Russia. The Norwegian government-owned Equinor energy company doubled its profits in 2022, compared to 2021, but postponed its plans for oil drilling to 2026 (Equinor, 2023). The outcome at COP28 was praised by the Norwegian authorities with reference to the fact that the Norwegian government had collected almost three times as much in taxes from oil companies in 2022, compared to 2021, according to the Norwegian Tax Administration (2023). Simultaneously, the industry association, Offshore Norway, released new, increasing investment figures.

The global oil market may be at the start of another multiyear boom cycle. Predictions indicate new licenses in Canada and the US and new exploration in South America and West Africa. Another aspect of the future fossil-fuel economy relates to the increasing costs of extraction because more inaccessible oil sources must gradually be used. In addition, Angola left the Organization of the Petroleum Exporting Countries (OPEC) oil cartel in 2023 after 16 years as a member, unhappy with the way the cartel managed the country's production quotas. Thus, OPEC consists of 12 countries since Qatar, Indonesia and Ecuador have also left the cartel for various reasons.

First, electrification is viewed by politicians as the solution for replacing fossil energy. This is not great news. The electric car is more than a century old and Swedish steel was produced with fossil-free charcoal long ago. A significant issue is that politicians claim, naively, that a decrease in carbon dioxide emissions could be achieved without negative consequences for society. Instead, new climate-smart 'green' technology will be the solution, together with rapid electrification. Then, business as usual may continue...

One example is a new 'green' industrial investment in northern Sweden. The HYBRIT initiative from the companies SSAB, LKAB and Vattenfall intends to create an entirely fossil-free value chain, with fossil-free pellets, fossil-free electricity and hydrogen (Vattenfall, 2022). It will require a huge increase in electricity to produce fossil-free iron. Also, natural gas will be needed as one component of the industrial process. This will lead to greenhouse gas emissions. The investment requires a large amount of electricity, which is not yet available. Moreover,

Swedish politicians seem to forget that this project is not the only one in Europe so there will be competition in the production of fossil-free steel.

The current electric system is not designed for large expansion in northern Sweden. A shortage of electricity is already expected in 2026–2027 according to an analysis by Svenska Kraftnät (2022). To meet demand in 2030, the electricity consumption in northern Sweden must be regulated, otherwise overloads threaten. In June 2023, the director of Finnish Energy concluded that Sweden has almost chronic restrictions on its electricity imports. The poor condition of the national grid of the state-owned Svenska Kraftnät public utility means that neither Finland nor Denmark or the European continent can supply future electricity to Swedes when there is a deficit in Sweden.

Second, serious politicians claim that their decisions should be research-based. This would justify consideration of the laws of thermodynamics. The first law stipulates that energy can neither be created nor destroyed, but only transformed from one form to another. Since the amount of energy in the universe is constant, the sum of energy supplied and used in an open system is equal to the change in internal energy. The second law postulates that the quality of energy, i.e. its working capacity, decreases every time an energy conversion occurs. If the energy is increased locally, a decrease occurs elsewhere in the universe. Energy decrease in the habitat is an irreversible process, which has been going on for a long time. Energy that does not carry out a job becomes waste or pollution. This principle also applies to a discussion on electrification.

Third, changes in energy systems are rarely complete. Different alternatives operate simultaneously as history shows (Rhodes, 2018). Any transformation from a dominant type of energy to another takes time. Forests and coal have gradually been replaced by oil and gas. Initially, the forests were an important source of charcoal in iron production, timber for homes and firewood. In the 1500s, massive quantities of oak timber were required for English and Swedish ship production. Burning coal became common in the 1600s because there was a shortage of forest. This required less work and was cheaper. In the 1700s, coal accounted for half of England's domestic

energy needs, but the proportion fell later during that century. Gradually, steam came into use (Rhodes, 2018).

The goal for carbon dioxide reduction to zero is set for 2050, which seems a brief time span from the perspective of history. The problem has recently been illustrated in the car industry. The German government requested an exemption from the rules of the European Commission to stop the sale of internal combustion engines after 2035. Instead, car engines running on so-called electrofuel (e-fuel) should be allowed, using a synthetic alternative to fossil fuel. An agreement was reached in March 2023 between the EU and Germany. Internal combustion engines powered by synthetic e-fuel can continue to be sold, even after the requirement that all new cars sold should be climate neutral by 2035. In late 2023, the UK government decided to postpone an earlier ban on the sale of new petrol and diesel cars from 2030 to 2035 (Humphries, 2023).

The issue of climate change is concentrated on the replacement of fossil fuels with wind power, bioenergy and nuclear power. This will be a momentous change for the whole of society. So far, there has been little political discussion about exactly how the transformation from fossil fuels to electrification will change society and the disadvantages of different renewable energy sources.

First, the expansion and increased efficiency of wind energy is strongly supported by the Swedish Energy Agency, most leading Swedish politicians and many climate activists. Wind energy is seen as the fastest way to increase electricity but it may not be easy. One challenge is the intellectual and mental process, which politicians have neglected to communicate to their fellow citizens. Many citizens oppose wind turbines close to their homes.

This problem led a French minister to decide, on March 9, 2024, that permits for onshore wind turbines and rules for the renewal of wind farms are illegal. The decision came after a legal appeal by the Fédération Environnement Durable and 15 associations (FED, 2024). It concerns the provisions of three successive versions of the noise measurement protocol which were intended to protect the health of all concerned residents. The decision not only affects current projects but also raises questions about the viability of ongoing projects and the future of existing French wind farms.

Second, it is reasonable to expect that the utilization rate for additional wind power in 2023–2026 would operate at the same level as the average in Sweden for 2010–2023. That was, however, only about 26%, which is in sharp contrast to the highest utilization rate measured globally of 57%, in windy New Zealand, close to the Roaring Forties of the Southern Hemisphere (B. Persson, personal communication, Bara, Sweden, 2024).

The government-owned China General Nuclear Power Group (CGN), controlled by the Communist Party, is the single largest owner of Swedish wind power with almost 7% of the wind power produced in the country. At least 1000 of the country's approximately 5200 wind farms are on the brink of bankruptcy due to financial difficulties. This was recently presented in a review by Jönköping International Business School in 2023, based on a database with annual reports from approximately 3500 major Swedish wind power companies, from 2010 onwards.

One example is the Markbygden Ett company, in northern Sweden, which is 75% owned by CGN. It is operating 179 wind turbines but has made economic losses for several years due to a very unprofitable electricity contract with the Norwegian company, Hydro Energi. That agreement means that Markbygden Ett has to sell a precise volume of electricity at a fixed price for 19 years. When the wind turbines failed to produce enough electricity, the company had to purchase on the open electricity market. Now, the company wants to terminate the agreement with Hydro Energi and sell electricity on the market during its restructuring. Markbygden Ett's decision on corporate restructuring in late 2023 was opposed by the Hydro Energi (Håkansson, 2024).

Globally, recent data shows that the renewable energy capacity added to energy systems grew by 50% in 2023, compared to the previous year (IEA, 2024a). In fact, solar power accounted for three-quarters of the global additions. The largest growth of solar and wind power took place in China but there were also increases in renewable energy capacity in Europe, the US and Brazil (IEA, 2024b).

There are major questions about how to produce food and how to live our lives in a new type of society, if a transition due to the climate issue is to be dealt with seriously, including sustainability. Among the likely effects are increased costs for transportation, leading to reduced consumption and economic growth. Transportation costs can be illustrated by an example from Indonesia. Rice is the staple food for most Indonesians, but annually varying imports are necessary. In recent years, the government has been limiting rice imports and instead supporting local rice farmers (Statista, 2023). Indonesia does not produce wheat domestically but is reliant on wheat imports to fulfil demand for wheat flour-based food for its current population of 275 million and as an ingredient for poultry, aquaculture, and livestock feed. This was originally based on an old trade deal on wheat with the USA, but Australia was the largest wheat supplier to Indonesia (46%) in 2022–2023 (Reidy, 2023). During that period the total consumption of wheat was estimated at 9.5 million t according to the Foreign Agricultural Service (FAS) of the US Department of Agriculture.

National food production will become more important for all countries. Declining global productivity and food production, together with demographic developments, will create increased demand for meat and fruit in low-income countries with a growing middle class. Import prices will increase with less access to various agricultural commodities. Ecological footprints must change, either by political will in an organized manner or by nature itself executing such a transformation. The timetable may vary.

With this knowledge, serious politicians would have been expected to take some quick action, even though these efforts may be unpopular. Many politicians and lobbyists, arguing for reduced greenhouse gas emissions, still fly by private jets to attend climate meetings. This applies both to the EU's elite and the world's wealthiest people (CE Delft, 2023). To underline the point, there has been a doubling of private flights within Europe according to Greenpeace and Eurocontrol (2024).

Another political decision would be to reduce the use of motor transport, and perhaps even motorsport events, which would be no new invention. As early as the 1970s, some streets were closed to car traffic in Bogota in Colombia, but in the last 4 years, more streets have been closed off and there are now more bike lanes. Another example is the 'Green Ring'; some 30 km of parks and cycling trails surrounding the Basque city of Vitoria-Gasteiz in northern Spain. It stems from a decision in the 1970s, stating that the city centre should be for pedestrians and cyclists. After 2006, superblocks were introduced and closed to through traffic in a city of about 250,000 inhabitants. About 70% of the city's areas are dedicated to pedestrians (Burgen, 2023).

At present, there are several car-free islands, for example, the Channel Island, Sark, with its horse-drawn vehicles, tractors with permits, electric vehicles for people with disabilities and bicycles. On other islands, there is legally restricted or no vehicle traffic but not petrol-driven cars. Emergency vehicles and trucks for collection of rubbish are allowed but horses and donkeys are used for transport on land and dhows are used at sea. Aside from Sark, examples around the world include Australia (two islands), Brazil (8), Croatia (7), Greece (Hydra), Germany (7), France (10), Indonesia (Siberut), Kenya (Lamu) and the United States (Machinac in Michigan state where cars have been forbidden since 1898).

Car-free Sundays could again be used as a simple contribution to reducing greenhouse gas emissions. This was done in Sweden during the Suez Crisis (1956) and the war between Egypt/Syria and Israel (1973) as a way of saving fuel when supplies were threatened. Another initiative could be a putting a gradual stop to using high-speed, non-essential Swedish A-tractors.

Increased Exploitation of Natural Resources

Mining

The idea of 'green growth' offers an opportunity for well-meaning politicians to continue with business as usual. All new 'green' technologies require continued – and increased – exploitation of the globe's natural resources and contribute to waste mountains. In addition, there will be greenhouse gas emissions. The electrification of transport requires increased availability of essential earth elements, such as lithium and cobalt, in addition to conventional mining of iron, copper, zinc, gold, etc. Without these resources, countries cannot build drones, electric cars, fighter jets and anti-aircraft missiles.

This will affect people and their environment, as exemplified when the Swedish Mining Inspectorate granted exploration permits in 2018 to different companies for 22,000 ha of productive land in the Scania province of south-eastern Sweden. The aim was to search for rare metals in an area where many people make their living. So far, no mining has been granted but this demonstrates the conflict between mining and long-term sustainable development in this region.

The European Commission has classified 34 minerals and metals as critical for society and industry. This is underlined by the fact that Europe produces only 3% of the world's metals but consumes 25% according to Swedish Geological Society (SGU, 2023). After a year with high energy costs, supply chain disruptions, and the US implementation of a large-scale Inflation Reduction Act (IRA), the EU's Critical Raw Materials Regulation was a step taken to revamp its reindustrialization and competitiveness investment plan in 2023 by setting targets for the production, refining and recycling of key raw materials for green and digital transitions. The demand for rare earth metals for wind turbines is expected to increase 4.5-fold by 2030. For lithium, the demand is estimated to increase 11-fold by 2030, and 57-fold by 2050. One objective for the EU is to reduce dependence 'on imports, often from quasi-monopolistic third country suppliers' (Bourgery-Gonse, 2023).

Extractive industries are working in many places in the world and many TNCs are active in mining. In the late Middle Ages, two-thirds of the world's gold supply was transported from West Africa to Genoa and Venice. But when the Europeans needed slaves for cotton and sugar production in the Americas, they no longer had to tackle the problems of operating in the interior of Africa. This changed when Cecil Rhodes started a diamond trade at the Kimberley mine in South Africa with funding from Rothschild & Co. He gained almost complete domination of the world diamond market and formed the De

Beers company in 1888. De Beers Group is still a dominant international mining corporation, operating in 35 countries. West Africa remains an important source for gold mining, as seen in Ghana. About 60% of its water bodies are now considered polluted, containing mercury and cyanide, largely due to illegal mining activities (Naadi and Lansah, 2021). About 35% of the gold is extracted by small-scale mining both in areas with large gold deposits and in forested regions with valuable cocoa farms. This extraction affects both the environment and humans.

China dominates the production of raw materials. It controls a large part of the extracting and refining processes of raw materials used for computers, LED lamps, semiconductors, batteries and solar cells. The Chinese have great control of the production of germanium (97%) and gallium (73%) according to a report from the Nordic Innovation (2023). The country decided to limit its exports of these minerals by 2023. China is also a big producer of tungsten (84%), antimony (87%) and indium (56%) explained by Lin *et al.* (2019). The Democratic Republic of Congo extracts some 60% of the world's cobalt, a metal needed in rechargeable battery electrodes (Lipton *et al.*, 2021). The global production of beryllium is dominated by the US (90%), whereas South Africa produces 70% of platinum and Russia some 46% of palladium (Statista, 2024).

Iron is one of the most common elements making up 4.5% of the Earth's crust and is widely distributed. Only exceptionally does the metal occur in its pure form in nature. The main ore minerals are usually oxides, such as hematite or magnetite. Ore prospectors normally seek a small part of the rock's content, but ores also include sulfur, phosphorus, arsenic and other heavy metals. In Sweden, ores can contain up to 35% arsenic. This places great environmental demands on mining stakeholders, implying conflicts between economic interests, environmental aspects and political strategies. In principle, a mining industry is unsustainable in the long term because mining companies, as a minimum, are required to dismantle mines and take safe care of all waste.

Since Sweden has one of the world's largest ore concentrations, it may be used to illustrate mining operations in relation to environmental aspects. Magnetite was reported in northern Sweden in the mid-1600s but large-scale mining began with the arrival of the railway. Established as a private company in 1890, Luossavaara-Kiirunavaara Aktiebolag (LKAB) has been 100% government-owned since the 1950s.

The main ore fields are the Kiruna mine and Malmberget, two of the world's largest underground mines. In Svappavaara, the ore in Leveäniemi mine is in an open pit. The company has subsidiaries for industrial minerals with processing plants, among others, in Finland, the UK, the Netherlands, Turkey and China. The expansion of the underground mines in Kiruna and Malmberget means that the towns must relocate. By 2035, Kiruna city will be phased out and a new centre created. Malmberget must move to Gällivare. This is called community transformation; a mine gives and a mine takes.

LKAB has deposits of minerals until 2060. The current goal is fossil-free steel and zero carbon dioxide emissions by 2045. In 2023, there were new discoveries of rare minerals, including graphite. In principle, current mining takes place despite the precautionary principle in the Swedish Environmental Code of 1998: 'damage or risk of damage must not be put into systems'. The ore operators are required to dispose of massive amounts of residual products, for instance, arsenic. These products are mixed with water (slurry) for disposal as close to the mine as possible and to settle there. An active open pit cannot be filled because more ore will be mined there.

At the start of a mine, an area is selected for landfill, and surrounded by dykes (ponds) and natural elevations. When the landfill is filled up, it must be elevated so new dams are built above the old ones at intervals of several years, since the dams expand over time. Large landfills can have an area of 10–15 km^2 and heights of tens of metres. For dam safety, water must be drained but a current trend is to reduce the amount of water in landfills.

Since the dams are permeated by both process water and annual precipitation, their residues, and other metals, are transported into the river valleys, to surrounding environments and soils, and into crop and forestry land. The fine-grained particles settle at a great distance from the point of discharge and the coarsest particles will end up closest to the ponds. Mining dams are supposed to keep sand contained, never to be removed, since the sand can spread uncontrollably. The sand holds minerals with lesser amounts of metal and often residues of various chemicals from the enrichment process. This means that mine dams must exist in perpetuity and be so stable that their safety can be guaranteed for a thousand years.

The Boliden company has a large stock of arsenic, but uncertainty prevails about how it should be handled. Its main products are zinc and copper concentrate, with some lead, gold and silver content. In principle, all should be decomposed, and unused rock is returned to solid rock. A possibility might be the use of microfungi, found 1000 m underground by researchers at the Swedish Museum of Natural History in LKAB's Kiruna mine. Some fungi species are believed to be able to purify the water pumped out of the mine (Nyström, 2020).

Another problem is the short-term nature of mining, exemplified by the Kallak mine in the Jokkmokk municipality, said to create 300 jobs over 14 years. But it was deemed to be of national interest according to the Geological Survey of Sweden (SGU). Usually, ores are used for short periods and can easily be replaced when the market changes. Then, it will be difficult to politically fulfil the sustainable development goals (SDGs) with current regulatory systems. The Sami Parliament has protested against mining for decades, claiming that the iron ore deposit areas are of national interest for reindeer herding. Such herding demonstrates sustainability, both for those who live in Kallak and their reindeer production, as practised for centuries.

In 2022, the Swedish government decided to grant a processing concession for the Swedish-registered Jokkmokk Mining, owned by the British Beowulf Mining company. The government also stated also that affected Sami communities should not be harmed. After mining is completed, the company should restore the land area so it can again be used for reindeer husbandry. The government decision had been appealed by Sirges Sami community and the Swedish Society for Nature Conservation, but was upheld in June 2024 by the Swedish Supreme Administrative Court. The Kallas mine was found to be of national public interest, as it previously had been by the SGU. Some 14 years of mining operations were considered more important than safeguarding continuous

long-term reindeer husbandry by Sami communities. The next step for the mining company is to apply for environmental permits. So far, it is unclear how this process will end but it hardly meets demands for sustainability.

Uranium for nuclear power is a special case. It occurs as a natural element in soil and in the bedrock of hundreds of minerals. It is more common than cadmium and mercury and is about as abundant as arsenic or molybdenum (Emsley, 2001). Uranium ore is mined in Namibia, South Africa, Kazakhstan, Uzbekistan and Russia but the largest mines are in Canada and Australia. After 2021, only two uranium mines remain in Australia after protests from Indigenous people and environmental organizations. Sweden's bedrock is rich in uranium and may hide as much as 15% of the world's uranium resources. Extraction has been prohibited since 2018 but is now being considered as a means of securing future Swedish energy supplies.

This raises the question of sustainability and its conflict with providing more electricity to secure continued growth and good living. The half-lives of uranium-235 and uranium-238 are 0.7 and 4.5 billion years, respectively. The Swedish Radiation Safety Authority has concluded it is about 710 million years. The waste is placed in sealed steel barrels in underground silos or trenches, leaving millions of tonnes of dangerous tailings that risk leaking into the environment. The wind can carry radioactive gases and polluted rainwater can reach the groundwater and enter the food chain, as it has done in, for example, Namibia. Residual products have caused health problems for the Navajo Indigenous people in the US due to uranium mining up to the 1980s (Ingram *et al.*, 2020). In 2003, one-third of the Navajo population lacks access to clean piped water due to contamination, a figure that was about 15% in 2020 (Environmental Protection Agency (EPA), 2024b). Canada struggles with contaminated water and poisoned fisheries from its mining operations.

From a scientific point of view it is interesting that some bacteria (*Shewanella putrefaciens*, *Geobacter metallireducens*) and some strains of *Burkholderia fungorum* use uranium for their growth (Min *et al.*, 2005). A lichen (*Trapelia involuta*) or the bacterium (*Citrobacter*) can absorb concentrations of uranium that are up to 300 times the level of their environment (Emsley, 2001).

Today's mining takes place both over and under the ground. Throughout the world, there are huge underground structures for mining and oilfields in the Middle East, the US, China, Nigeria and other countries, as well as under the sea. It has been estimated that 50 million km of tunnels and holes have been drilled for underground natural resources (Macfarlane, 2019). These industries have major economic, environmental, social and cultural impacts, both at the national level and on the population in the locality.

Rising demand for metals has led to discussions about deep-sea mining. It is a process of retrieving mineral deposits from the seabed below 200 m. This depth covers about two-thirds of the total seafloor and contains an extensive array of geological features. These include abyssal plains, which are found 3500–6500 m below the sea's surface, volcanic underwater mountains known as seamounts, hydrothermal vents with bursting water heated by volcanic activity, and deep trenches such as the Mariana Trench (IUCN, 2022a). Relevant metals are copper, nickel, aluminium, manganese, zinc, lithium and cobalt, all important for technologies like smartphones, wind turbines, solar panels and batteries (Economist Impact, 2023). One active company is Moana Minerals Ltd, which is attempting to source these metals from seafloor nodules found near the 15 islands of the Cook Islands in the South Pacific Ocean. This may be less problematic since the Cook Islands is a self-governing country in association with New Zealand (Moana Minerals Ltd, 2024).

By mid-2022, the International Seabed Authority (ISA), which regulates activities on the seabed beyond national jurisdiction, had issued 31 contracts to explore deep-sea mineral deposits. This included some 1.5 million km^2 of international seabed. In mid-2021, the Government of Nauru notified the ISA about its intention to start deep sea mining and there are plans in Norway as of 2023. Mining in international waters could commence in 2026 but this raises questions about how to mine ores, since these international areas belong to the world as a whole and not to individual governments (IUCN, 2022b).

Another concern is that current research has suggested that mining in the deep sea could destroy marine biodiversity and ecosystems and risk sustainability. In many remote places far below the sea's surface there might be unknown species that are uniquely adapted to conditions of lack of sunlight and high pressure. Many of these species are unknown to science, with gaps in the knowledge of marine biodiversity and ecosystems making it difficult to assess the potential impact of deep-sea mining in the marine environment.

Urbanization and loss of farmland

Expansion of agriculture is usually the main driver of forest loss, but urbanization can cause indirect loss of forest. Between 1992 and 2015, the world may have lost up to up to 35 million ha of forest to urbanization (van Vliet, 2019). Although urban regions across the world have grown rapidly in recent decades there are few studies on the impacts of this urban expansion. Global natural dryland habitats cover 40% of global land area and provide habitats for 28% of endangered species (Ren et al., 2022). Global urban expansion from 1992 to 2016 has resulted in an average 0.8% loss of dryland habitat quality but the indirect impacts were 10–15 times greater. Globally, almost 60% of threatened species were affected by dryland urban expansion (Ren et al., 2022). According to analysis based on the land cover data of the European Space Agency's Climate Change Initiative, some 22 million ha of shrubland may have been lost (Plummer et al., 2017).

In 2021, the world's total agricultural land was 4.8 billion ha, one-third of the global land area. Cropland covered 1.6 billion ha and permanent meadows and pastures 3.2 billion ha (FAO, 2023). Since 2000, cropland area has grown by 6%, while permanent meadows and pastures have decreased by 5%. Since 1961, arable land worldwide has decreased by one-third. This reduction is due to soil erosion, reforestation and desertification, where global climate change has played some role.

In 2021, global arable land covered 1.4 billion ha, mainly temporary crops such as wheat, rice and maize. Some 183 million ha were permanent crops, an area which has grown by 40% since 2000. By comparison, land area for fallow and meadows had declined. In 2021, the world average cropland area per person was 0.2 ha per capita, a decrease of 18% since 2000 (FAO, 2023).

Another reason for the declining arable area is that the world's urban has areas have spread over productive soils. Estimates show that around 450 million ha are currently cultivated in and around the world's largest cities. By 2030, it is estimated that just over 2% of today's global cultivated area will disappear because of urbanization, especially in Africa and Asia. This means an estimated loss of between 1.6 to 3.3 million ha of prime farmland since 2000, according to the United Nations Convention to Combat Desertification (UNCCD) (2019).

According to UN projections, China's largest city, Shanghai, had some 24 million inhabitants in 2020. It is estimated to quadruple its population over the next 50 years. Likewise, the population of Beijing has tripled over the last 30 years. Chinese urbanization has risen from 19% in 1979 and is expected to reach about 80% by 2050 (World Bank, 2018). Rural–urban migration accounts for the movement of over 200 million residents to Chinese cities, but another 200 million rural residents have become *in situ* urbanized via administrative reclassification (Kan and Chen, 2022). Between 1987 and 2000 there was land expansion in China's north-east and north-west regions. This meant a small net increase in cultivated land, but this area declined by 1.2% between 1995 and 2000. The major causes of changing land use are ecological land use conversion, non-agricultural construction and readjustment of the land use structure (Huang *et al.*, 2005).

According to UNCCD (2019) there were 28 megacities ten years ago, home to 453 million people. In the 1970s, Schumacher (1973) warned that a city's population should not rise above 500,000, albeit an idealistic concept at that time. The current political thinking is to encourage urban growth, sometimes leading to megacities. By 2030, 13 new ones are expected to emerge in the less-developed regions. This trend is seldom noted in strategic discussion on future food security and given little attention in most agricultural research.

One example is Greater Cairo in Egypt, with nearly 20 million people, which is expected to double in size in the next few decades. Thus, a new city was announced in 2015 to become Egypt´s new administrative and financial capital, being part of a larger initiative called Egypt Vision 2030. One official reason for the initiative was to relieve congestion in an already crowded Cairo. The new city is located 45 km east of Cairo in a largely undeveloped area halfway to the city of Suez. It is expected to house a population of almost 7 million people on 700 km².

The move of more than 30,000 Egyptian government employees has been held up due the COVID-19 pandemic and construction delays. By May 2023, 14 ministries and government entities had been relocated to the New Administrative Capital according to Ahram Online (2023). To some extent, this project is financed by Chinese capital and some Emirati developers but it is also funded by the Egyptian government through loans and high-interest bonds. Foreign investors have shown little interest in Egypt since the military has a strong command of the economy.

Another example is from Indonesia. In 2019, a 10-year plan was announced to transfer all government offices to a new capital city, Nusantara. It is to be inaugurated in August 2024, replacing Jakarta with its 10 million people. The new capital will be situated in the province of East Kalimantan on the island of Borneo, on an expected area of 2560 km², relatively free from volcanoes, earthquakes and tsunamis. The project has been estimated to cost US$35 billion and construction began in August 2022. In addition, the government allocated US$40 billion to save Jakarta from sinking in the next decade. Parts of it have already begun to sink by about 25 cm/year.

These two examples show there might be a limit to current megacities. The construction of a new one will be expensive with environmental consequences and raises the question of sustainability. With an already fragile economy, the Egyptian project, with an estimated cost of US$59 billion and additional costs of arms purchasing, has led to a quadrupling of the national debt over a decade (Walsh and Lee, 2022). The Indonesian government was planning to provide some 20% of the funding for Nusantara with the rest coming from domestic and foreign investors. However, so far investors have been reluctant due to Indonesia's political uncertainty and its lack of investment in infrastructure.

The trend of urbanization also indirectly affects biodiversity, and places demands on cultivation opportunities in urban areas that were previously unthinkable. Competition will arise with large-scale solar cells to provide electricity. Parking spaces will become expensive and so will food. Working from home and still receiving a salary has become possible after the COVID-19 pandemic.

So far, only a minority of high-income earners have preferred to leave metropolitan environments. A recent study by the US Department of Agriculture found, however, that as the urban population grows, residents have discovered that in addition to congestion and housing shortages, they have had increasingly limited access to supermarkets, groceries and health stores. Thus, they can no longer profit from their lower prices.

In India, loss of agricultural land is occurring more often around smaller cities than around areas suitable for agriculture close to urban expansion of large cities (Pandey and Seto, 2015). The north-eastern states experienced the least loss of agricultural land. Urban conversion of agricultural land was concentrated in states with high rates of economic growth and good agricultural productivity. Since 2006, the amount of agricultural land converted to urban areas has been increasing steadily. At the same time, India has brought 9.8 million ha of degraded land under restoration since 2011 (Aggarwal, 2018).

In Ethiopia, urbanization leads to higher demand for agricultural land in the peri-urban areas of the country (Ayele and Tarekegn, 2020). In contrast, only about 8% of land is arable in Zimbabwe according to FAO. That country has lost more than 6 million ha of forest land since 2000. In 2021, the country's top exports were gold, nickel mattes, raw tobacco and diamonds, a radical change with the past. Prior to independence in 1980, there was a commonly held view that Zimbabwe was the bread basket of Africa. This narrative gives an impression that Zimbabwe lost its 'bread basket' status after President Robert Mugabe introduced land reform, which resulted in the decline of Zimbabwe's agricultural output. But there is limited evidence to suggest that the country was

a dominant player in Africa's food production of staple foods prior to that period (Sihlobo, 2017). Since the FAO started recording African agricultural statistics beginning in 1961, maize and wheat has never surpassed a 10% share of Africa's production.

Although the land reform initially meant an agricultural decline, more recent studies have shown that – based on a 30-year experience – the productivity of small producers has grown slowly with output escalating. The land reform has recast land-based social relations in important ways, with the poor gaining more than previously believed (Moyo, 2011).

Later, high inflation led farmers to invest in their cattle, which keep their value when the local currency decreases. As in earlier times, cattle can later be traded for other services and products, for example, a bicycle, solar cells or for transport to the hospital. Recently, Zimbabwe registered its largest wheat harvest, 375,000t in 2022. It made the country self-sufficient in wheat (Gbane, 2023). The harvest in 2022 was 13% higher than the previous year, breaking a half-century-old record, saving up 300 million Zimbabwe dollars in import costs.

Loss of farmland also occurs in the US which has 10% of the world's arable area. Only 17% of American land is ideal for farming so this development is critical. Between 1992 and 2012, some 12.5 million ha of farmland were lost in the US (Hunter et al., 2022). Some of this was America's best agricultural land with suitable weather for food production. Expansion of cities and suburbs was the cause of most of the farmland loss. In the UK, almost 810,000 ha of grassland have been lost to urban development and new woods between 1910 and 2015, according to satellite analysis by the UK Centre of Ecology and Hydrology, 2020 (Beament, 2020).

In Norway, the total agricultural area in 2018 was approximately 0.986 million ha, of which about 0.806 million ha was arable land. Between 2001 and 2016, the total agricultural area in Norway decreased by 6%, a decline mainly attributed to changes in land use (NIBIO, 2020). In Sweden, almost 190,000 ha of arable land have disappeared over the past 25 years for housing, golf courses and roads, according to the Swedish Board of Agriculture and SCB (2023). Total arable land in Sweden was reported as 2.5 million ha in 2021. Nowadays, large logistics centres are built on Sweden's most fertile arable land despite the number of lorries needing to be reduced to save the climate according to both local and national politicians. Every year, around 600 ha of agricultural land are exploited in Sweden. It is up to each individual municipality to decide and strike a balance.

The Pågen bakery company was recently allowed to build a giant bakery in southern Sweden on almost 60 ha of land, which is classified as one of the world's best agricultural areas. Future food is viewed as less important by local politicians, preferring to focus on the 200 jobs to be generated. Nonetheless, the bakery still needs grain to be produced – somewhere. In 2013, the Swedish Board of Agriculture claimed that agricultural land is being exploited without careful consideration. Better legislation is needed since agricultural land must be seen as a strategic resource, linked to future national security.

In this perspective of diminishing arable land, the EU Commission proposed a Nature Restoration Act in 2023. It seems short-sighted, bureaucratic and dubious with regard to future agriculture and sustainability. It relates to decisions made in 2022 at the meeting of the Kunming–Montréal Global Biodiversity Framework. It aims to arrange continued and sustainable restoration of resilient and biodiverse nature and to save biodiversity.

Land and Water Grabbing

Land grabbing as a process

Land grabbing goes back a long way to territorial wars, precolonial land seizures by foreign rulers and dispossession of native peoples in Australia and North America. Land grabbing reappeared internationally when global food prices increased in 2007–2008. Private and public actors and sovereign governments acquired land resources in low-income countries with good water supply for capital accumulation to secure their own food security. The major actors were Saudi Arabia, China, and South Korea but private investors were also involved.

Land grabbing is a process through which governments or private investors from developed or growing economies buy or lease a vast amount of land in low-income countries. The term 'land grabs' was defined in the Tirana Declaration (2011) by the International Land Coalition, consisting of 116 organizations from community groups to the World Bank (International Land Coalition, 2011).

Three major causes of land grabbing may be identified: food security, energy security and pursuit of economic gain (Poderati, 2022). Also, water can be of interest and mining is an instance of water grabbing. The pollution of streams and rivers with mine tailings illustrates how mining can affect the local political economy. Corruption enables land grabbing in several ways. It can be state officials accepting bribes from a company to gain access to land or it can be institutionalized. Due to skewed decision-making in government, confidential business or political elites, the national laws to seize land can be circumvented without consequences.

In general, foreign-owned companies consider certain points when deciding upon land investments. One is sector, such as food for export or for food security, land for biofuel or water for hydropower and electricity. Size is important. The larger the size of the investment, the more security investors will seek in their acquisitions. Legal aspects are a third point. In many East Asian countries, foreigners are not allowed to own land. For example, in Thailand and Indonesia they must lease from private citizens. In sub-Saharan Africa, a long-term lease from the government is the only option since land is held by the government or on behalf of the people. Lease contracts allow for protection against expropriation risks and currency devaluations.

The scale of land grabbing

Over the years, there have been a range of estimates of land grabbing that differ in quality. Large-scale land deals were reported to have risen by 20 million ha between 2005 and 2009 according to the International Food Policy Research Institute (von Braun and Meinzen-Dick, 2009), by 45 million ha since 2007–2008 according to the World Bank (2010) and by 227 million ha since 2000 according to Oxfam

(2011). Often, land areas targeted for large-scale investment have been described as idle, marginal or degraded land, unpopulated and unproductive. The World Bank sent an early positive signal of a large global reserve of potential land of between 445 million and 1.7 billion ha (Oxfam, 2011).

In contrast, Brautigam (2015) claimed that early land deals by Chinese investors had not always been implemented. This included the 3 million ha of farmland in the Democratic Republic of Congo for oil palm plantations, and the lease of 2 million ha in Zambia for a biofuel project. The analysis of and visits to some of the 60 large-scale deals announced during 1987–2014 showed that only 38 of the land acquisitions had led to a transfer of just 240,000 ha.

Other organizations have also been concerned with issues of land grabbing. The international non-profit organization GRAIN (2016) estimated that there were 500 deals covering 30 million ha in 70 countries, with half of the deals in Africa. Some of the large projects had gone bankrupt, flopped or were stopped by community resistance, sometimes with increasing violence. This was highlighted by Jurkevics (2022), reporting land grabs of approximately 56 million ha dispossessing and displacing inhabitants during 2008–2010.

Other smaller actors, for example, Global Witness, have also been documenting land grabbing with a focus on its negative effects on people in various countries. According to Global Witness (2014), about 12% of Papua New Guinea had been annexed by timber and palm-oil companies, using a leasing system intended for small-scale agriculture without government action. Land grabs in Myanmar have devastated livelihoods after many conflicts in a long civil war. In 2012, the Myanmar government established the Vacant, Fallow and Virgin Lands Management Law and the Farmland Law, aimed at managing the use and distribution of farmland. The Farmland Law established the Farm Management Bodies (FMBs), which were replicated at state and local levels, replacing community bodies and are responsible for 'guidance and control' on key land issues, ranging from disputes and transfer of rights to land registration (Henley, 2014). These laws and their administrative bodies have been criticised for encouraging land grabbing, and do not acknowledge customary land rights, which dominate in south-eastern Myanmar (Glatz, 2014). More emphasis has been placed on official land ownership acquired through applications, a system that is flawed with corruption and loosely respected. The FMBs lack adequate mechanisms for appealing over land disputes through a judicial process (Henley, 2014), making it difficult for displaced farmers to legally protest against their loss of land. Nonetheless, there have still been many complaints by farmers. There is, however, an urgent need for a unified national land policy to reduce land grabbing from the farmers.

During 2001–2018, Cambodia lost one-quarter of its tree cover according to Global Forest Watch (2019), using satellite data from the University of Maryland. More than 800,000 Cambodians are estimated to have been affected according to Global Witness, the International Federation for Human Rights and the Climate Counsel. Many were illegally forced off their land, both by the government and various corporations involved in timber and other activities. There were forced evictions, murders, violence and imprisonment of land activists (Surma, 2021). It led to a complaint against the Cambodian government submitted by an international law firm to the International Criminal Court (ICC) to consider land grabbing as a crime against humanity.

In a recent study on land grabbing, 128 case studies from 124 articles out of 252 peer-reviewed articles published since 2007 were analysed by Yang and He (2021). The findings revealed gaps in the existing literature in terms of conceptualization, methodology and research area. More interdisciplinary, holistic research is required to allow for broader regional/temporal contexts to get more evidence-based data.

A review of articles on land grabbing during the last decade showed changes since 2010 (Wolford et al., 2024). The land deals can be of diverse types and more 'green' environmental deals have been signed. Modern technology has made it easier for investors to use cryptocurrency instead of pension and hedge funds. Past efforts to regulate land grabbing through new institutional mechanisms have been futile. Future global regulations must involve new political powers, such as China, Russia, India, Brazil and the Gulf states.

Major actors in land grabbing

China's strategic expansion relates not only to maritime claims and infrastructure exports. The country is also a major trading partner with Africa. Over 10,000 Chinese-owned firms are operating throughout Africa according to Shepard (2019).

Agricultural land has been a priority for Chinese food security. According to Land Matrix, a European land monitoring organization, from 2011 to 2020, Chinese companies have gained control of 6.5 million ha of purchased or leased land around the world, devoted to agriculture, forestry and mining (Chiba *et al.*, 2021). This is about three times more than other foreign companies, for example, 1.6 million ha by British companies, 860,000 ha held by American companies, and 420,000 ha controlled by Japanese companies.

China has also acquired land in industrialized countries. According to a 2021 report from the US Department of Agriculture (USDA, 2021), China owned approximately 155,400 ha of American agricultural land. Half of this land area was owned by 85 Chinese investors; individuals, companies or the government. As a rule, buyers of land must complete an USDA form, indicating if they are purchasing the land for themselves, or for a government or some other entity. Lawmakers are concerned over the delays in reporting as purchasing land is also considered as a threat. The federal oversight system on reporting is considered weak and enforcement is minimal.

The previously American company, Smithfield Foods, with some 51,800 ha of land was bought by a Chinese firm in 2013. Four years later Syngenta was bought by a Chinese firm, which now owns, or leases a total of 2430 ha in the US (PBS News, 2014). NBC News reported purchases by Chinese entities (11 in 35 states) between 2022 and mid-2023 (Strickler and Moeder, 2023).

Saudi Arabian companies are also active in land grabbing. They control rice farms in Ethiopia, Sudan and the Philippines, cattle ranches in California and Arizona in the United States, wheat fields in Ukraine and Poland, ranches in Argentina and Brazil, and shrimp producers in Mauritania (Cooke, 2016). One specific example is the privately owned Saudi Star Agriculture Development Plc. In 2011, it acquired 10,000 ha of land along the Alwero River in the Gambela region of Ethiopia. It was a 60-year lease, free of land rent, which attracted the company to plan another acquisition of 500,000 ha in the same and other regions for the production of a projected 1 million t of rice, maize, teff, sugarcane and oilseed.

The company plans to dam the Alwero River for the irrigation of rice for export. Farms along the riverbank originally produced maize for local consumption using shifting cultivation. Land was lost since forests have been cleared. Since Gambela lacks a formal system of land tenure and property rights, the protests of the villages had insignificant effect. Impacted communities were forced to relocate, being told that clearing their ancestral lands was justified because it was actually 'government land'. Fieldwork by the Oakland Institute revealed that the Ethiopian government has handed over blanket discretion over Gambela´s water and resources to Saudi Star. The lease contracts were concluded without community consultation and they were kept confidential (Oakland Institute, 2011).

In general, most farms owned by Arab companies worldwide are in low-income countries. For instance, Qatar's wealth fund has holdings in Latin America and Africa (Spagat and Batrawy, 2016). Saudi Arabia prefers investing in high-income countries with political stability, exemplified by the US. The Almarai Company bought about 700 ha of US farmland in 2016, which doubled its holdings in California with a good water supply from the Colorado River, together with previous purchases of land in Arizona (Spagat and Batrawy, 2016).

Despite drought conditions, these areas are attractive to water-seeking companies because there are strong legal protections for agriculture and they have good water rights. The intention is to grow alfalfa to feed dairy cows in Saudi Arabia (Spagat and Batrawy, 2016). This is a policy change since the country had previously attempted to grow food crops at home. The priority is food security which requires land and

Table 7.1. South Korea's acquisitions of agricultural land in foreign countries as of contracts signed between 2005 and 2016 (Statista, 2016).

Country	Hectares
Sudan	690,000
Paraguay	240,000
Indonesia	171,400
Cambodia	67,813
Papa New Guinea	56,000
Laos	16,023
Mozambique	15,000
Philippines	11,300
Russia	10,000
India	688
Mongolia	270

resources overseas to guarantee a long-term supply. Then, alfalfa need not be grown at home.

The same principle applies to the American company, Al Dahra ACX Global Inc., owned by Al Dahra Agriculture Company of the United Arab Emirates. It is growing both alfalfa and hay in California and Arizona (Spagat and Batrawy, 2016). Worsening drought has led to a debate about whether Arizona should be doing more to protect its groundwater resources. The state attorney general of Arizona stressed in 2023 that most Arizonans are protesting that their state was allowing foreign-owned companies to use their water for free to grow alfalfa and send it home to Saudi Arabia (Naishadam, 2023).

Companies from South Korea are among the new important actors in land grabbing (Table 7.1). In addition, some Western European companies and the Nordic countries have accumulated land in Eastern Europe, especially in the 'Black Earth' area of Russia and Ukraine.

Land grabbing for biofuels

Climate change and discussions about the need for biofuels to replace fossil-fuel energy have led both governments and foreign companies to look at increased oil palm cultivation. This amounts to more than 4 million ha in southern and central Africa since 2005. The Singapore-based company, Olam International Ltd., has planted 100,000 ha of various crops in Nigeria, Mozambique, Ghana and Côte d'Ivoire, and has plans for oil palm cultivation in Gabon. The Singapore-based company, Wilmar International Ltd., has already planted a similar acreage in Ghana, Côte d'Ivoire and Uganda. It plans for strategic partnerships in oil palm cultivation in Gabon, China, Papua New Guinea and Tanzania (Wilmar, 2023).

Another example is the Malaysian company, Sime Darby, which produces rubber and palm oil from plantations on almost 1 million ha in about 20 countries through 200 subsidiaries, for example, in Malaysia, Indonesia and Liberia. Both Malaysia and Indonesia are major producers of palm oil, accounting for almost one-third of world production of 17 oils and fats in 2016 (Kushairi Din, 2017). In Malaysia, there were 5.4 million ha of oil palm plantations: the majority as private estates (61.2%). Felda Global Ventures Holdings Berhad (FGV) controls over 850,000 ha of land in Malaysia, including approximately 500,000 ha that it leases and manages for smallholders (FGV, 2023). In Indonesia, Golden Agri-Resources Ltd. (GAR), with a corporate office in Singapore, manages more than 532,000 ha of oil palm plantations in Indonesia and Liberia, including smallholder farmers (Golden Agri-Resources, 2023). The company is one of the largest oil palm plantation companies in the world and has set aside some 80,000 ha for conservation.

The European Socfin/Bolloré conglomerate controls about 400,000 ha of land concessions through a large network of holdings and subsidiaries. It operates in ten countries, in West Africa, Indonesia and Cambodia. Half of the land area is used for industrial plantations of oil palms and rubber trees, and sales and marketing of oil palm seeds (GRAIN, 2020). Since palm oil in Africa is normally produced by small-scale farmers, the large plantations cause conflict between the firm and local farmers. This frequently leads to land loss and displacement of the small producers.

Water grabbing and development

Water grabbing relates to hydropower development, mining and the control of water for development. The latter is demonstrated in the Chinese use and increased control of the Mekong River, with controversy over transboundary water issues and China's upstream hydropower developments. There are 11 large hydropower dams on the main stream of the Lancang River which have a significant environmental impact. Already in the 1990s, China had begun sharing agricultural development experience with Myanmar and Laos. It offered neighbouring Mekong countries special treatment through the ASEAN–China free trade agreement. To stimulate legal trade, China provided financing and technical support for the expansion and dredging of Mekong River sections in Myanmar and Laos. This led to improved connectivity, not only through hydropower grids but also railways, highways, and telecommunication networks, in a region traditionally linked across mountains and by waterways. By October 2023, China, Laos, Myanmar, and Thailand had conducted 134 law enforcement activities on the Mekong River (Gong, 2023). The Thai government has not agreed to a joint patrol entering the country's waters. The development cooperation often showed that closer security cooperation and border controls were required. Chinese overseas investors and businesspeople have taken advantage of the positive image of the Belt and Road Initiative (BRI) for private and even illicit activities (Gong, 2023) Also, regional criminal networks have become an issue through illegal wildlife trade in the borderlands of the Golden Triangle (northern areas of Laos, Myanmar and Thailand). Chinese crime groups seem to be major players, connected to other crime groups in Laos and Myanmar (van Uhm and Wong, 2021).

The Mekong River flows through China, Myanmar, Laos, Thailand, Cambodia and Vietnam. It sustains almost 300 million people with its water influencing both rice crops in times of drought or flooding, and the fishing industry in Laos. One problem for collective decision-making is the river's seasonality between the flooding season and the low-flow season (Gao, 2021). Most of the lower Mekong countries flourish on the seasonality of the river with good fisheries. But China prefers both a stable river flow, a stable neighbourhood and a secure continental shipping route.

The Mekong region is the subject of political tensions between neighbouring states, which are at various stages of economic development. Since 2000, China has promoted infrastructure investment for development in the region, through the BRI. The China-led Lancang–Mekong Cooperation (LMC) framework was established in 2016 as an attempt to manage the problems of the region. One priority was rural development and poverty reduction. Another one was hydropower dams. China has financed, developed and constructed over 53 dams in Myanmar, Laos, Cambodia, Vietnam and Thailand (Gao, 2021). This has resulted in environmental destruction and accusations of forced evictions and land-clearing across the region. In addition, China has assisted with investment in railways, highways, telecommunication networks and financial aid in economic and trade zones.

This strong reliance on China means it can exert its power to control the water of the Mekong River. China is challenged and often accused of infringing on the other countries' water resources through its upstream hydraulic projects, with social and environmental impacts (Liebman, 2005).

Since the lower Mekong countries have experienced recent high economic growth, there is a demand for more electrical power. At the same time, there are increasing concerns that the debts of these countries to China will be problematic. Laos may be forced to sell state assets to China as repayments for loans. Such a trend has been evident in other countries. When the China Harbour Engineering Company (CHEC) developed a port in southern Sri Lanka, the country was unable to repay the loans. Instead, it was forced to hand over the port to the CHEC on a 99-year lease. Officials estimate it will take about 25 years to complete the project (Ethirajan, 2022). Another example is the expressway in Colombo, Sri Lanka which will be owned by the CHEC after its completion and after collecting the tolls for 18 years.

Water grabbing for electricity and economic development

Examples of water grabbing of much older origin are from the Belo Monte project in Brazil and the Sardar Sarovar Dam in India. They were planned for increased electricity production and led to massive expulsions of people, and the flooding of farm and grazing land, fields and forests. In these cases, land and water were not grabbed but utilized by the government for domestic energy production to guarantee higher economic growth. This is now a priority for many governments globally, searching for alternative energy sources to fossil oils. The Swedish government plans to double its electricity production by 2045 to maintain current economic standards. The experiences of these projects illustrate the conflict between huge demands for fossil-free energy for societies and their consequences for the environment, biodiversity and people. The latter may be affected by new hydropower dams, nuclear plants and a vast number of large wind power parks. Since increased electricity production will be expensive there will be lessons to learn for current politicians.

Both dam projects are several decades old. The Sardar Sarovar Dam was officially founded in 1961, a few years prior to the start of the Green Revolution, to provide water and electricity to four Indian states in dry areas. Project development started in 1979, funded by a loan from the World Bank with a focus on increased irrigation.

Construction of the dam began in 1987, but the work was stalled in 1995 by the Supreme Court of India over deep concerns about the displacement of many people. In 1994, the World Bank was requested to withdraw by the Government of India. The state governments were unable to comply with the loan's environmental and other requirements (World Bank, 1995). There were also protests by people affected by the irrigation work. The project was revived in the early 2000s with a lower dam height, which was increased in 2006 after several cases in the Supreme Court. Eleven years later the dam was inaugurated with a height of 139 m, being part of the Narmada Valley Project, a series of large irrigation and hydroelectricity multi-purpose dams on the Narmada River.

Since the late 1980s, the dam has been very controversial, met protests and caused the displacement of many people. Its environmental impact and net costs and benefits have been much debated. Indian activists claimed in 2019 that at least 178 villages in Madhya Pradesh were submerged after the dam had reached its maximum capacity (Satheesh, 2019). Now, almost 2 million ha in drought-prone areas

have been irrigated. In 2021, there was water for irrigation in summertime; for the first time according to *The Indian Express* (Raja, 2021).

The Supreme Court ruled compensation should be provided but about 28,000 of the project-affected families have yet to receive it. The government and the activists differed on the total number of families. Some recent findings suggest that those displaced are considered better off than their former forest neighbours in their ownership of a range of assets, such as TVs, mobile phones, vehicles and access to schools, hospitals and agricultural markets. Resettlement was found to have helped vulnerable groups more than the less vulnerable. The fear that resettlement will destroy the lives and lifestyles of tribal people has been exaggerated (Swaminathan *et al.*, 2021).

The original plan for the Belo Monte project in Brazil dates from 1975 during the military dictatorship. The government policy was to construct hydroelectric dams to guarantee national energy security. The plans met initial controversy, but were revitalized in the late 1990s. and redesigned in the 2000s. They still met opposition, both nationally and internationally, with questions about the project´s economic viability, the efficiency of the dams and the serious impacts on the region's people and environment. From the beginning, the government was reluctant to offer transparency to the people who would be affected by the six planned dams. Due to the protests, Belo Monte finally became the only dam option, being redesigned between 1989 and 2002 with a reduced reservoir surface area. It lies in the northern part of the Xingu River in the state of Pará.

In two environmental impact assessments (2002 and 2008), the Brazilian environmental agency (IBAMA) had stated the adverse effects on fauna, flora, the water supply, fish migration routes and the temporary disruption of the water supply in the Xingu riverbed for seven months. IBAMA granted the final license to construct the dam in mid-2011 after several court cases and renewed strong protests by the tribes of the region (the Second Encounter of the Peoples of the Xingu). The consortium agreed to pay costs to address social and environmental problems. The power station was completed in late 2019.

The Belo Monte Dam is a complex of dams, numerous dykes and canals to supply water to two different power stations. The main dam, Pimental, has a reservoir of 359 km² and diverts much of the Xingu's water flow through a 17 km canal to a secondary reservoir and hydroelectric station. The diversion reduced the river's flow of water and stopped the annual of flooding, reducing a fish fauna of some 600 fish species (Moutinho, 2023). It also meant the dam would only be producing 10% of its hydropower capacity between July and October. Initially, its overall capacity factor was estimated to be about 39%. The average capacity factor of hydroelectric power plants is between 30–80%, higher than that of wind power. That is why Brazil's Energy Expansion Plan calls for three more large dams in Amazonia by 2029. Hydroelectric power produces over 66% of Brazil's electrical energy (IEA, 2022).

Now, Belo Monte generates less than half its installed capacity (Moutinho, 2023). According to Andritz Hydro (2021), Belo Monte provides renewable energy for about 60 million people. As claimed long ago, the project would only make financial sense – and be sustainable – if the five original upstream dams were built to guarantee a year-round flow of water (Fearnside, 2006). That would directly or indirectly affect many more people from the Indigenous communities.

Now, two dams jointly flood 51,600 ha, of which 38,000 ha are located on the floodplain above the Pimental dam (Killeen, 2021). It is a major difference to the initial plan of the reservoir, flooding 400 km² of forest. Collaborative research with the community and partially funded by the company, Norte Energia, have found a 29% decline in the number of species and a 9% drop in the abundance of all fish in the Big Bend (Volta Grande) stretch of the river (Moutinho, 2023). There may be an impact on the numerous giant Amazon River turtles during the their breeding season. Norte Energia agreed in 2022 to pay reparations to about 2000 local fishermen. Now, a Canadian mining company wants to build a large gold mine, which is opposed by the Indigenous communities.

The project has had a negative impact on the livelihoods of Indigenous communities and displaced many people, in addition to its adverse effects on the environment. It has been strongly criticized by the Indigenous communities and numerous environmental organizations in both Brazil and internationally. The Brazilian government has been criticized by the UN Human Rights Council. The International Labour Organization (ILO) has pointed out that the Brazilian state was in violation of the Indigenous and Tribal Peoples Convention, 1989 No.169.

Belo Monte is ranked as one of the largest hydropower plant in the world. The dam is funded by the Brazilian Development Bank with some contributions from private investors in the mining and construction industry. Due to a lack of interest from private foreign investors, the government must rely on Brazilian pension funds, credits and public money. This implies a heavy burden on future Brazilian taxpayers.

Some effects of land and water grabbing

The overall trend of grabbing land and natural resources is threatening. Emerging and low-income countries may come under foreign control and experience long-term security implications, especially in terms of food security. Land grabbing is, however, a much broader issue, affecting climate change, the environment, biodiversity, risks of new pandemics and capital concentrations. Investments in countries are often based on loans which may put the recipients in a debt-trap.

In most cases, land grabbing leads to protests, violence, the displacement of people and even killings. In 2021, some 200 people were murdered for their commitment in environmental campaigns (Hines, 2023). The following year 177 people were murdered. In African and Latin American countries, ensuring Indigenous peoples' collective rights to lands, territories and natural resources is critical. The Indigenous people are more vulnerable because there is lack of political recognition of their traditional access to their ancestral lands and natural resources (Poderati, 2022).

In theory, agriculture is one of the promising avenues for Africa, where it is a large contributor to economic activity. Some 60% of the world's uncultivated arable land is in Africa. When it comes to food and agriculture, China is a big importer of food. Its imports were worth US$13 billion in 2000. By 2020, their worth had risen to US$161 billion but Africa only accounted for 2.6% of that total (Miriri and Bavier, 2022). According to the Agricultural

Business Chamber of South Africa, China may be missing out on ways to increase imports from Africa. The country prefers bilateral deals and even separate protocols for each agricultural product. This is very time-consuming compared to broad trade deals. Another issue is that African nations now claim they must boost exports to China since they cannot afford more Chinese loans.

National governments have the ultimate power in controlling foreign buyers or lessees of their land and natural resources, and to ensure that buyers and lessees use their land and resources in a way that supports long-term sustainability – a requirement that can currently be classified as almost non-existent. The international community also has an important task in identifying better control mechanisms for this global threat and measurements of its impact. This calls for more up-to-date national and international research. It may require financing of serious actors involved in studying land grabbing and its effects on people and their rights.

Declining Biodiversity

Plants appeared when land was colonized some 450 mya. Higher oxygen content (up to 35%) during the Devonian and Carboniferous periods led to exceptional growth. The more complex the organisms on Earth, the faster they seem to die out although the general lifetime for complex organisms is about four million years (Gould, 1993). In fact, since the beginning of evolution 3.8 billion years ago, 99% of all species that have lived, no longer exist. Five major biological events have meant mass extinctions when species have disappeared (Annex III). Such mass extinctions are characterized by the loss of at least 75% of species within a geologically short period of time, without human involvement.

In recent times, the most dominant wave of global plant germplasm exchange occurred between the Age of Exploration (15th–17th centuries) and the Industrial Revolution (1760–1840) (Naithani, 2021). During this period, a few early plant collections were conducted with some systematic cataloguing and classification of both flora and fauna in the new

continents. Plants were transported in bulk and large plantations were established with new cash crops, significantly changing their environments.

It was, however not the first exchange of plants with effects on global biodiversity. Prior to this, the Silk Road had contributed to exchanges of both plants and other goods. But the spread of plants had started around 3000 BCE, when Polynesian seafarers began sailing between South-east Asia, Africa and South America. Later, Polynesian ancestors of the Māori brought sweet potato (kúmara) as a food plant, arriving in New Zealand in the 13th century (McLauchlan, 2018).

It is assumed that at least 84 cultivated plants travelled from South America to Asia and Africa, for example, maize, amaranth, cashews, pineapples, custard apples, peanuts, pumpkins, gourds, arrowroot, guava, sunflowers, basil and water hyssop (Naithani, 2021). Hemp and another 15 plants have come from Asia to Africa and South America. Closer examination of all printed sources on precolonial West Africa may provide more details as to which crops the Europeans introduced, whence, where and when.

Currently, the Holocene extinction is claimed as ongoing. It is a result of human activity, population growth and overconsumption of the Earth's natural resources. The global biodiversity assessment by the Intergovernmental Science-Policy Platform on Biodiversity and Ecosystem Services (IPBES) asserted in 2019 that out of an estimated 8 million species, about 1 million plant and animal species are currently threatened with extinction.

Tropical rainforests cover only about 6% of the land surface but contain more than half of the world´s animal life and about two-thirds of the flowering plants. The loss of tropical primary forest totalled 41 million ha in 2022 according to the World Resources Institute (WRI, 2023a). The tropics lost 10% more primary rainforest in 2022 than in 2021, according to data from the University of Maryland; most losses took place in Brazil and the Democratic Republic of the Congo (Jong, 2023). Rainforests in Borneo have declined by 55% since 2000 partly due to palm oil production which used to be concentrated to East Asia (Moate, 2023). This is an illustration of how the largest commercial production of an economic crop is often far removed from its centre of origin.

Globalization plays a key role in allowing alien species to enter a country's borders. In the ballast of cargo ships, it is estimated that 10,000 alien species can be spread through so-called global swarming. The sand flea is an early illustration. It is native to Brazil but arrived in a shipload at Ambriz, in modern-day Angola, on the west coast of Africa in 1872. It spread north along the coast, reached the Congo, and then spread quickly along the western coast of Africa to Sierra Leone and the former colony of Portuguese Guinea.

Research has shown that older versions of the potato leaf mold (*Phytophtera infestans*) came via the ships from Peru that transported guano to England and Ireland. It has been argued that the crew of the ships had brought infected potatoes for human consumption during the long crossing. In August 2023, the first case of African swine pest was discovered in wild boars in Sweden. It reached Europe in 2007 and the EU in 2014. Widespread infection drastically threatens pig production.

Although the IPBES has made an estimate, it is not well known exactly how many species there are on the globe. Wilson (2010) suggested between 3–200 million species. Other sources have claimed that up to 97% of the world's plant and animal species were still to be discovered. For example, bacteria in large numbers have been around for millions of years. So have acarids but they were only discovered in 1965 (Bodanis, 2003). Almost 100,000 fungi have been discovered but the Earth may harbour some 1.8 million species (Leakey and Lewin, 1996). Irrespective of these rough estimates there is a real need for the scientific community to intensify the search for new species, with strong political support. There is a need for more botanic collections of both seed and new species.

Some efforts have already been made. The All Species Foundation was an organization launched in 2000, aiming to catalogue all species on Earth by 2025. It was seen as a step forward in expanding, modernizing and digitizing taxonomy. The Foundation started with a large grant from the Schlinger Foundation but faced difficulty in finding continued funding and, after a period of dormancy, finally ended its work in 2007 (Berger, 2005).

In a study by the Arizona State University, almost 17,000 new species of plants and animals were found in 2006. Insects accounted for more than half the number and there were also 2000 plants and 1000 vertebrates. In 2019 alone, 1942 plants and 1886 fungi were found, hitherto unknown to science. Several of them can become food, drink, medicines or fibre (ASU, 2006).

In 2023, scientists at the Natural History Museum in London and the California Academy of Sciences reported the discovery of 968 new species (Strickland, 2023). In the same year, Swedish researchers claimed they had identified, with the help of DNA technology, about 2000 new insects to add to Swedish fauna (van Klink *et al.*, 2024). This was considered as a warning of increased climate change. Several of the species were suspected to have come from warmer regions.

New environments may be interesting places for several types of advanced biotechnological research. Bacteria can live and have adapted to many strange places such as soda-rich lakes, inside mountains, ice-cold water, and deep in oil wells. One bacterium is immune to radioactivity (*Deinococcus radiodurans*). In Australia, *Micrococcus radiophilus* can live in concentrated sulphuric acid. The sea probably has many new species. Therefore, the High Seas Treaty reached in early 2023 can be important in aiming to safeguard 30% of the high seas by 2030. The Treaty, signed by 193 countries, shields marine biodiversity from overfishing, deep-sea mining and shipping traffic, but the issue is its implementation through practical action.

The Clarion-Clipperton Zone in the Pacific Ocean, between Hawaii and Mexico, is twice the size of India. It is some 3500–5500 m deep with large mineral resources (manganese, copper, cobalt and nickel), attracting many companies. In previous biological studies of the area around 5500 species have been registered, most of them new to science (Rabone *et al.*, 2023).

Biodiversity is changing and species are lost, which is problematic but follows a natural historical pattern. Species can disappear quickly, gradually or by extinctions in the long-term. Today, the political agenda requests that all threatened species should be preserved. This is an ambitious task, considering historical experience where biodiversity has constantly changed. It raises the question of its efficiency and what other efforts would be more effective.

The Quechua people in Peru cultivate and eat a variety of foods. For centuries, they have undertaken their own conservation and adaptation efforts. Since Inca times, they have cultivated various crops extending across different altitudes and domesticated the potato (*Solanum* spp.) from several wild potato species in the Andean region. Five Indigenous Quechua farming communities have been cultivating around 1300 different native varieties of potatoes over the centuries, safeguarding this genetic resource for future generations; a feature of sustainability. In 1997, this led to the creation of the Parque de la Papa close to Cuzco at an altitude of 3400 m. It comprises 12,000 ha and is run by the Quechua communities.

The importance of variability in climatic conditions for biodiversity was first noted by the Russian botanist Nikolai Vavilov (1887–1941). From 1924 to 1935 he was the director of the Lenin All-Union Academy of Agricultural Sciences at Leningrad. During this period, he organized more than 100 botanical–agronomic expeditions to many countries, with visits to traditional farmers' fields, rich in plant diversity. As a botanist, Vavilov collected more than 380,000 crop samples: an impressive effort forming the basis of an understanding of the concept of centres of diversity (Annex IV).

Vavilov concluded early that genetic diversity was lost over time so new crop varieties were needed. Seed banking would help minimize such losses and serve as a model for safely conserving seeds *ex situ* for plants from different regions and environments. This gradually led to the creation of Vavilov's gene bank in 1940. Later, other countries followed suit, for example, the US National Seed Storage Laboratory in the late 1950s, the International Maize and Wheat Improvement Center (CIMMYT) and Consultative Group for International Agricultural Research (CGIAR) centres in the 1960s, India (1986) and China (2013). Today, there are many international crop accessions available, among others within the One CGIAR (Chapter 8).

There are currently instances where politicians can act quickly. The herring stock in the central Baltic Sea is reported to have decreased by over 80% compared to the 1970s. The larger herring has been replaced by spigg and cod has disappeared in several places (Baltic Sea Brief, 2021). It has taken over 20 years to get an agreement for sustainable management with international goals to govern the fisheries policy within the EU. The problem is that large trawlers catch the most fish. In 2021, 20 industrial trawlers caught 96% of all Swedish catches in the Baltic Sea (Svensson, 2022). Ruling politicians have the power to ban these large trawlers.

Another approach may be to investigate new food alternatives. Until the early 1900s, langoustines were an unwanted bycatch on the Swedish west coast. In this region, the fish catches have varied over hundreds of years and so have the food traditions. The langoustine is a close relative of lobster, but smaller and slimmer. In the past, they were thrown overboard because there was no demand from Swedish consumers. However, in the 1930s local fishers found some outlets for them. The substantial change came after the World War II when they became popular and they still are.

When foreign species appear in new environments, they usually turn invasive. It is not surprising. Existing flora and fauna lack the power to compete with intruders in an environment established over a long time. The invasive species are a significant threat to many native habitats, and bring significant costs to agriculture, forestry, the environment and recreation. The tenth session of IPBES in 2023 focused on invasive alien species and their control. It was concluded that both people and nature are seriously threatened by their increase all over the world. They are considered a threat to both the economy and food and water security. Details are found in the Global Register of Introduced and Invasive Species (GRIIS) with validated and verified national checklists of alien and invasive alien species at the country, territory and associated island level (GBIF, 2020).

Although several countries, in particular the US and Australia, are extremely cautious about importing animals, soil and plant material, it seems insufficient. Within the EU, the emphasis is on the long-term, undesirable consequences of freedom of movement – about animals and plant material. Most EU countries have abolished their internal border controls, and instead the EU's external borders will be responsible for customs and border control, which has proved to be insufficient. Currently, it is common that homecoming travellers to Sweden bring food with them from their home countries or from holidays, such as meat, fruit, etc. This is not allowed for passengers travelling outside the EU due to the risks of spreading diseases and invasive species.

According to the EU (2022), the list of invasive species should be amended with 22 new species, totalling 88. In the US the term 'invasive species' can refer to introduced/naturalized species, feral species or introduced diseases. There are approximately 50,000 foreign species, and the number is increasing, causing major environmental damage and economic losses estimated in the mid to late 1990s at almost US$120 billion per year (Pimentel *et al.*, 2005).

Another specific example is Australia, where humans played a significant role through new species arriving with the European explorers and settlers. One species is the European honey bee, introduced almost 200 years ago to pollinate food plants. Now, they have become an aggressive species, winning in competition with native species for dwelling hollows in trees. About 3% of Australians have become allergic to honey bee venom and sometimes the stings even kill people. In 1788, 49 pigs were shipped to Sydney as domestic livestock. In the 1880s, they ran wild and spread across the northern and eastern parts of the country, deserts excluded. Today, feral pigs destroy the land by eating almost anything and affecting watercourses and swamps. The effects have been dramatic and have caused considerable costs over the past 60 years. Feral cats are on top of the list, followed by rabbits, fire ants, annual ryegrass and pigs (Khan, 2021).

In the UK, there are about 20 non-native invasive species, for example, Japanese knotweed, the Asian hornet and the American bullfrog. In Sweden, one recent discovery (2022) was the aggressive killer shrimp (*Dikerogammarus villosus*). It is an amphipod crustacean originally found in the Ponto-Caspian region of Eastern Europe.

The Kunming–Montréal Global Biodiversity Framework has four overarching goals to be achieved by 2050, specified in 23 targets. The framework focuses on ecosystem and species health, including efforts to halt human-induced species extinction and the

sustainable use of biodiversity. Moreover, there should be equitable sharing of the benefits from genetic resources and their implementation and responsibility for funding the biodiversity finance gap of US$700 billion per year. Among the targets to be achieved by 2030 are the 30% conservation of land and sea, 30% restoration of degraded ecosystems, halving the introduction of invasive species and a US$500 billion/year reduction in harmful subsidies (UNEP, 2022).

The EU set a goal to take action that covers at least 20% of the EU's land and water surface by 2030. All ecosystems need to be restored by 2050 according to the proposal, which means that 16% of forest and agricultural land in Sweden alone would disappear. Three billion trees will be planted in the EU and large forested areas can be subject to several types of restoration. Watercourses will be restored so that water can flow freely and some 30% of arable land created from drained peatlands will be restored to wetlands. The EU Council and the European Parliament reached a compromise in late 2023, which the European Parliament approved in early 2024 by a narrow margin. It must, however, be formally approved by the EU Council of Ministers. Considering future food security in Europe, the proposed law will hardly guarantee long-term sustainability.

The major task of the COP16 at Cali, Colombia was to lay out a funding plan as a follow up to the Kunming-Montreal Global Biodiversity Framework. It failed due to splits between rich and poor countries, especially over increased funding. The COP16 summit agreed only on a new permanent body to represent Indigenous peoples' interest in biodiversity and a new fund. It will serve as a fee for the use of DNA biodiversity in digital databases (Einhorn, 2024). It was stipulated to be 1% of the profits of companies or 0.1% of their revenues from pharmaceuticals, cosmetics and biotechnology. Half the funds are to be directed to Indigenous people but they fund it voluntary. When the Convention on Biological Diversity was adopted in 1992, world society had promised to financially compensate middle- and low-income countries that were rich in biodiversity. This has not yet materialized so the new promise seems unrealistic.

Increasing Antibiotic Resistance

Antibiotics are lifesaving but can be abused. When antibiotics are overdosed or used excessively, bacteria develop resistance and antibiotic resistance occurs. All bacteria can mutate. Many deaths occur due to common, previously treatable infections because the bacteria that cause them are becoming resistant to treatment. Infections such as pneumonia, tuberculosis, gonorrhoea and salmonellosis are becoming harder to treat. The antibiotics used to treat them are becoming less effective (Gregory, 2022).

The World Health Organization (WHO) considers antibiotic resistance to be one of the biggest threats to human health. Resistance occurs naturally, but misuse of antibiotics in both humans and animals is accelerating the process (WHO, 2015). Antibiotic resistance leads to longer hospital stays, higher medical costs and increased mortality. Given the ease and frequency with which people can travel globally, necessary efforts are required from politicians. Beyond the medical and ethical issues of using antibiotics to increase growth in the food industry there is an economic dimension. The World Bank has warned that by 2050, drug-resistant infections could cause global economic damage on par with the 2008 fiscal crisis.

According to *The Lancet* (2022), more people died in 2019 as a direct result of antimicrobial resistance infections (AMR) than of HIV/AIDS and malaria. The analysis covered more than 200 countries and territories. It showed that AMR was solely responsible for an estimated 1.27 million deaths worldwide and associated with an estimated 4.95 million deaths (Gregory, 2022). Regionally, deaths caused directly by AMR were estimated to be highest in sub-Saharan Africa and South Asia. If this trend continues, more people will die from resistant microorganisms in 2050 than the number dying from cancer today.

Globally, about 70% of all antibiotics are used on animal livestock and only 30% in human health care according to the Axfoundation (2023). Globally, antimicrobial usage is projected, based on current trends, to increase by 8.0% by 2030 (Mulchandani et al., 2023). The European Centre for Disease Prevention and Control (ECDC) has estimated

that around 35,000 people die annually in the EU, Iceland and Norway from infections that can no longer be treated with antibiotics (WHO, 2022). Annually, some 1.2 million deaths occur globally due to infections caused by resistant bacteria; a figure expected to reach some 19 million by 2050.

Antibiotics are routinely fed to livestock, poultry and fish on Western industrial farms to increase the weight of the animals and thereby the profit. It can even be done to compensate for the poor condition of the animals. In the US, antibiotics have been used in livestock farming since the 1950s, but the US Food and Drug Administration (FDA) started monitoring the quantity of antimicrobial drugs sold for livestock use in 2008. Most antibiotics sold in the US are for livestock use (80%) but there are large variations between other countries. There is an elevated risk of developing multi-resistant bacteria and the chance of polluting the environment is high (Larsson and Flach, 2022).

Large livestock farms may produce at least 2–3 million chickens per year or house around 700–800 milking cows. They live close together and any disease not only has great effect on the animals but also on humans. Such farms must be biosecured and hermetically closed with antiseptic procedures. The animals are vulnerable since a bacterium can easily mutate and infect all of them, ending up with the slaughter of all those infected. The system can be considered safe and economic for the producer but less than acceptable for the animals. They are forced to spend all their life in an environment far from what they originally were used to.

The potential of aquaculture has led to the rapid development of the shrimp aquaculture industry, particularly in South-east Asia. Frequent diseases have become a major risk factor in their production, which has declined as the use of antibiotics has proved ineffective. Important shrimp bacterial diseases include hepatopancreatic necrosis and necrotizing hepatopancreatitis. A parasitic disease is crayfish plague and all of these diseases are evolving into new types due to resistance (Yu et al., 2022).

Antibiotics reach the environment via excretions from humans, domestic animals, inadequate disposal of unused drugs, environmental contamination in aquaculture and from waste from factories producing of antibiotics.

With low production costs, China and India have become the world's largest producers of antibiotics and there are reports showing excessive emissions of antibiotic residues from manufacturing (Larsson and Flach, 2022).

More than half of the antibiotics given to animals are the same or like those used in human medicine. The use of antibiotics in the food industry and for healthy animals should be stopped as the WHO has recommended. The risk of developing multi-resistant bacteria and polluting the environment is too high. There is a great need to change how antibiotics are prescribed and used. In agriculture, antibiotics should only be given to animals under veterinary supervision. Animals should be vaccinated to reduce the need for antibiotics and alternatives to antibiotics should be used whenever available. They should not be used for growth promotion or to prevent diseases in healthy animals. Instead, farm biosecurity must be improved.

The health industry invests in the research and development of new antibiotics, vaccines, diagnostics and other tools. There are some new antibiotics in development but none of them are expected to be effective against the most dangerous forms of antibiotic-resistant bacteria according to current information from the WHO (WHO, n.d.). Since the food industry uses as much antibiotics as health care, there are obviously major challenges for consumers in choosing the right food. Swedish food is safe because the use of antibiotics in animal husbandry is the lowest in the EU. It was forbidden for animals in 1986. Any routine admixture of antibiotics in animal feed has been banned throughout the EU since 2006. From 2022, EU rules have not allowed antibiotics for humans to be given to animals.

The WHO has referred to the spread of multidrug-resistant bacteria as a silent pandemic. In 2015, this led researchers, at the universities of Warwick and Nottingham to form the interdisciplinary group Ancientbiotics which has a great interest in medieval pharmacology. Traditionally, medicinal plants have been the basis for many of the pharmaceutical industry's most important achievements. For example, both aspirin from white willow (*Salix alba*) and moose grass (*Filipendula ulmaria*) contain salicylic acid. Foxglove flower (*Digitalis purpurea*) has long been used for circulatory problems and heart failure. The plant contains a substance that increases the power of the heartbeat.

Severe acute respiratory syndrome coronavirus (SARS-CoV-2) is the strain of coronavirus that caused the COVID-19 pandemic in 2019. It demonstrated how devastating a biological weapon could be to the world if there are scientific mistakes. Globally, some 3000 laboratories are working with pathogens like SARS. In spite of high security, one of those scientists may by mistake take a wrong decision, leading to the spread of another dangerous agent.

Increasing Migration

Globalization leads to increased international travel and increased migration. When the United Nations High Commissioner for Refugees (UNHCR) was founded in 1951, the number of refugees in the world was estimated at just over 1 million. Then, refugees came from countries that had been involved in World War II, different to the situation of today. The refugee problem became a political issue on the international agenda in connection with war and crisis in the Middle East. There were no preparations by the UN, the EU, or individual countries at the beginning of the Syrian crisis in 2011. It was not until mid-2018 that most UN member states endorsed a global migration agreement with 23 formulated goals, the Global Compact for Safe, Orderly and Regular Migration (UN, 2022). Its intention was to reduce irregular, disorderly and illegal migration.

In 2019, the UN Department of Economic and Social Affairs (UNDESA) estimated the number of international migrants worldwide to be 272 million (defined as persons outside their country of origin for 12 months or more). Half of them were estimated to be economically active. This estimate was an increase of 3% compared to 2017 and 12.7% compared to 2013 (ILO, 2021).

In 2022, the global estimate was around 281 million international migrants, living in a country other than their country of birth (IOM, 2022). That figure was more than three times the estimated number in 1970. The economic situation in a range of low-income countries has not improved since 2000 and after the COVID-19 pandemic. In 2022, there were almost 1 million applications for asylum within the European Union.

In 2019, Europe hosted the largest number of international migrants (82 million), followed by North America (59 million) and northern and western Africa (49 million) (United Nations, 2019). In 2021, the US Border Patrol reported more than 1.6 million encounters with migrants along the US–Mexico border. That was more than quadruple the number of the prior fiscal year and the highest annual total on record (Gramlich and Scheller, 2021).

There are several factors contributing to people having to move from their environments, leading to more migrants. The United Nations Environment Programme (UNEP) has warned that water and land shortages have begun to cause social disruption in the semi-arid Sahel region of Africa. Degraded land in South Sudan has forced 2 million people to move. Nomads in northern Nigeria have experienced more drought than usual and moved south with cattle herds to find new pastures. This had led to serious conflict over land resources with settled farmers in central Nigeria and neighbouring countries. In Asia, there is a growing concern that China is holding back water at dams, resulting in drought impacting the Mekong River region, which in turn causes migration because of environmental degradation. This growing trend includes people who are seeking better employment opportunities despite half a century of development aid.

By the end of 2022, the number of people displaced by war, persecution, violence and human rights abuse was about 108 million; an increase of some 19 million compared to 1 year earlier according to UNHCR (UNHCR, 2023). The conflict in Sudan increased the global total to an estimated 110 million by mid-2023. Half of all refugees come from Syria, Afghanistan and Ukraine and most of the refugees (76%) are hosted by low- and middle-income countries. Turkey has the largest number of refugees in the world (3.6 million) (Frontline, 2023).

A feature of current trade is that it includes migrants who are trafficked across borders at great expense by a large network of smugglers. It has become a commercial activity, a reminder of old times. In 1562, Sir John Hawkins undertook the first English slave expedition. Over 200 years, an estimated 10–12 million slaves were shipped out of Africa to North America, the Caribbean and Brazil. British ships carried one-third of all slaves shipped across the Atlantic until the abolition of the trade by the UK parliament in 1807 (Richardson, 2022). Scandinavian countries participated as well.

Slavery has a long history and was not uncommon in Europe during Phoenician and Roman times. Slaves were used for services and the Romans also used them for agriculture (William Jr, 2013). During three historical periods, the slave trade expanded significantly in the western Indian Ocean and Red Sea region: at the turn of the common era (about 1st century CE), the 10th–13th centuries and the 19th century (Perry, 2021). Muslim societies impacted slavery in areas such as the west coast of India, East Africa, Yemen and Arabia, Ethiopia, Nubia and Egypt. Merchants generally trafficked small numbers of enslaved people as part of larger mixed cargoes of luxury goods and other commodities. A large number of slaves were imported to the Muslim-ruled area of the Iberian Peninsula (711–1492), From there Muslim and Jewish merchants operated to market slaves to other parts of the world (Constable, 1996). The East African slave trade from the Indian Ocean began after Arab and Swahili traders took control of the Swahili Coast and sea routes during the 9th century.

Some African peoples, such as the Yao, Makua, and Marava, were involved in slavery. They were fighting against each other, so prisoners of war were sold, mostly locally. When Zanzibar became part of Oman in 1698 after the defeat of the Portuguese in Mombasa, a new slave trade was started by an Arab elite on the island. Merchants from Oman settled on Zanzibar and expanded the overall trade, including with traders from the Indian subcontinent. The Zanzibar archipelago developed as the centre of the East African slave trade in addition to its trade in ivory and cloves. Many East Africans were sold as slaves by Arabs to the Middle East and other places, via the Sahara Desert and the Indian Ocean. The treaty signed in 1873 to make the slave trade in the archipelago illegal was not effective until slavery was abolished in East Africa in 1909 (Fröhlich, 2019).

Even today, slavery persists in parts of the world. Slave labour is invisible when people sew clothes, work in mineral mines or are sold as wives. Still, international firms depend on raw materials from countries where the use of slave labour is suspected. Few companies can completely control the entire production chain, even though international agreements prohibit the holding of anyone in slavery or slave-like conditions. The UN has estimated that slavery affects 115 out of 198 countries as they may risk using slave labour in their production. In 2021, some 50 million people were living in modern slavery. The definition of 'modern slavery' includes forced labour and forced marriage, trafficking in human beings and the exploitation of children. The most recent global estimate is 28 million in forced labour and 22 million in forced marriages (International Labour Organization (ILO) *et al.*, 2022). Compared to 2016 global estimates, 10 million more people were in modern slavery in 2021. North Korea, Eritrea and Mauritania stand out in a negative sense. North Korea is the only country in the world where slavery has not yet been outlawed. In 2023, some 6000 people from Thailand had applied for working permits for berry picking in Sweden (Swedish Migration Agency, 2023). Low salaries and poor working conditions made the Swedish Minister for Gender Equality characterize this as imported slavery.

At the end of 2023, an asylum and migration settlement was reached within the EU. Asylum seekers and migrants will be received at the EU's external border for health and ID checks, and biometric facial registration. Those who have weak grounds for asylum should be fast-tracked to asylum centres near the EU's external border or at airports. The previous rule that the first country of reception should take care of asylum seekers was abandoned as all member states must share the responsibility.

Increasing Population

Globally, the world's population increased from two to four billion people between 1930 and 1974. Since then, it has almost doubled with most people living in Asia (Table 7.2).

About half of the world's countries will double their population in 35 years (Table 7.3). Globally, the population will continue to grow in Africa, India and some countries in South America, but growth is more doubtful in China. According to the UN's forecast, there will be 11 billion people on the planet by 2100 compared to just over 8 billion today. The populations in Europe and North America have been growing at an annual rate of less than 1% since the mid-1960s, reaching a growth rate close to zero in 2020 and 2021. In contrast, the annual growth rate of the population of sub-Saharan Africa peaked at 3% in 1978 and remained above 2.8% during the 1980s. Since then, sub-Saharan Africa has been the region with the fastest-growing population. The population of Africa's cities is projected to double to 1.5 billion people by 2050 (IPCC, 2022).

In early 2023, India passed China and became the country in the world with the largest population, according to recent statistics from the UN; 1428 million and 1425 million, respectively (UN, 2023). Half of the Indian population is under the age of 30, and is expected to be the fastest-growing population in the coming years (Worldometer, 2023).

Table 7.2. Population of the world and its major areas 1750–1999 (in millions) (United Nations (1999)).

Area	1750	1800	1850	1900	1950	1999
World	791	978	1262	1650	2521	5978
Africa	106	107	111	133	221	767
Asia	502	635	809	947	1402	3634
Europe	163	203	276	408	547	729
Latin America and the Caribbean	16	24	38	74	167	511
North America	2	7	26	82	172	307
Oceania	2	2	2	6	13	30

Table 7.3. Population prospects of the world and the sustainable development goal (SDG) regions in 2022, 2030 and 2050 according to the medium scenario by the United Nations (2022) (in millions).

Region	2022	2030	2050
World	7942	8512	9687
Sub-Saharan Africa	1152	1401	2094
Northern Africa and Western Asia	549	617	771
Central and South Asia	2075	2248	2575
East and South-east Asia	2342	2372	2317
Latin America and the Caribbean	658	695	749
Europe and North America	1120	1129	1125
Australia/New Zealand	31	34	38
Oceania	14	15	20

The UN expects poverty reduction and economic and social development even in today's low-income countries. There has been some early speculation that the global birth rate may decline after about 2050, in contrast to the UN's assessment. If 2.1 children are born per woman, the population decreases (UNFPA, 2023), although this estimate was made before the COVID-19 pandemic. Mass epidemics have previously played a fatal role in limiting the number of people.

Maybe some politicians see the COVID-19 pandemic and other pandemics as a regulatory factor. According to the WHO, there have been 767 million confirmed cases of COVID-19, including almost 7 million deaths, as of mid-2023. Less than 2% of the population in low-income countries were vaccinated. The overwhelming majority of politicians have claimed that once poverty has been eradicated and development has taken place in all countries, population growth will not be a major problem. A conventional view is that increasing urbanization and the declining importance of religion would contribute to women wanting to have fewer children. If living conditions improve, most families desire fewer children.

China tested child restriction with its one-child policy spanning 1979–2015. According to the Communist Party, the population of China in 2015 was 400 million less than it would otherwise have been compared to 1970. The number of births per Chinese woman decreased from 2.6 to 1.6 in 30 years. There was a large surplus of men and lower numbers of young Chinese had increasingly large numbers of older people to support. Since death rates and child mortality decreased, a greater number of people survived longer and the number of women of childbearing age increased. Chinese economists and demographers warn of the serious risk of both an aging and a declining population. It would prevent China's plans to become a great power and Chinese domestic food production would still be a problem.

Population control is not new to history. Plato and Aristotle had already claimed that Greek cities should practice birth control. Aristotle believed that an excessive population in Greek cities would lead to social disruption and end in poverty. Since then, the population issue has often returned for debate. Thomas Malthus (1798) dystopia was prevented, among other things, by the introduction of agricultural technology, leading to increased productivity with improved animal husbandry and higher yields.

During the 1960s, there was talk of child restriction, the growing population, and even a population bomb. The Club of Rome's predicted the end of the world, focusing on the number of people in relation to the capacity of natural resources and the environment to support human activity. It was an initial thought towards a concept of sustainability, as the environment started to degrade. Again, a new agricultural technology contributed to higher food production. It was the Green Revolution, though another warning of a population explosion appeared (Ehrlich and Ehrlich, 1990). The issue of overpopulation recurred during the economic crisis of 2008 with a growing awareness of the expanding degradation of the environment. But leading experts and politicians insisted that the Earth had the capacity to produce food for 10 billion people without significant environmental problems.

Today, the situation is different to the mid-19th century. The number of people is much higher and there is no new agricultural technology, ready and available, which could lead to a rapid change in countries which cannot produce sufficient food for their people. It is unclear whether a new biotechnological or digital revolution can contribute to a solution in those countries. In contrast, the baby-bust economy, caused by a reduction in births, is now the political focal point. Before the end of this century the number of people on the planet may shrink for the first time since the Black Death in the 14th century. The root cause is not a surge in deaths, but a slump in births (The Economist, 2023). Across most of the world the fertility rate, the average number of births per woman, is declining. Although the trend may be familiar in certain countries, its extent and its consequences are not.

In 2000, the world's fertility rate was 2.7 births per woman, comfortably above the 'replacement rate' of 2.1, at which a population is stable but declining. The largest 15 countries by gross domestic product (GDP) all have a fertility rate below the replacement rate. That includes the United States and much of the developed world, but also China and India (The Economist, 2023). In 2021, the US population expanded at its slowest rate in history. Much of its population growth came from immigration (The New York Times, 2022). On the other hand, the Amish population is one of the fastest growing populations. The Amish people live in 32 US states with the largest populations in Pennsylvania and Ohio. The fertility rate is high, and the average family had seven children in the 1970s and 5.3 children per woman in the 2010s (Hurst and McConnell, 2010).

A new study in *The Lancet* (2024) predicts that the populations of three-quarters of the world's countries will decline by 2050. It is a global projection by the University of Washington's Institute for Health Metrics and Evaluation (IHME). Only in six countries is the birth rate projected to exceed a fertility rate of 2.1 children per woman. A trend below 2.1 children per woman is projected to expand to most countries by 2100, assuming the new findings are reliable.

The WHO has claimed the results of the study should be interpreted with caution due to limitations in the methods used. Another concern may be unreliable data from many low-income countries. Over the years, the fertility rate has varied. Nevertheless, if these predictions are true, it would be beneficial for natural resources and the future food security of planet Earth. It will, however, take a long time. On the other hand, countries affected by the decline may take early political action to prevent it, threatening their future economic development.

Moreover, previous predictions on population growth have been insufficient. It would be naive to believe that the Earth, with sustainability, can accommodate all human beings who desire today's level of Western-type development, without any long-term consequences. The increasing trend of refugees and migration to other countries will continue. Considering climate change, increasing greenhouse gases and a decrease in global natural resources, it seems relevant that the international community should have an 18th goal to supplement the 17 SDGs, stating that each country should aim for balanced population growth.

Increasing Consumption and Consumerism

In half a century, global consumption has tripled. If ideal annual growth, considered to be 4% continues, the global economy will be more than thirty times larger by the turn of the next century. That is far from sustainable. Today's capitalism neglects consequences to the environment and nature. Sir David Attenborough expressed that humans are a scourge to the world and the well-known physicist Stephen Hawking argued that we need a new world to live on. It is unrealistic to think that the current approach of the modern Western consumer society can be spread to the entire world. A growing global population with higher demands will further increase consumption.

Planet Earth has reached a stage where it is not possible, with sustainability, to raise the material standard of the current world population to a level characterizing a well-developed Western country. The World Wide Fund for Nature (WWF) has observed that we would need 4.2 globes if everyone lived like Swedes. Another measurement is the Earth Overshoot Day. It refers to the day when a kind of annual budget of global natural resource consumption would show an end date. This day has been calculated since 1970 when it fell on 30 December in Sweden. In 2024, it was on 21 April.

The dilemma for politicians is that they must constantly work for economic growth so that consumption can increase and please their citizens, in turn generating more waste and pollution of the environment. This calls for an innovative approach to measuring development. Political leadership must act in the interests of a long-term perspective on consumption. Not everything is necessary, or should be available immediately, or must be material things.

The classical measurement of growth is the GDP. It is the value of total annual consumption of goods and services, gross capital formation and exports minus imports. Primarily, it measures economic activity without considering negative environmental effects or depletion of natural resources. Emissions from international aviation and shipping are not included in any country or region total. There is no international agreement that these emissions should be allocated to the country of origin or destination. To calculate consumption-based emissions, detailed trade data between countries and the emissions intensity across many industries and sectors in each country is required (Global Carbon Budget, 2023). Furthermore, there is insufficient high-quality, high-resolution data prior to 1990 to produce these calculations on consumption-based emissions. Greenhouse gas emissions from consumption represent a significant portion of total emissions, but are often not included in official statistics reported to the UN. Such omission can lead to a poor understanding of a country´s environmental impact (Onakuse, 2021). There is a need for more comprehensive metrics, for example, the Gross Ecosystem Product (GEP) for the inclusion of environmental factors and the contributions of the natural ecosystem to human well-being.

It was a positive signal when all Swedish political parties on the Environmental Objectives Committee in early 2022 agreed that Sweden should become the first country in the world with consumption-based environmental goals. These goals should be included in the climate policy framework, together with international air travel. The politicians did not, however, indicate how the objectives are to be achieved. Instead, the task was given to the Environmental Objectives Committee which has refrained from a proposal, so far. The EU's climate package for 2030, with updated rules for emissions trading, included shipping and air traffic and was accepted by the European Parliament in spring 2023.

Other measurements have a similar focus to GDP. The Human Development Index (HDI) uses per capita income as part of the calculation, but also considers life expectancy and education level. The Gini coefficient is a unit of measurement of inequality in terms of the income distribution of a population. It is usually calculated based on disposable relative income after taxes and transfers. But income dispersion is significantly reduced if the redistributive effects of tax-financed welfare are considered. High economic inequality can lead to social crises, especially in times of unrest.

Reduced consumption of various products is necessary. This may include reducing air travel by introducing taxes. One quick and easy solution would be for a government to increase the value added tax on unsustainable products and human behaviour, to indicate the urgency. This will give governments more resources, for example, for easy and rapid access to health care, social services, and pensions for more older people. Other solutions would be to prevent the availability of easy-to-get and quick credit for consumption, and higher taxes on income from capital.

In principle, traditional self-sufficiency was a way of life with significant durability. Since it is still common in societies of the low-income countries, it should be valuable to draw lessons from it. One example would be to learn from how life was lived for a long time on the island of Bali in Indonesia. Bali is known for its old traditions and gentle people, based on the old philosophy of Tri Hita Karana. The concept implied that life should be lived in harmony with other people, the environment and with spirituality. Such a life was characterized by sustainability and well-being for its inhabitants but has recently been replaced by Western modernity and consumerism through tourism.

Today, local people on Bali dislike the tourists, who are welcomed by the government for their money. After the COVID-19 pandemic, some tourists have started to behave as if they own the island and do whatever they wish. Foreign visitors, sometimes push aside local people at the beach and reside in new luxury villas that compete with the rice fields. According to Deutsche Welle there are some 3000 digital nomads in Indonesia, most of them on Bali: a new form of colonialism. Several hundred local people have protested in writing against careless driving, noisy parties, drug use, drunkenness and indecent behaviour (Matthes, 2022). But worldwide, digital nomads have had government support by being granted special visas, such as by the governments in Thailand and Barbados. In early 2023, the Indonesian authorities proposed a ban on tourists using mopeds, launched a campaign to educate foreign visitors in local customs, and proposed an increase in tourist tax. A special task force of local officials has been set up to watch the activities of foreign visitors.

Increasing Waste Mountains

Increased consumption leads to more waste. According to the World Bank, the world generates 2 billion t of municipal solid waste annually (Kaza *et al.*, 2018). Conservatively, about one-third is not managed in an environmentally safe manner. Despite high-income countries accounting for only 16% of the world's population they generate about one-third of global waste. The total is expected to grow to 3.4 billion t by 2050 with a positive correlation between waste generation and income level (World Bank, n.d.).

Waste must be minimized in all sectors and circular production reintroduced. Although circular production can contribute it is not the whole solution. Only 1% of what is handed in for recycling in Sweden today becomes new clothes. The service life of products can be extended and not just guided by fashion trends for annual renewal. Nor is it justified to replace a fully functional one-year-old car or mobile phone with a newly manufactured model. All production and environmental costs must be reflected in the price tag which would not be popular.

One-third of all food produced globally by weight is lost or wasted between farm and fork; more than 1 billion t (UNEP, 2021). At the same time, one of ten people globally remain malnourished today and 800 million people are hungry. If current trends persist, food loss and waste will double by 2050, although there is an objective in the SDGs to halve food waste and reduce food losses by 2030. According to the United Nations Development Programme (UNDP), food waste occurs at the same level in middle-income countries as it does in high-income countries (Goodwin, 2023). According to the Swedish Food Agency, 35 million lunch portions of food were thrown away in Swedish schools in 2022.

Plastic constitutes the third-highest source of waste globally. The total volume of plastic waste is growing in line with increases in the global population and per capita consumption. Since 2017, Malaysia has been the world's largest importer of plastic waste and has more than 1300 plastic manufacturers (Chen *et al.*, 2021). Global recycling in 2019 was, however, only 24% according to the World Bank (2019). Estimates by UNEP in 2018 indicated that only 21% of global plastic waste was recovered. Incineration and recycling constituted 12% and 9%, respectively. The remainder of the plastic is disposed of in landfills or burnt. Plastic pollution could be reduced by 80% by 2040 if countries and companies change policy and markets, using existing technologies, according to a report by UNEP for World Environment Day 2023 (UNEP, 2023).

The current political focus on electrification implies that electronic waste will increase significantly in the future. In 2010, global electronic waste was around 34 million t according to the UN Global E-Waste Monitor and was estimated to reach 74 million t by 2030 (GESP, 2024). Only some 17% is recycled, such as gold, silver and platinum, in contrast to the goal set by the EU of 65% recycling after 2019.

The UN reports annual waste of US$9.5 billion of essential metals in electronic debris. It is thrown away instead of being recycled (Baldé *et al.*, 2023). In Europe, the recycling rate for electronic waste is 55% but the global average drops to around 17% and almost zero in parts of South America, Asia and Africa. Research from the United Nations University has indicated that about 60,000t of e-waste has been sent to Nigeria annually (Vidal, 2013). About three-quarters comes from EU countries, the United States and China. About one-fifth of the waste was nonfunctional. It was argued that few countries understood the scale and there was little tracking of e-waste. This delivery to Nigeria has continued despite, since 2002, it being illegal to import and export e-waste to and from Nigeria under the Basel Convention. Recently, an ambitious project was launched in Lagos, aiming to reform the electronics sector and attempting to stop the toxic toll of improper management of e-waste in Nigeria (UNEP, 2019). This may reduce environmental problems.

References

Aggarwal, M. (2018) India has brought 9.8 million hectares of degraded land under restoration since 2011. *Mongabay*. Available at: https://india.mongabay.com/2018/09/india-restored-9-8-million-hectares-of-degraded-land-since-2011/ (accessed 11 July 2024).

Ahram Online (2023) 14 ministries and government entities have relocated to the New Administrative Capital. Presidential statement, Al-Ahram, Cairo. Available at: https://english.ahram.org.eg/News/50 0121.aspx (accessed 7 September 2024).

Andritz Hydro (2021) Green Energy for 60 million people. *Hydro News*, Andritz Hydro, Belo Monte, Brazil. Available at: https://www.andritz.com/hydro-en/hydronews/hn34/belo-monte-brazil (accessed 7 September 2024).

ASU (2006) State of Observed Species. A report by Arizona State University's (ASU) International Institute for Species Exploration, the International Commission on Zoological Nomenclature, the International Plant Names Index, and Thompson Scientific, Tempe, AZ.

Axfoundation (2023) Antibiotika inom livsmedelsproduktion – för en ansvarsfull antibiotikaanvändning. Axfoundation, Stockholm. Available at: https://www.axfoundation.se/projekt/antibiotika-inom-livsm edelsproduktion-antibiotikaresistens#:~:text=in%20your%20browser.-,Problem,en%20omfattande %20anv%C3%A4ndning%20av%20antibiotika (accessed 7 September 2024).

Ayele, A. and Tarekegn, K. (2020) The impact of urbanization expansion on agricultural land in Ethiopia: A review. *Environmental & Socio-Economic Studies* 8(4), 73–80. Available at: https://doi.org/10.2478/environ-2020-0024 (accessed 11 July 2024).

Baldé, C.P., Yamamoto, T. and Forti, V. (2023) Datasets on invisible E-waste supporting the International E-waste Day 2023. United Nations Institute for Training and Research, (UNITAR), Bonn, Germany.

Baltic Sea Brief (2021) A collapse of herring could threaten the Baltic Sea ecosystem. *Baltic Sea Brief 35*, Baltic Waters, Stockholm. Available at: https://balticwaters.org/en/baltic-sea-brief-35/ (accessed 7 September 2024).

Beament, E. (2020) Nearly two million acres of British grassland lost to urban development and woods in 25 years, analysis shows. *The Independent*. Available at: https://www.independent.co.uk/climate-change/news/uk-grassland-lost-farm-pasture-playing-fields-urban-development-a9609986.html (accessed 6 July 2024).

Benny, J. (2023) QatarEnergy signs second major natural gas supply deal with China. *The National*. Available at: https://www.thenationalnews.com/business/energy/2023/06/20/qatarenergy-signs-second-major-natural-gas-supply-deal-with-china/ (accessed 7 September 2024).

Berger, J.K. (2005) Mission possible: ALL species foundation and the call for discovery. In: *Proceedings of the California Academy of Sciences* 56, Supplement I, (10), pp. 114–118. Available at: https://res earcharchive.calacademy.org/research/scipubs/pdfs/v56/proccas_v56_n10_Suppl.pdf (accessed 7 September 2024).

Bodanis, D. (2003) *The Secret House: The Extraordinary Science of an Ordinary Day*. Berkley Books, New York.

Bourgery-Gonse, T. (2023) EU unveils Critical Raw Materials Act, aiming to lessen dependence on China. *Euractiv*. Available at: https://www.euractiv.com/section/economy-jobs/news/eu-unveils-critical-raw-materials-act-aiming-to-lessen-dependence-on-china/ (accessed 6 July 2024).

Brautigam, D. (2015) *Will Africa Feed China?* Oxford University Press, Oxford, UK.

Bryson, B. (2003) *A Short History of Nearly Everything*. Broadway Books, New York.

Burgen, S. (2023) 'People are proud of the green spirit of ours': How a small Spanish city rejected cars. *The Guardian*. Available at: https://www.theguardian.com/environment/2023/dec/20/vitoria-gasteiz -spanish-city-rejected-cars (accessed 11 July 2024).

CE Delft (2023) CO2 emissions of private aviation in Europe. Publication commissioned by Greenpeace CE. CE Delft, Delft, Netherlands.

Chen, H.L., Nath, T.K., Chong, S., Foo, V., Gibbins, C. *et al.* (2021) The plastic waste problem in Malaysia: Management, recycling and disposal of local and global plastic waste. *SN Applied Sciences* 3, 437. Available at: https://doi.org/10.1007/s42452-021-04234-y (accessed 11 July 2024).

Chiba, D., Watanabe, S. and Nitta, Y. (2021) Chinese companies corralling land around world. *Nikkei Asia*. DOI: https://asia.nikkei.com/Spotlight/Datawatch/Chinese-companies-corraling-land-around-world.

Clean Development Mechanism (CDM) (n.d.) Frequently asked questions. Available at: https://cdm.unfccc. int/faq/index.html (accessed 6 July 2024).

Constable, O.R. (1996) *Trade and Traders in Muslim Spain: The Commercial Realignment of the Iberian Peninsula 900-1500*. Cambridge University Press, Cambridge, UK.

Cooke, K. (2016) Saudi agricultural investment abroad: Land grab or benign strategy? *Middle East Eye*. Available at: https://www.kurdistanagriculture.org/2016/10/saudi-agricultural-investment-abroad .html (accessed 7 July 2024).

Crützen, P.J. and Ramanathan, V. (2000) The ascent of atmospheric sciences. *Science* 290(5490), 299–304. Available at: https://doi.org/10.1126/science.290.5490.299 (accessed 12 July 2024).

Ditlevsen, P. and Ditlevsen, S. (2023) Warning of a forthcoming collapse of the Atlantic meridional overturning circulation. *Nature Communications* 14, 4254. Available at: https://doi.org/10.1038/s41467 -023-39810-w (accessed 12 July 2024).

Economist Impact (2023) Five things you need to know about deep-sea mining. Available at: https://impa ct.economist.com/sustainability/ecosystems-resources/five-things-you-need-to-know-about-deep -sea-mining (accessed 6 July 2024).

Ehrlich, P. and Ehrlich, A.H. (1990) *The Population Explosion*. Frederick Muller Ltd, UK.

Einhorn, C. (2024) Global summit on nature adopts a novel way to pay for compensation. In: *16th Conference of the Parties to the Convention on Biological Diversity (COP16)*, Cali, Colombia. New York Times, November 2, 2024. Available at: https://www.nytimes.com/2024/11/02/climate/cop16-c ali-colombia-nature-biodiversity.html

Emsley, J. (2001) *Nature's Building Blocks: An A-to-Z Guide to the Elements*. Oxford University Press, Oxford, UK.

Environmental Protection Agency (EPA) (2024a) Sources of greenhouse gas emissions. EPA, Washington D.C. Available at: https://www.epa.gov/ghgemissions/sources-greenhouse-gas-emissions (accessed 9 July 2024).

Environmental Protection Agency (EPA) (2024b) Providing Safe Drinking Water in Areas with Abandoned Uranium Mines. United States Environmental Protection Agency (EPA). Available at: https://www. epa.gov/navajo-nation-uranium-cleanup/safe-drinking-water (accessed 7 September 2024).

Equinor (2023) Equinor fourth quarter 2022 and year end results. Equinor ASA, Stavanger, Norway. Available at: https://www.equinor.com/news/equinor-fourth-quarter-2022-and-year-end-results (accessed 7 September 2024).

Ethirajan, A. (2022) Colombo port city: A new Dubai or a Chinese enclave? *BBC News*. Available at: https ://www.bbc.com/news/world-asia-59993386 (accessed 7 September 2024).

Eurocontrol (2024) European aviation overview 2023. Available at: https://www.eurocontrol.int/publication /eurocontrol-european-aviation-overview (accessed 7 September 2024).

Fahey, D.W., Doherty, S.J., Hibbard, K.A., Romanou, A. and Taylor, P.C. (2017) Physical drivers of climate change. In: Wuebbles, D.J., Fahey, D.W., Hibbard, K.A., Dokken, D.J., Stewart, B.C. *et al.* (eds) *Climate Science Special Report: Fourth National Climate Assessment*, Vol. I. US Global Research Program, Washington D.C.

Fearnside, P.M. (2006) Dams in the Amazon: Belo Monte and Brazil's hydroelectric development of the Xingu River Basin. *Environmental Management* 38(1), 16–27.

Fédération Environnement Durable (FED) (2024) French Council of State annuls wind turbine permits, major impact on energy future. FED, Paris. Available at: https://environnementdurable.org/documen ts/CP-%20Conseil-Etat-annulation-normesEN-DEF.pdf (accessed 6 July 2024).

Felda Global Ventures (FGV) (2023) Progressing Sustainably. Annual Integrated Report 2023. FGV Holdings Berhad, Kuala Lumpur, Malaysia. Available at: https://www.fgvholdings.com/investor-relat ions/annual-reports-presentations/ (accessed 7 September 2024).

Food and Agriculture Organization (FAO) (2023) Land statistics and indicators 2000–2021 Global, regional and country trends. FAOSTAT Analytical Brief 71. FAO, Rome. Available at: https://www.fao.org/stat istics/highlights-archive/highlights-detail/land-statistics-and-indicators-(2000-2021).-global--region al-and-country-trends/en (accessed 7 September 2024).

Friedrich, J., Ge, M., Pickens, A. and Vigna, L. (2023) *This Interactive Chart Shows Changes in the World's Top 10 Emitters*. Insights, World Resources Institute. Available at: https://www.wri.org/insights/inter active-chart-shows-changes-worlds-top-10-emitters (accessed 5 July 2024).

Fröhlich, S. (2019) East Africa's forgotten slave trade. *Deutsche Welle*. Available at: https://www.dw.com/ en/east-africas-forgotten-slave-trade/a-50126759 (accessed 12 July 2024).

Frontline (2023) Conflicts in Sudan, Ukraine aggravate global refugee crisis: UN report. *Frontline*. Available at: https://frontline.thehindu.com/news/world-refugee-day-record-110-million-people-now-forcibly -displaced-worldwide-rise-in-number-an-indictment-of-the-world-says-un-refugee-agency-unhcr- report/article66986107.ece (accessed 8 July 2024).

Gao, C. (2021) China and the Mekong River disputes: Can a new framework bring new compromises? *China Focus*. Available at: https://chinafocus.ucsd.edu/2021/04/09/china-and-the-mekong-river-dis putes-can-a-new-framework-bring-new-compromises/ (accessed 7 July 2024).

Gbane, N.C. (2023) Zimbabwe: New African agricultural power house. *AfricaNews*. Available at: https:// www.africanews.com/2023/01/08/zimbabwe-new-african-agricultural-power-house/ (accessed 7 September 2024).

GESP (2024) *The Global E-Waste Monitor 2024*. The Global E-waste Statistics Partnership (GESP) managed by United Nations Institute for Training and Research (UNITAR) and International Telecommunication Union (ITU), Geneva, Switzerland. Available at: https://www.itu.int/en/ITU-D/ Environment/Pages/Publications/The-Global-E-waste-Monitor-2024.aspx (accessed 7 September 2024).

Glatz, A.K. (2014) Myanmar: Comprehensive solutions needed for recent and long-term IDPs alike. The Internal Displacement Monitoring Centre (IDMC). Available at: https://api.internal-displacement. org/sites/default/files/publications/documents/201407-ap-myanmar-overview-en.pdf (accessed 7 September 2024).

Global Biodiversity Information Facility (GBIF) (2020) Global Register of Introduced and Invasive Species GRIIS – Luxembourg. Invasive Specialist Species Group (ISSG), Rome. Available at: https://www. gbif.org/dataset/a811c07f-206d-46b8-ad59-93af4e2ce7c0 (accessed 8 July 2024).

Global Carbon Budget (2023) Population based on various sources (2023) – with major processing by Our World in Data. 'Per capita consumption-based CO_2 emissions – Global Carbon Project.' Available at: https://ourworldindata.org/grapher/consumption-CO2-per-capita (accessed 7 September 2024).

Global Forest Watch (2019) What's Happening in Cambodia's Forests? Global Forest Watch Blog. Available at: https://www.globalforestwatch.org/blog/forest-insights/whats-happening-in-cambodi as-forests/ (accessed 7 September 2024).

Global Witness (2014) The people and forests of Papua New Guinea under threat: The government's failed response to the largest land grab in modern history. Global Witness.

Golden Agri-Resources (2023) Our business: What we do. Golden Agri-Resources, Singapore. Available at: https://www.goldenagri.com.sg/about-us/our-business/ (accessed 7 September 2024).

Gong, X. (2023) The Mekong Region is a test of China's global development and security model. *Carnegie China*. Available at: https://carnegieendowment.org/research/2023/12/the-mekong-region-is-a-test -of-chinas-global-development-and-security-model?lang=en (accessed 12 July 2024).

Goodwin, L. (2023) The global benefits of reducing food loss and waste, and how to do it. World Resources Institute. Available at: https://www.wri.org/insights/reducing-food-loss-and-food-waste (accessed 8 July 2024).

Gould, S.J. (1993) *Eight Little Piggies: Reflections in Natural History*. W.W. Norton & Company.

GRAIN (2016) The global farmland grabs in 2016: How big, how bad? GRAIN, Barcelona. Available at: https://grain.org/en/article/5492-the-global-farmland-grab-in-2016-how-big-how-bad (accessed 12 July 2024).

GRAIN (2020) Socfin/Bolloré plantation activities in 2018: Who benefits? GRAIN, Barcelona. Available at: https://grain.org/en/article/6443-unravelling-the-socfin-bollore-plantations-thanks-to-profundo (accessed 7 July 2024).

Gramlich, J. and Scheller, A. (2021) What'S happening at the U.S.-Mexico border in 7 charts. *Pew Research Center*. Available at: https://www.pewresearch.org/short-reads/2021/11/09/whats-happening-at-the-u-s-mexico-border-in-7-charts/ (accessed 8 July 2024).

Gregory, A. (2022) Antimicrobial resistance now a leading cause of death worldwide, study finds. *The Guardian*. Available at: https://www.theguardian.com/society/2022/jan/20/antimicrobial-resistance-antibiotic-resistant-bacterial-infections-deaths-lancet-study (accessed 8 July 2024).

Griffin, P. (2017) *The Carbon Majors Database: CDP Carbon Majors Report 2017*. CDP Worldwide. Available at: https://cdn.cdp.net/cdp-production/cms/reports/documents/000/002/327/original/Carbon-Majors-Report-2017.pdf (accessed 27 June 2024).

Håkansson, L. (2024) Sweden's largest wind farm faces bankruptcy. *Arctic Business Journal*. Available at: https://www.arctictoday.com/swedens-largest-wind-farm-faces-bankruptcy/ (accessed 6 July 2024).

Henley, G. (2014) Case study on land in Burma. Overseas Development Institute (ODI) with the assistance of the UK Department for International Development (DFID). Available at: http://dx.doi.org/10.12774/eod_hd.march2014.henley (accessed 7 September 2024).

Hines, A. (2023) Decade of defiance: Ten years of reporting land and environmental activism worldwide. *Global Witness*. Available at: https://www.globalwitness.org/en/campaigns/environmental-activists/decade-defiance/ (accessed 12 July 2024).

Huang, J., Zhu, L., Deng, X. and Rozelle, S. (2005) Cultivated land changes in China: The impacts of urbanization and industrialization. *Proceedings of SPIE* 5884, 135–149. Available at: https://doi.org/10.1117/12.613882 (accessed 12 July 2024).

Humphries, A. (2023) Rishi sunak pushes back ban on new petrol and diesel cars to 2035. *BBC News*. Available at: https://www.bbc.com/news/live/uk-66863110 (accessed 6 July 2024).

Hunter, M., Sorensen, A., Nogeire-McRae, T., Beck, S., Shutts, S. *et al.* (2022) *Farms Under Threat 2040: Choosing an Abundant Future*. American Farmland Trust, Washington, D.C.

Hurst, C.E. and McConnell, D. (2010) *An Amish Paradox: Diversity and Change in the World's Largest Amish Community*. John Hopkins University Press, Baltimore, MA.

Ingram, J.C., Jones, L., Credo, J. and Rock, T. (2020) Uranium and arsenic unregulated water issues on navajo lands. *Journal of Vacuum Science and Technology A* 38(3), 031003. Available at: https://doi.org/10.1116/1.5142283 (accessed 7 September 2024).

Intergovernmental Panel on Climate Change (IPCC) (2022) *Summary for Policymakers*. Cambridge University Press, Cambridge, UK and New York, NY, USA, pp. 3–33. Available at: https://doi.org/10.1017/9781009325844.001 (accessed 5 July 2024).

Intergovernmental Panel on Climate Change (IPCC) (2023) *AR6 Synthesis Report*. IPCC, Geneva, Switzerland. Available at: https://www.ipcc.ch/report/ar6/syr/resources/spm-headline-statements/ (accessed 5 July 2024).

Intergovernmental Panel on Climate Change IPCC Pachauri, R.K. and Meyer, L.A. (eds). (2024) *Climate Change 2014: Synthesis Report. Contribution of Working Groups I, II and III to the Fifth Assessment Report of the Intergovernmental Panel on Climate Change*. IPCC, Geneva, Switzerland.

International Energy Agency (IEA) (2022) *Global Energy Review: CO2 Emissions in 2021*. IEA, Paris. Available at: https://www.iea.org/reports/global-energy-review-co2-emissions-in-2021-2 (accessed 12 July 2024).

International Energy Agency (IEA) (2023) *World Energy Outlook 2023*. Licence: CC BY 4.0 (report); CC BY NC SA 4.0 (report); CC BY NC SA 4.0 (Annex A). IEA, Paris. Available at: https://www.iea.org/reports/world-energy-outlook-2023 (accessed 7 September 2024).

International Energy Agency (IEA) (2024a) Renewables 2023. IEA, Paris. Available at: https://www.iea.org/reports/renewables-2023 (accessed 12 July 2024).

International Energy Agency (IEA) (2024b) Massive expansion of renewable power opens door to achieving global tripling goal set at COP28. IEA, Paris. Available at: https://www.iea.org/news/massive-expansion-of-renewable-power-opens-door-to-achieving-global-tripling-goal-set-at-cop28 (accessed 6 July 2024).

International Labour Organization (ILO) (2021) *Global Estimates on International Migrant Workers – Results and Methodology. Executive Summary, 30 June*. ILO, Geneva, Switzerland.

International Labour Organization (ILO), International Organization for Migration (IOM), and Walk Free (2022) Global Estimates of Modern Slavery Forced Labour and Forced Marriage. ILO, IMO and Walk Free Geneva, Switzerland. Available at: https://www.ilo.org/publications/major-publications/global-estimates-modern-slavery-forced-labour-and-forced-marriage (accessed 7 September 2024).

International Land Coalition (2011) *Tirana Declaration*. Global Assembly, Tirana, Albania. Available at: https://www.landcoalition.org/about-us/aom2011/tirana-declaration

International Organization for Migration (IOM) (2022) *World Migration Report*. IOM, Geneva, Switzerland.

International Union for Conservation of Nature and Natural Resources (IUCN) (2022a) Deep-sea mining IUCN, Gland, Switzerland. Available at: https://www.iucn.org/resources/issues-brief/deep-sea-mining (accessed 6 July 2024).

International Union for Conservation of Nature and Natural Resources (IUCN) (2022b) Deep-sea mining, IUCN, Gland, Switzerland. Available at: https://www.iucn.org/sites/default/files/2022-07/iucn-issues-brief_dsm_update_final.pdf (accessed 6 July 2024).

Jong, H.N. (2023) New data show 10% increase in primary tropical forest lost in 2022. *Mongabay*. Available at: https://news.mongabay.com/2023/06/new-data-show-10-increase-in-primary-tropical-forest-loss-in-2022/ (accessed 7 July 2024).

Jurkevics, A. (2022) Land grabbing and the perplexities of territorial sovereignty. *Political Theory* 50(1), 32–58. Available at: https://doi.org/10.1177/00905917211008591 (accessed 12 July 2024).

Kan, K. and Chen, J. (2022) Rural urbanization in China: Administrative restructuring and the livelihoods of urbanized rural residents. *Journal of Contemporary China* 31(136), 626–643. DOI: 10.1080/10670564.2021.1985841. (accessed 12 July 2024).

Kaza, S., Yao, L.C., Bhada-Tata, P. and Van Woerden, F. (2018) *What A Waste 2.0: A Global Snapshot of Solid Waste Management to 2050*. World Bank, Washington, D.C. Available at: http://hdl.handle.net/10986/30317 (accessed 12 July 2024).

Khan, J. (2021) Invasive species have cost Australia $390 billion in the past 60 years, study shows. *ABC News*. Available at: https://www.abc.net.au/news/science/2021-07-30/invasive-species-cost-billions-australia/100333710 (accessed 12 July 2024).

Killeen, T.J. (2021) *A Perfect Storm in the Amazon*, 2nd edn. The White Horse, Winwick, UK.

Kushairi Din, A. (2017) Malaysian oil palm industry performance 2016 and prospects for 2017. Ministry of Plantation Industries & Commodities, Kuala Lumpur, Malaysia.

Larsson, D.G.J. and Flach, C.F. (2022) Antibiotic resistance in the environment. *Nature Reviews Microbiology* 20, 257–269. Available at: https://doi.org/10.1038/s41579-021-00649-x (accessed 8 July 2024).

Leakey, R.E. and Lewin, R. (1996) *The Sixth Extinction: Patterns of Life and the Future of Humankind*. Knopf Doubleday Publishing Group, New York.

Liebman, A. (2005) Trickle-down hegemony? China's "peaceful rise" and dam building on the Mekong. *Contemporary Southeast Asia* 27(2), 281–304. Available at: https://www.jstor.org/stable/25798737 (accessed 10 July 2024).

Lin, S., Mao, J., Chen, W. and Shi, K. (2019) Indium in Mainland China: Insights into use, trade, and efficiency from the substance flow analysis. *Resources, Conservation and Recycling* 149, 312–321. Available at: https://doi.org/10.1016/j.resconrec.2019.05.028 (accessed 7 September 2024).

Lipton, E., Searcey, D. and Forsythe, M. (2021) Race to the future: What to know about the frantic quest for cobalt. *The New York Times*. Available at: https://www.nytimes.com/2021/11/20/world/china-congo-cobalt-explained.html (accessed 7 September 2024).

Ljungqvist, C.F. (2017) *The Climate and Humans for 12 000 Years* (in Swedish). Dialogos, Stockholm.

Macfarlane, R. (2019) *Underland: A Deep Time Journey*. Penguin, UK.

Makhijani, A. (1992) *Climate Change and Transnational Corporations: Analysis and Trends, United Nations Centre on Transnational Corporations, Environment Series no.2*. United Nations, New York.

Malthus, T.R. (1798) *An Essay on the Principle of Population*. J. Johnson, London.

Matthes, G. (2022) Bali lures digital nomads despite controversy. *Deutsche Welle*. Available at: https://www.dw.com/en/bali-lures-digital-nomads-despite-controversy/a-63669834 (accessed 7 September 2024).

McLauchlan, G. (2018) Māori horticulture: Growing kūmara and other crops the traditional way. *NZ Gardener*.

Min, M., Xu, H., Chen, J. and Fayek, M. (2005) Evidence of Uranium biomineralization in sandstone-hosted roll-front uranium deposits, Northwestern China. *Ore Geology Reviews* 26(3–4), 198–206. Available at: https://doi.org/10.1016/j.oregeorev.2004.10.003 (accessed 12 July 2024).

Miriri, D. and Bavier, J. (2022) Insight: Africa's dream of feeding China hits hard reality. *Reuters*. Available at: https://www.reuters.com/world/africa/africas-dream-feeding-china-hits-hard-reality-2022-06-28/ (accessed 12 July 2024).

Moana Minerals Ltd (2024) The Cook Islands Nodule Project – Unpacked. Moana Minerals, Cook Islands. Available at: https://www.moanaminerals.com/ (accessed 6 July 2024).

Moate, M. (2023) What causes deforestation in Borneo and how do we stop it? Earth Org, Hong Kong. Available at: https://earth.org/deforestation-in-borneo/ (accessed 7 September 2024).

Moutinho, S. (2023) A river's pulse. *Science* 379, 6627. Available at: https://doi.org/10.1126/science.adg 5424 (accessed 7 July 2024).

Moyo, S. (2011) Three decades of agrarian reform in Zimbabwe. *The Journal of Peasant Studies* 38(3), 493–531. Available at: https://doi.org/10.1080/03066150.2011.583642 (accessed 7 September 2024).

Mulchandani, R., Wang, Y., Gilbert, M. and Van Boeckel, T.P. (2023) Global trends in antimicrobial use in food-producing animals: 2020 to 2030. *PLOS Global Public Health* 3(2), e0001305. Available at: https://doi.org/10.1371/journal.pgph.0001305 (accessed 7 September 2024).

Naadi, T. and Lansah, S. (2021) *The Illegal Gold Mines Killing Rivers and Livelihoods in Ghana*. BBC. Available at: https://www.bbc.com/news/world-africa.58119653 (accessed 11 August 2024).

Naishadam, S. (2023) In Arizona, fresh scrutiny of Saudi-owned farm's water use. *Associated Press*. Available at: https://apnews.com/article/water-foreign-farms-arizona-drought-saudi-arabia-2fe3ea1 fad43b14ca118cf85196f3e9a (accessed 7 July 2024).

Naithani, S. (2021) *History and Science of Cultivated Plants*. Open Educational Resources, Oregon State University, Corvallis, OR. Available at: https://open.oregonstate.education/cultivatedplants/ (accessed 12 July 2024).

National Academy of Sciences (2020) *Climate Change: Evidence and Causes: Update 2020*. The National Academies Press, Washington, D.C.

National Oceanic and Atmospheric Administration (NOAA) (2022) Antarctic ozone hole slightly smaller in 2022. NOAA, Silver Spring, MD. Available at: https://www.noaa.gov/news-release/antarctic-ozone-hole-slightly-smaller-in-2022 (accessed 1 September 2024).

NIBIO (2020) *Norwegian Agriculture. Status and Trends 2019*. Norsk Institutt for Bioekonomi, NIBIO POP, 6, 8. Available at: https://nibio.brage.unit.no/nibio-xmlui/bitstream/handle/11250/2643268/NIBIO_ POP_2020_6_8.pdf?sequence=4 (accessed 7 September 2024).

Nilsen, E. (2022) Biden is blamed for downturn in new oil drilling, but fossil fuel companies are the ones hitting pause. *CNN*. Available at: https://edition.cnn.com/2022/10/09/politics/us-oil-drilling-opec-gas-prices-biden-climate/index.html (accessed 6 July 2024).

Nordic Innovation (2023) *Nordic Innovation Annual Report 2023*. Nordic Innovation, Oslo., Norway.

Norwegian Tax Administration. (2023) Tax revenue from the oil companies, Trondheim, Norway.

Nyström, J. (2020) *Planet of the Mushrooms (in Swedish)*. Bonniers Fakta. Livonia Print, Lithuania.

Oakland Institute (2011) Understanding land investment deals in Africa: Saudi Star in Ethiopia. Oakland Institute, Oakland, CA. Available at: https://www.oaklandinstitute.org/sites/oaklandinstitute.org/files /OI_SaudiStar_Brief.pdf (accessed 7 July 2024).

Onakuse, S. (2021) GDP ignores the environment: Why it's time for a more sustainable growth metric. The Conversation, The Conversation Trust (UK) Ltd, London. Available at: https://theconversation .com/gdp-ignores-the-environment-why-its-time-for-a-more-sustainable-growth-metric-170820 (accessed 7 September 2024).

Oxfam (2011) Land and power: The growing scandal surrounding the new wave of investments in land. 151 Oxfam Briefing Paper, Oxfam Australia, Carlton, Australia. Available at: https://policy-practice. oxfam.org/resources/land-and-power-the-growing-scandal-surrounding-the-new-wave-of-investm ents-in-l-142858/ (accessed 7 September 2024).

Pandey, B. and Seto, K.C. (2015) Urbanization and agricultural land loss in India: Comparing satellite estimates with census data. *Journal of Environmental Management* 148, 53–66. Available at: https:// doi.org/10.1016/j.jenvman.2014.05.014 (accessed 12 July 2024).

PBS News (2014) Who's behind the chinese takeover of world's biggest pork producer?. *PBS News*. Available at: https://www.pbs.org/newshour/show/whos-behind-chinese-takeover-worlds-biggest -pork-producer (accessed 7 July 2024).

Perry, C. (2021) Slavery and the slave trade in the western indian Ocean world. In: Perry, C., Eltis, D., Engerman, S.L. and Richardson, D. (eds) *The Cambridge World History of Slavery: Volume 2, AD 500–AD 1420*. Cambridge University Press, Cambridge, UK, pp. 123–152.

Pimentel, D., Zuniga, R. and Morrison, D. (2005) Update on the environmental and economic costs associated with alien-invasive species in the United States. *Ecological Economics* 52(3), 273–288. Available at: https://doi.org/10.1016/j.ecolecon.2004.10.002 (accessed 12 July 2024).

Plummer, S., Lecomte, P. and Doherty, M. (2017) The ESA climate change initiative (CCI): A European contribution to the generation of the Global Climate Observing System. *Remote Sensing of Environment* 203, 2–8. Available at: https://doi.org/10.1016/j.rse.2017.07.014 (accessed 7 September 2024).

Poderati, G. (2022) Human rights aspects and soil governance: A special focus on land grabbing. *Soil Security* 6, 100042. Available at: https://doi.org/10.1016/j.soisec.2022.100042 (accessed 12 July 2024).

Poynting, M. and Stallard, E. (2024) What are el niño and la niña, and how do they change the weather?. *BBC News*. Available at: https://www.bbc.com/news/science-environment-64192508 (accessed 5 July 2024).

Qian, H., Zhu, X., Huang, S., Linquist, B., Kuzyakov, Y. *et al.* (2023) Greenhouse gas emissions and mitigation in rice agriculture. *Nature Reviews Earth & Environment* 4, 716–732. Available at: https://doi.org/10.1038/s43017-023-00482-1 (accessed 7 September 2024).

Rabone, M., Wiethase, J.H., Simon-Lledo, E., Emery, A.M., Jones, D.O.B. *et al.* (2023) How many meta-zoan species live in the world's largest mineral exploration region? *Current Biology* 33, 2383–2396. Available at: https://doi.org/10.1016/j.cub.2023.04.05 (accessed 12 July 2024).

Raja, A. (2021) Explained: How Sardar Sarovar Dam is providing irrigation water in summer for the first time in history. Available at: https://indianexpress.com/article/explained/explained-how-sardar-saro var-dam-is-providing-irrigation-water-in-summer-for-the-first-time-in-history-7350969/ (accessed 7 September 2024).

Reidy, J. (2023) Indonesia wheat imports decline in 2022–23. WORLD-GRAIN.com. Available at: https ://www.world-grain.com/articles/18864-indonesia-wheat-imports-decline-in-2022-23 (accessed 7 September 2024).

Ren, Q., He, C., Huang, Q., Shi, P., Zhang, D. *et al.* (2022) Impacts of urban expansion on natural habitats in global drylands. *Nature Sustainability* 5, 869–878. Available at: https://doi.org/10.1038/s41893-02 2-00930-8 (accessed 7 September 2024).

Rhodes, R. (2018) *Energy. A Human History*. Simon & Schuster, New York.

Richardson, D. (2022) *Principles and Agents: The British Slave Trade and Its Abolition*. Yale University Press, New Haven, CT.

Riley, T. (2017) Just 100 companies responsible for 71% of global emissions, study says. *The Guardian*. Available at: https://www.theguardian.com/sustainable-business/2017/jul/10/100-fossil-fuel-compa nies-investors-responsible-71-global-emissions-cdp-study-climate-change (accessed 5 July 2024).

Satheesh, S. (2019) Hundreds of India villages under water as Narmada dam level rises. *Aljazeera*. Available at: https://www.aljazeera.com/news/2019/9/23/hundreds-of-india-villages-under-water-as-narmada-dam-level-rises (accessed 7 July 2024).

SCB (2023) Land use in sweden in 2021. Statistics Sweden (SCB), Stockholm.

Schumacher, E.F. (1973) *Small Is Beautiful. A Study of Economics as if People Mattered*. Blond & Briggs, UK.

SGU (2023) Critical and strategic raw materials. Geological Survey of Sweden, (SGU). Available at: https: //www.sgu.se/en/mineral-resources/critical-raw-materials/ (accessed 7 September 2024).

Shepard, W. (2019) What China is really up to in Africa. *Forbes*. Available at: https://www.forbes. com/sites/wadeshepard/2019/10/03/what-china-is-really-up-to-in-africa/ (accessed 7 September 2024).

Sihlobo, W. (2017) Was Zimbabwe ever the breadbasket of Africa? Africa Check Foundation, Kenya Ikigai Westlands, Nairobi, Kenya. Available at: https://agbiz.co.za/uploads/AgbizNews17/171201_Zimbab we-bread-basket-Africa-before-land-reform-program.pdf (accessed 7 September 2024).

Spagat, E. and Batrawy, A. (2016) Saudi land purchases fuel debate over US water rights. *Associated Press*. Available at: https://apnews.com/general-news-07508055f70c486e90871737d8f4558d (accessed 7 July 2024).

Statista (2023) Rice market in Indonesia: Statistics and facts. Value of rice import Indonesia 2019-2023. Available at: https://www.statista.com/topics/12200/rice-market-in-indonesia/#statisticChapter (accessed 7 September 2024).

Statista (2024) Palladium mine production worldwide 2023, by country. Available at: https://www.statista. com/statistics/273647/global-mine-production-of-palladium/ (accessed 7 September 2024).

Statista (2016) *Republic of Korea's Acquisitions of Agricultural Land Abroad 2016*. Statista Research Department, Jan 12, 2016. Available at: https://www.statista.com/statistics/381160/korean-acquisit ions-of-agricultural-land-abroad/

Statista (2022) *Number of operational nuclear reactors worldwide from 1954 to 2022*. Statista Research Department. June 28 2024. Available at: https://www.statista.com

Strickland, A. (2023) A legless lizard and hundreds of other new species were discovered in 2023. *CNN*. Available at: https://edition.cnn.com/2023/12/29/world/new-species-2023-scn/index.html (accessed 7 September 2024).

Strickler, L. and Moeder, N. (2023) Is China really buying up U.S. farmland? Here's what we found. *NBC News* Available at. Available at: https://www.nbcnews.com/news/investigations/how-much-us-farm land-china-own-rcna99274 (accessed 7 September 2024).

Surma, K. (2021) Lands grabs and other destructive environmental practices in Cambodia test the International Criminal Court. *Inside Climate News*. Available at: https://insideclimatenews.org/new s/23032021/land-grab-deforestation-international-criminal-court-cambodia-ecocide/ (accessed 12 July 2024).

Svenska Kraftnät (2022) *Market Analysis 2022. Analysis of the Power System 2023-2027*. Nr 2022/3235, (in Swedish). Svenska Kraftnät (SVK), Sundbyberg, Sweden.

Svensson, A. (2022) Hur många industritrålare finns det i Sverige? *Svenssons Nyheter – Njord*, Zaramis Media. Available at: https://fiske.zaramis.se/2022/07/19/hur-manga-industritralare-finns-det-i-sveri ge/ (accessed 7 September 2024).

Swaminathan, S., Anklesaria, A. and Neeraj, K. (2021) Are Resettled Oustees from the Sardar Sarovar Dam Project Better Off Today than their Former Neighbors Who Were Not Ousted?. National Bureau of Economic Research, NBER Working Paper No. 24423, Cambridge, MA.

Swedish Migration Agency (2023) Cirka 5 400 tillstånd beviljade för nästan 6 000 ansökande bärplockare. Swedish Migration Agency.

The Economist (2023) Global fertility has collapsed, with profound economic consequences. *The Economist*. Available at: https://www.economist.com/leaders/2023/06/01/global-fertility-has-collap sed-with-profound-economic-consequences (accessed 8 July 2023).

The Lancet (2022) Global burden of bacterial antimicrobial resistance in 2019: A systematic analysis. *The Lancet* 399(10325), 629–655. Available at: https://doi.org/10.1016/S0140-6736(21)02724-0 (accessed 12 July 2024).

The Lancet (2024) Global fertility in 204 countries and territories, 1950-2021, with forecasts to 2100: A comprehensive demographic analysis for the Global Burden of Disease Study 2021. *The Lancet*. Available at: https://doi.org/10.1016/S0140-6736(24)00550-6 (accessed 12 July 2024).

The New York Times (2022) Amid slowdown, immigration is driving U.S. population growth. *The New York Times*. Available at: https://www.nytimes.com/2022/02/05/us/immigration-census-population.html (accessed 7 September 2024).

United Nations (2019) The number of international migrants reaches 272 million, continuing an upward trend in all world regions, says UN. UN, New York. Available at: https://www.un.org/development/de sa/en/news/population/international-migrant-stock-2019.html (accessed 8 July 2024).

United Nations (1999) World Population Monitoring 1999. Population Growth, Structure and Distribution. Based on data from the World Population Prospects: the 1998 Revision. Vol.1. United Nations Population Division, Economic and Social Affairs, United nations, New York. Available at: https:// www.un.org/development/desa/pd/sites/www.un.org.development.desa.pd/files/files/documents/2 020/Jan/un_2000_world_population_monitoring_1999_-_population_growth_structure_and_distribu tion.pdf

United Nations (UN) (2022) *Global Compact for Migration*. UN, New York. Available at: https://www.un.or g/en/migration2022/global-compact-for-migration (accessed 8 July 2024).

United Nations (UN) Food Systems (2023) *Making Food Systems Work for People and Planet: UN Food Systems Summit +2, Report of the Secretary-General*. Available at: https://www.unfoodsystemshub .org/docs/unfoodsystemslibraries/stocktaking-moment/un-secretary-general/sgreport_en_rgb_upd ated_compressed.pdf?sfvrsn=560b6fa6_33 (accessed 5 July 2024).

United Nations Convention to Combat Desertification (UNCCD) (2019) The Land in Numbers 2019. Risks and Opportunities. UNCCD, Bonn, Germany.

United Nations Environment Programme (UNEP) (2019) Nigeria turns the tide on electronic waste. UN Environment Programme, Nairobi, Kenya. Available at: https://www.unep.org/news-and-stories/pre ss-release/nigeria-turns-tide-electronic-waste (accessed 7 September 2024).

United Nations Environment Programme (UNEP) (2021) *UNEP Food Waste Index Report 2021*. UNEP, Nairobi. Available at: https://www.unep.org/resources/report/unep-food-waste-index-report-2021 (accessed 11 July 2024).

United Nations Environment Programme (UNEP) (2022) *Kunming-Montreal Global Biodiversity Framework*. UNEP, Nairobi. Available at: https://www.unep.org/resources/kunming-montreal-global-biodiversity -framework (accessed 8 July 2024).

United Nations Environment Programme (UNEP) (2023) UN roadmap outlines solutions to cut global plastic pollution. UNEP, Nairobi. Available at: https://www.unep.org/cep/news/blogpost/un-roadma p-outlines-solutions-cut-global-plastic-pollution (accessed 8 July 2024).

United Nations Framework Convention on Climate Change (UNFCCC) (2023) COP 28: What was achieved and what happens next? UNFCCC. Available at: https://unfccc.int/cop28/5-key-takeawa ys (accessed 6 July 2024).

United Nations High Commissioner for Refugees (UNHCR) (2023) UNCHR calls for concerted action as forced displacement hits new record in 2022. UNNHCR, Geneva, Switzerland. Available at: https:// www.unhcr.org/news/press-releases/unhcr-calls-concerted-action-forced-displacement-hits-new-r ecord-2022 (accessed 8 July 2024).

United Nations Population Division (2022) UN Population Division Data Portal.

United Nations Population Fund (UNFPA) (2023) The problem with too few. UNFPA. New York. Available at: https://www.unfpa.org/swp2023/too-few (accessed 7 September 2024).

United States Department of Agriculture (USDA) (2021) *Foreign Holdings of U.S. Agricultural Land Through December 31, 2021*. Farm Service Agency and the Farm Production and Conservation Business Center, U.S. Department of Agriculture, Washington DC. Available at: https://www.fsa.usd a.gov/Assets/USDA-FSA-Public/usdafiles/EPAS/PDF/2021_afida_annual_report_through_12_31_2 021.pdf (accessed 7 September 2024).

van Uhm, D.P. and Wong, R.W.Y. (2021) Chinese organized crime and the illegal wildlife trade: Diversification and outsourcing in the golden triangle. *Trends in Organized Crime* 24, 486–505. Available at: https:// doi.org/10.1007/s12117-021-09408-z (accessed 7 September 2024).

van Vliet, J. (2019) Direct and indirect loss of natural area from urban expansion. *Nature* 2, 755–763. Available at: https://doi.org/10.1038/s41893-019-0340-0 (accessed 12 July 2024).

van Klink, R., Sheard, J.K., Høye, T.T., Roslin, T., Do Nascimento, L.A. *et al.* (2024) Towards a toolkit for global insect biodiversity monitoring. *Philosophical Transactions of the Royal Society B* 379, 1904. Available at: https://doi.org/10.1098/rstb.2023.0101 (accessed 7 2024).September

van Westen, R.M., Kliphuis, M. and Dijkstra, H.A. (2024) Physics-based early warning signal shows that AMOC is on tipping course. *Science Advances* 10(6). Available at: https://doi.org/10.1126/sciadv.ad k1189 (accessed 12 July 2024).

Vattenfall (2022) HYBRIT: A unique, underground, fossil-free hydrogen gas storage facility is being inau- gurated in Lulea. Vattenfall, Stockholm. Available at: https://group.vattenfall.com/press-and-media /pressreleases/2022/hybrit-a-unique-underground-fossil-free-hydrogen-gas-storage-facility-is-bein g-inaugurated-in-lulea (accessed 6 July 2024).

Vidal, J. (2013) Toxic 'e-waste' dumped in poor nations, says United Nations. *The Guardian*. Available at: https://www.theguardian.com/global-development/2013/dec/14/toxic-ewaste-illegal-dumping-dev eloping-countries (accessed 7 September 2024).

von Braun, J. and Meinzen-Dick, R. (2009) Land grabbing' by foreign investors in developing countries: Risks and opportunities. IFPRI Policy Brief. IFPRI, Washington, DC.

Walsh, D. and Lee, V. (2022) A new capital rises in Egypt, but at what price? *The New York Times*. Available at: https://www.nytimes.com/2022/10/08/world/middleeast/egypt-new-administrative-ca pital.html (accessed 12 2024).July

Western, L.M., Vollmer, M.K., Krummel, P.B., Adcock, K.E., Crotwell, M. *et al.* (2023) Global increase of ozone-depleting chlorofluorocarbons from 2010 to 2020. *Nature Geoscience* 16, 309–313. Available at: https://doi.org/10,1038/s41561-023-01147-w (accessed 12 July 2024).

William Jr, D.P. (2013) *Slavery in the Medieval and Early Modern Iberia*, The Middle Age Series. University of Pennsylvania Press.

Williams, S. and Montaigne, F. (2001) *Surviving Galeras*. Houghton Mifflin Company, Boston, and New York.

Wilmar (2023) Annual Report 2023. From farm to table: Growing our future. Wilmar International Limited, Singapore. Available at: https://www.wilmar-international.com/annualreport2023/documents/Wilma r-International-Limited-Annual-Report-2023.pdf (accessed 7 September 2024).

Wilson, E.O. (2010) *The Diversity of Life*. Harvard University Press, Cambridge, MA.

Wolford, W., White, B., Scoones, I., Hall, R., Edelman, M. *et al.* (2024) Global land deals: What has been done, what has changed, and what's next?. *The Journal of Peasant Studies*. Available at: https://doi.org/10.1080/03066150.2024.2325685 (accessed 12 July 2024).

World Bank (1995) *Annex:2 government cancellation of bank loan. project completion report: india: narmada river development – gujarat: sardar sarovar dam and power project: annexes to part I, and part III*. P. 148 (pages 23 of 61 of Annex 2). World Bank, Washington D.C.

World Bank (2010) Rising global interest in farmland: can it yield sustainable and equitable benefits?. World Bank, Washington D.C.

World Bank (2018) Urban population (% of total population) – China. World Bank, Washington D.C. Available at: https://data.worldbank.org/indicator/SP.URB.TOTL.IN.ZS?locations=CN (accessed 10 July 2024).

World Bank (2019) Solid Waste Management. World Bank Group. Washington DC. Available at: https://www.worldbank.org/en/topic/urbandevelopment/brief/solid-waste-management (accessed 7 September 2024).

World Bank (n.d.) Trends in solid waste management. World Bank, Washington D.C. Available at: https://datatopics.worldbank.org/what-a-waste/trends_in_solid_waste_management.html (accessed 8 July 2024).

World Health Organization (WHO) (2015) Antibiotic Resistance: Multi-country Public Awareness Survey. WHO, Geneva, Switzerland. Available at: https://iris.who.int/bitstream/handle/10665/194460/9789241509817_eng.pdf;sequence=1 (accessed 8 July 2024).

World Health Organization (WHO) (n.d.) Antibiotic Resistance. WHO, Geneva, Switzerland. Available at: https://www.afro.who.int/health-topics/antimicrobial-resistance (accessed 8 July 2024).

World Health Organization (WHO) (2022) *WHO/ECDC Report: Antimicrobial Resistance Remains a Threat to Health in European Region*. WHO, Geneva, Switzerland. Available at: https://www.who.int/europe/news/item/26-01-2022-who-ecdc-report-antimicrobial-resistance-remains-threat-to-health-in-european-region (accessed 8 July 2024).

World Meteorological Organization (WMO) (2023) COP28 concludes with historic agreement to try to tackle the climate crisis. WMO. Available at: https://wmo.int/news/media-centre/cop28-concludes-historic-agreement-try-tackle-climate-crisis (accessed 6 July 2024).

Worldometer (2023) India Demographics 2023. Available at: https://www.worldometers.info/demographics/india-demographics/ (accessed 7 September 2024).

World Resources Institute (WRI) (2023a) Getting the Transition Right for People, Nature and Climate: WRI Strategic Plan for 2023–2027. WRI, Washington D.C. Available at: https://www.wri.org/strategic-plan (accessed 12 July 2024).

World Resources Institute (WRI) (2023b) Fossil fuels face a reckoning at the COP28 climate summit. WRI. Available at: https://www.wri.org/news/statement-fossil-fuels-face-reckoning-cop28-climate-summit (accessed 6 July 2024).

Yang, B. and He, J. (2021) Global land grabbing: A critical review of case studies across the world. *Land* 10(3), 324. Available at: https://doi.org/10.3390/land10030324 (accessed 7 September 2023).

Yu, Y.-B., Choi, J.-H., Kang, J.-C., Kim, H.-J. and Kim, J.-H. (2022) Shrimp bacterial and parasitic disease listed in the OIE: A review. *Microbial Pathogenesis* 166, 105545. Available at: https://doi.org/10.1016/j.micpath.2022.105545 (accessed 12 July 2024).

Zhang, Z., Guan, D., Wang, R., Meng, J., Zheng, H. *et al.* (2020) Embodied carbon emissions in the supply chains of multinational enterprises. *Nature Climate Change* 10, 1096–1101. Available at: https://doi.org/10.1038/s41558-020-0895-9 (accessed 12 July 2024).

8 Towards New Agricultural Technology

Abstract

Brief texts provide examples of agricultural innovations during past agricultural revolutions and technical steps in agricultural development during the agricultural revolution of the Middle Ages, the Columbian exchange, the British and European agricultural revolution, agricultural innovations during the two World Wars and the rise of new technology with the Green Revolution. Specific examples of agricultural innovations are highlighted from very early innovators without the benefit of organized research, and from colonial times. This historical information is supplemented by examples of new technology from recent agricultural innovation by the private sector, for instance, in biotechnology or creating lab-grown meat. Moreover, there is a brief discussion of public sector agricultural research and technology developed in the interests of sustainability by international organizations and aid donors, focusing on the Consultative Group of International Agricultural Research plus its transformation into One CGIAR, and a general discussion of agricultural universities and their work towards sustainability.

Past Agricultural Revolutions and Major Technical Steps in Agricultural Development

History shows that the concept of sustainability is not a modern idea but has indirectly been part of a process of designing and developing new agricultural technology in line with agroecology. This has taken place over generations which ought to be instructive when discussing future approaches. In early history, agriculture often developed by specific, simple innovations that were achieved by trial and error. Then agriculture could make progress and produce more food. There were minimal negative consequences. The innovations fitted into nature and farmers adjusted them into their agroecology.

This historical perspective is also justified since agroecology is now gaining prominence as a research area and means of generating solutions. It combines local and scientific knowledge, focusing on the interactions between plants, animals, humans and the environment. One example is the Transformative Partnership Platform on Agroecological Approaches to Building Resilience of Livelihoods and Landscapes (Agroecology TPP) recently developed by CIFOR and ICRAF and includes several other research partners (CIFOR-ICRAF, n.d.). It is partly based on recommendations from a regional symposium on agroecology in 2015 by the Food and Agriculture Organization (FAO). This FAO web-based platform shares relevant knowledge on agroecology, identified as an integration of science, practices and social processes. The platform provides updates on current states of knowledge on these three aspects (FAO, n.d.).

The first agricultural revolutions

Agriculture began independently in various parts of the world leading to early civilizations. They included Sumer in Mesopotamia, ancient Egypt and Sudan, the Indus Valley civilization, ancient China and ancient Greece. In the Old World and the New World, at least 11 distinct regions were involved as independent centres of origin for the domestication of both plants and animals, and even inventions of early agricultural technology. These centres are briefly presented in Annex IV with their major agricultural characteristics.

The Middle Ages agricultural revolution

During the Middle Ages, further improvements were made in agriculture. Monasteries were important centres for collecting knowledge related to forestry and agriculture. The manorial system allowed large European landowners to control their land and its labourers as peasants or serfs. By 900 CE, developments in iron smelting led to the creation of agricultural implements such as ploughs, hand tools, horseshoes and harnesses. Old metal remains have shown products made of iron ore from lakes and marshes in northern Sweden. These are exemplified by an axe and knives of old iron-made steel, which included several layers of iron. These are the oldest examples of so-called temperature-treated steel objects that have been found in Europe. Recent research has shown that this process started before the Middle Ages when hunting and trapping groups needed iron, forming an integral part of the hunter–gatherer subsistence economy in Northern Fennoscandia (Bennerhag *et al.*, 2021). This took place during the Iron Age, contemporary with the Roman Empire.

The carruca heavy plough was an improvement on the earlier scratch plough due to a breakthrough around 1000 CE (Andersen *et al.*, 2016). In northern Africa, people had no metallurgical skills but in most of Africa south of the equator, the beginnings of farming and of ironworking took place at about the same time. Thus, there was sharp contrast between the metal-using farming people and their stone-tool-using hunter–gatherer neighbours and immediate predecessors (Phillipson, 2005).

During the Middle Ages, the watermills the Romans had introduced were improved, as well as the windmills. They were used for grinding grain into flour, cutting wood and processing flax and wool.

European farmers moved from a two-field to a three-field crop rotation, resulting in somewhat increased productivity. Crops included wheat, rye, barley and oats. From the 13th century onward, peas, beans and vetches became common as fodder crops for animals and for their properties in nitrogen-fixation fertilizing. Crop yields increased and stayed steady right up to the 18th century (Campbell and Overton, 1993).

Between the 8th and the 14th centuries, the Islamic world underwent an agricultural transformation described as the Arab agricultural revolution (Watson, 1983). Across the Islamic world, this revolution changed the economy, population levels and distribution, vegetation cover, agricultural production, urban growth, the distribution of the labour force, and diet. There were many factors for such a transformation, such as the large diffusion of crops and plants along Muslim trade routes over much of the Old World. Moreover, the Muslims introduced summer irrigation to Europe and initiated a form of plantation system for sugarcane cultivation, using slaves (Janick, 2008).

Among the major crops introduced to the Muslim-ruled area of the Iberian Peninsula were sugarcane, rice and cotton, together with the associated techniques for their cultivation and their inclusion in the cuisine. Other crops included citrus, fruit and nut trees. Vegetables included eggplant, spinach, and chard together with spices such as cumin, coriander, nutmeg and cinnamon. Intensive irrigation, based on Roman technology, crop rotation and agricultural manuals was adopted.

The Columbian exchange and the Age of Exploration

After 1492, there was a global exchange of crops and livestock breeds that were previously only grown locally. Maize, potatoes, sweet potatoes, tomatoes and manioc were the most important of a large number of crops that were introduced from the New World to the Old World. Maize and manioc (cassava) had been introduced from Brazil into African countries by early Portuguese traders, becoming staple foods. The potato was brought to Spain from South America in the mid-1500s. Moving in the opposite direction, varieties of wheat, barley, rice and turnips were brought to the New World. Horses, cattle, sheep and goats were unknown in the Americas prior to the arrival of European settlers. The exchange of important crops across the Atlantic Ocean had a lasting effect on many cultures (Crosby, 2003).

The British and European agricultural revolution

Between the 17th and the mid-19th centuries, agricultural productivity increased in Britain, particularly for wheat yields. It was a gradual process when agricultural output grew faster than the population. It has been debated whether it started in Britain or if the Dutch reduced the agricultural workforce more quickly up to 1650. Historians have disputed whether there was one single revolution or several steps of agricultural developments. Recently, it was argued that statements about 'the Agricultural Revolution' are difficult to sustain (Jones, 2016).

The agricultural development was a complex process in a feudal Europe under large landowners, the aristocracy and the Church, the land being worked by peasants and sharecroppers. Slowly, land ownership shifted so that more farmers could acquire land of their own. This allowed individual farmers to make technical changes and adopt innovations. Gradually, farm size increased and so did production and the population. Farmers spent more time and labour on their farming with improved farm tools and machinery. This freed up a large percentage of the workforce, which in turn could drive the Industrial Revolution.

A crucial development step was trading between growing markets. It required merchants, some forms of credit and knowledge of the markets of supply, demand and pricing. This led to a national food market, beginning with wheat. Agricultural commerce was aided by the expansion of different means of transportation. The arrival of railways significantly reduced land transport.

Increasing populations required more food. In the Netherlands, this was achieved by land reclamation from a constantly changing coastline, eroded by wind and water. The early inhabitants had built primitive dikes to protect their settlements from the sea and new land was claimed from this natural process. In the 14th century, smaller land strips were reclaimed by filling with sand or other land materials. The use of windmills for pumping water in the 15th century allowed for the draining of water, followed by the creation of a large system of polders to drain the wetlands to be reclaimed and used for agriculture.

A major innovative process was the improvement of the plough, illustrating a long-term development. A light, wheelless plough with iron blades, drawn by oxen, was used by Roman times. It could turn the topsoil of the

Mediterranean regions but failed to turn the heavier soils of northern Europe. It was followed by a wheeled plough, initially drawn by oxen and later by horses. The next step was the mouldboard plough. It was initially developed by the Dutch, who had acquired a Chinese plough, through the activities of the Dutch East India Company in the early 1620s.

The heavy Chinese iron mouldboard plough had been developed during the Han dynasty in the 1st and 2nd centuries or even before the 3rd century BCE (Temple, 1986). The ploughshare of cast iron was shaped like a V, with the blade carving into the ground. Two arms turned the earth away from the blade, reducing friction. The Dutch improvements meant the plough could be pulled by one or two oxen instead of the six or eight needed for the heavy-wheeled northern European plough. Dutch contractors brought their improved plough to Britain for wet soils on moors.

Ploughs were steadily improved in Britain, from Joseph Foljambe's patented Rotherham iron plough in 1730 to James Small's improved 'Scots plough' in 1763. By 1789, Ransomes, Sims and Jefferies produced 86 plough models for different soils (Barlow, 2003). The latest version was considered the cheapest and best plough. It spread to Scotland, France, and America. Improved ploughs led to the clearing of northern European forests. But the mouldboard was too weak to manage the American black prairie soils. A new type was invented in the mid-19th century by John Deere, who designed the all-steel, one-piece share and mouldboard in 1837. Later, assorted designs followed for tractors.

In Sweden, Carl Stjernsvärd of Engeltofta farm consulted his Scottish friend for agricultural advice. In 1803, this led to the arrival of Scottish blacksmiths and mechanics at the farm, located in the southern Scania province. They started production of an improved plough and other farming tools in an agricultural factory. During the first ten years, more than 700 new ploughs were produced and put into use (Hammarlund, 1956).

As a result of the Columbian exchange with the Americas, maize was grown in Spain after 1525. It gradually spread to France as a major food in rural areas by the beginning of the 18th century. In northern Europe, maize became popular as animal fodder. Potatoes took time to be adopted and cultivated, despite yielding about three times more calories than wheat or barley. They could be grown in nutrient-poor soils and became a staple crop throughout Europe by the late-18th century, adding variety to the European diet. Nonetheless, dependence on the potato was risky, as exemplified by the Great Famine in Ireland in the 1840s, causing one million deaths and another million people to emigrate.

In 1747, a method for making sugar from sugarbeets was discovered. Sugarbeets required less labour and fuel than sugarcane. They could easily be grown in the colder climate of Europe. In 1812, Benjamin Dalesart discovered a method of extracting sugar on an industrial scale, giving impetus to the cultivation of sugarbeet.

For a long time, the depletion of nutrients was the main problem in sustaining agriculture, mostly nitrogen. Thus, productive farmland was often left in fallow to allow the soil to generate. The Dutch had started a four-field rotation system which the British popularized in the 18th century. Different crops were planted so that they could take up various types and quantities of nutrients from the soil as the plants grew. Later, turnips and nitrogen-fixing clover were planted instead of fallow.

This Norfolk four-course crop rotation system of wheat, turnips, barley and clover opened with clover as a fodder and grazing crop. Clover was important for nitrogen fixing. That fodder crop allowed for more livestock, whose manure was later added to the soil to increase its fertility. In addition, guano with a high content of nitrogen, phosphorus and potassium was primarily imported as fertilizer from Peru during the 1800s.

The introduction of new crops in a new environment is always problematic since they may be attacked by – for them – new diseases and pests. Native to the Rocky Mountains in the United States, the Colorado potato beetle started to destroy potato crops in Nebraska in the late 1850s and spread rapidly across America and then to Europe. In the 1870s, several Western European countries banned imports of American potatoes to avoid infestation, but this was not effective. The beetle has the potential to spread to the temperate areas of the rest of the world. An arsenic compound was introduced to fight the Colorado potato beetle. In the 1880s, a French researcher

discovered a solution of copper sulphate and lime that killed potato blight, a first step to modern pesticides.

In Sweden, the farming system was based on permanent meadows harvested for hay and grazing of large, forested areas of understory with food to eat. Farmers used infields with one or two crops, such as barley, rye, oats or wheat depending on the climatic region. Yields were low and sometimes peas or turnips were planted (Rundgren, 2023). After the 1850s, the infields were given animal manure which increased crop productivity. Usually, fields had to be fallowed every second or third year until this system was replaced by the four-course crop rotation system.

The seed drill originated from China and had been introduced into Italy in the mid-16th century. In fact, Chinese farmers had used a seed drill since the 2nd century BCE and row-planted crops, facilitating more efficient weeding and harvesting. In Europe, some agricultural machinery was invented, such as Jethro Tull's version of the seed drill in 1701 but it had little impact. Most agricultural machinery was not improved until the second half of the 19th century. Powered farm machinery began in 1812 with Richard Trevithick's stationary steam engine, used for driving a threshing machine. In 1801, he had already started to build a full-size steam road locomotive (Puffing Devil).

In the mid-18th century, some breeding of sheep to stabilize certain qualities to select for large and long wool started in Britain. Organized selective breeding of plants and livestock first began in the late 19th century. In the 1840s, scientific investigations had already started at the Rothamsted Experimental Station on the impact on crop yield of inorganic and organic fertilizers. Artificial fertilizer manufacturing factories were founded. Sodium nitrate from deposits in Chile, was imported to Britain, together with guano. In 1842, coprolites were discovered outside Felixstowe in Suffolk in the UK by John Stevens Henslow. His findings showed their potential as a source of phosphate after treatment with sulphuric acid. After patenting the process, mining started on an industrial scale for their use as fertilizer.

Agricultural changes and innovations during the two World Wars

The first commercially successful, gasoline-powered, general-purpose tractor was constructed in 1901 by Dan Albone (1860–1906). However, the main development was when the 1923 International Harvester Farmall tractor began to replace draft animals. After that, a range of self-propelled mechanical combines, planters, transplanters and other equipment were produced.

Another major breakthrough came with the Haber-Bosch method for synthesizing ammonium nitrate, first patented by the German, Fritz Haber. In 1910, Carl Bosch, while working for the BASF company, commercialized the process.

After a period of increasing globalization of agricultural markets, World War I seriously limited farming in the nations at war. This resulted in sharp increases in agricultural prices and farm incomes in countries outside the war zones. During the 1920s, farm prices for most goods fell drastically during the Great Depression, causing a lot of unemployment in agricultural regions. Most countries tried to protect the large farmers by raising tariffs and setting quotas on farm imports. Mixed farming was promoted in research by UK universities from the 1920s, but few significant agricultural innovations were developed during this period. During the next decade, marketing boards were set up in the United Kingdom for milk, potatoes and for sugarbeet as a new crop. In the 1930s, all governments regulated agriculture, for example, by subsidies, setting minimum prices, purchasing surpluses or limiting output. Frequently, new regulations caused other problems, requiring other regulatory amendments (Fishback, 2022).

Starting in the 1920s in the Soviet Union, collective farming was widely practiced, and it was also adopted by the Eastern Bloc countries, China and Vietnam. A consequence of the drive for collective farming was the Soviet famine of 1932–1933 (Iordachi and Bauerkamper, 2014). By the mid-1930s, much of the arable land in Britain had reverted to pasture or scrubland. Despite some politicians arguing against government intervention, a special Food and Supply Sub-Committee was established and

proved quite successful. At this time, domestic agriculture supplied almost 90% of Sweden's food consumption with one-third of the population engaged in farming (Olsson, 1978).

In 1939, Germany had become 83% self-sufficient in basic crops. During World War II, Denmark exported food to Germany. Swedish farmers were given strong government support and the country earned income from the export of iron ore to Germany. Most European farmers outside the war zones made profits because of high food prices.

Energy was not a major problem to Swedish agriculture during World War II. Horses were still in use as the main drought power. In 1941, gas-powered tractors came into operation, usually powered by home-produced gas or wood from the farm. Tractors that could not be converted to gas were run on domestically produced motyl (a mixture of gasoline and liquor used by military vehicles). Other agricultural vehicles were also run on motyl. Electricity consumption was low. Hay and the grain harvest were dried outdoors and there were few combine harvesters. Natural products were used for plant protection and mercury, for seed treatment, was the only chemical used. Livestock production faced problems due to a shortage of imported concentrated feed, which was replaced by leaves, reeds, etc. The Swedish government gave priority to milk production which was never rationed during the war.

After World War II, the use of synthetic fertilizers increased rapidly and so did the use pesticides. An agricultural extension service had been developed in the United States. In the Soviet Union, Joseph Stalin's vision of a Soviet socialist society led to private property being forcibly seized and peasants being organized in collectives. The resistance of the peasant population was treated ruthlessly since it was assumed that farmers, through their work for the state, would become socialists.

In China, the Great Leap Forward (1958–1962) was a five-year economic plan executed by Mao Zedong and the Chinese Communist Party. The goal was to modernize the country's agricultural sector using communist economic ideologies. Private plot farming was abolished, and farmers were forced to work on collective farms where all production, resource allocation and food distribution was controlled by the Communist Party. Large-scale irrigation projects, with little input from trained engineers, were initiated. Untested new agricultural techniques were introduced around the country. A nationwide campaign to exterminate sparrows, believed to be a major pest on grain crops, resulted in massive locust swarms in the absence of natural predation by the sparrows. Grain production fell sharply, resulting in the Great Chinese Famine (1959–1961) (Investopedia, 2024).

The launching of the Great Leap Forward had other effects. The unrealistic figures from the regions were sent to the government, e.g. the Communist Party, even though people were starving. These figures were the fake news of that time, but the official reporting attracted great international attention. For example, propaganda about a Chinese agricultural miracle was well received in Tanzania which adopted its policy on collective farming. Swedish development assistance officials were eager to support such efforts in Tanzania in the early 1970s. While millions of Chinese starved, the country remained a net exporter of grain. Mao had refused offers of international food relief so as to indicate to the outside world that his plans were a success (Peng, 1987).

The strong belief in collective agriculture transformed Tanzania over the course of a decade from being one of Africa's food exporters to an importer of food. Ninety percent of the country's farmers lived on collective farms in 1979 but accounted for only 5% of the country's agricultural production. The fake news from China helped to reshape reality, giving way to the written word and lack of critical scrutiny by aid administrators.

The Green Revolution

The Green Revolution was a package of initiatives in the areas of research, development and technology transfer, expanding from high-income to low-income countries in the 1960s. The package increased food crop productivity growth over a period of 50 years in low-income countries (Pingali, 2012). New high-yielding crop varieties increased yields by 44% between 1965 and 2010, increased income and reduced

population growth, saving 1 million people from starvation (Gollin *et al.*, 2021). The Green Revolution began in Asia and Central and South America. New, improved seed, initially of rice, maize and wheat, had been developed at the International Rice Research Institute (IRRI), founded in 1960, and the International Maize and Wheat Improvement Center (CIMMYT), founded in 1966. They became partners in a network of international agricultural research institutes within the Consultative Group of International Agricultural Research (CGIAR) in 1971. Its focus was on agriculture in low-income countries. New, improved seed varieties were included as one element of a technology package. This was an innovative approach rather than the conventional introduction of one single innovation. This package was based upon, among other things:

- rapid advances in science and technology with new high-yielding varieties and hybrids of various food crops (initially maize from CIMMYT and rice from IRRI). They responded well to irrigation and fertilizer use.
- good agricultural service with synthetic fertilizers and pesticides and access to credit and extension advice.
- supporting policies on land reform, infrastructure, marketing and affordability, and the expansion of irrigation.
- rural challenging work of the agricultural population to increase productivity. The positive effects of the technical solutions led to rapid improvements in global food production.

The Green Revolution significantly increased yields in Asia, initially, but this approach now seems insufficient. Yields have levelled off. The genetic 'yield potential' for wheat has increased, but not for rice, and this also applies to maize. Furthermore, herbicide-resistant weeds appear within a couple of decades, and insects become resistant to insecticides even quicker. In India, the innovative technology was initially applied in areas where the land was owned by the farmer. It enabled investments in increased irrigation, access to electricity for water pumps, marketing and the support of an expanded advisory system.

The Green Revolution imposed a scientific package on nature which gave satisfactory results. The future may be more doubtful. The political will is lacking for costly investments in irrigation facilities and there might be a shortage of water. The Green Revolution delivered useful calories through the production of wheat, maize and rice in the low-income countries. However, this led to a nutrition problem since traditional crops were displaced, smallholders were left behind and the modern type of agriculture led to environmental problems. Also, farming has become less profitable, farmers are getting old and younger generations are sceptical about becoming farmers. The future requires science to better understand nature to obtain sustainability. Not even the designed technical package is sustainable in the future. There is a need to revisit the concept of agroecology, a concept forgotten in current agricultural approaches focusing on technology only. Initially, farmers had control of both their farmland and the surrounding ecosystem services, but this is less so now.

Agricultural Innovations: From Early Days to Current Approaches

Early agricultural inventors

Agricultural innovations began thousands of years ago. In shifting cultivation, planters just used a stick. In Māori horticulture, the kō was used and the Andean foot plough was used by the Inca people. In north-west Scotland, they used the 'cas-chrom bent foot', a crooked spade. It went out of use in the Hebrides as late as in the early part of the 20th century.

Around 3000 BCE, the first seed-ploughs appeared in pictographs from Uruk in Mesopotamia, one of the first cities in the world. They were found on seals from around 2300 BCE according to the British Museum. Lal (2001) refers to archaeological evidence of an animal-drawn, wooden plough used in the Indus Valley civilization at around 2500 BCE in Harappa, central Punjab. The Sumerians had taught the Harappans how to utilize this wooden plough. In the late 2nd century, by the end of the Han dynasty, Chinese heavy ploughs had been developed that had iron ploughshares and mouldboards (Greenberger, 2006).

Archaeological records from around 2300 BCE show that wheeled carts were in use in the Indus Valley civilization. For agricultural purposes, the Chinese had designed a hydraulic-powered trip hammer by the 1st century BCE (Needham, 1986). They began using the square-pallet chain pump to lift water from a lower to a higher elevation to fill irrigation canals. The system was powered by a waterwheel or by oxen pulling a system of mechanical wheels.

Along the Nile in Egypt, agriculture was practiced with large-scale irrigation from around 10,000 BCE up to 4000 BCE. A similar development took place in the Indus Valley. Also, the Aztecs in Mesoamerica had developed irrigation systems and formed terraced hillsides, fertilized their soils and developed artificial islands, known as 'floating gardens'. Between 400 BCE and 900 CE, the Maya used extensive canals and managed to farm the swampland on the Yucatán Peninsula (Mascarelli, 2010).

During the Han dynasty, there was a nationwide granary system in Chinese agriculture. In India, crops were planted in two or six rows and granaries were used for storage. The people of the Inca Empire stored their food surpluses in buildings called qullqas: an effective means for maintaining food security (Mumford, 2012).

There are also old examples of more complex engineering, for example, the rice terraces in China and Bali. In the Philippines, the Ifugao ethnic group has occupied the mountains in remote areas of the Cordillera mountain range on the island of Luzon. They have a complex system of stone and mud walls, created to grow rice on terraces on the steep slopes by careful carving of the natural contours of hills and mountains. There are terraced pond fields integrated with an intricate irrigation system using water harvested from the mountain-top forests. Nowadays, this Hungduan area is a farming system with rich biodiversity and a globally important example of agricultural heritage. There are some 280 plant species with 58 endemic and four threatened species (Baguilat, 2011). There are also 93 animal species with 10 endemic and four threatened species. According to archaeological evidence, this system has been in use, almost unchanged, for some 2000 years. It demonstrates both an innovative strength and a blend of social and cultural balance in a political environment. It also demonstrates the concept of real sustainability with the use of fossil-free energy and little or no environmental degradation.

Another example is the underground channels that supplied water to towns in North Africa. The water channels in Morocco (khettaras) and Algeria (foggara) were created several centuries ago in the Tafilalt oases of south-eastern Morocco (Lightfoot, 1996). The channels were designed to transport water down the slopes by gravity to irrigate fields in the oases. The khettaras consisted of gently inclined horizontal tunnels dug into a sloping terrain. When the horizontal tunnel hit the water table, gravity caused water to simply flow downhill in the channel towards outlets at the base of the slope. The water lay below the surface, but the access shafts used for construction and maintenance were above ground (Earth Observatory, 2016). Different civilizations established their own regulations for the management and mobilization of the water. The users of the system had to abide by rules formulated to protect, promote and care, to enhance the sustainability of the system (Beraaouz et al., 2022).

Tafilalt was an economic, cultural and spiritual bridge between sub-Saharan Africa, North Africa and Europe. Caravans met at the Sijilmassa trading centre. It became a source for cultural and spiritual influence for Fez, Marrakech and Tangier. In the 1970s, the water table was lowered by widespread use of diesel pumps and vertical wells. This caused some khetteras to dry up. The groundwater table has dropped significantly, reducing the cultivation of date palm in the Tafilalt oasis. Droughts have made the situation more severe, and the maintenance of the system has declined. Traditional knowledge about this type of sustainability is being lost.

Early inventors used trial and error for problem-solving; an approach also applied for local problems in Western farming communities up to the early 1920s. Local blacksmiths played a key role in working out practical solutions to their satisfaction, but the final models might be different. Alpern (1992) has underlined that the early introduction of iron bars made it possible for African blacksmiths to make more farm tools and steel-bladed machetes for planting. The introduction of firearms meant that farmers could protect crops from wild animals and hunt more wild game. Also, fishhooks were brought to Africa. Initially, such actions were made in harmony with nature and through a biological eye, so there were few, if any, negative consequences. When looking towards future sustainability we must consider this basic concept to avoid environmental degradation. This calls for biological-based technologies with minimum or preferably no waste.

At the start of the Atlantic slave trade in the 16th century there were craftsmen in Africa using iron, but Europe had the advantage. Shipbuilding and the broader economic context of the time, including profits from slavery, were used to fund technological advances, for example, the invention of the steam engine. James Watt has been mentioned as having had connections with transatlantic slavery (Mullen, 2020). Although the steam engine was not directly funded by Caribbean plantation owners, it was used to improve efficiency on plantations, for example, in Jamaica, St. Lucia, Martinique and Cuba (Tann, 1997). The diffusion of steam power to the plantation economy of the Caribbean was on a significant scale. According to Tann (1997), 148 Fawcett & Littledale engines were ordered between 1813 and 1825 and 119 Boulton & Watt engines between 1803 and 1825.

Examples of new agricultural developments during the colonial period

There were few initiatives of a permanent kind to establish specific institutions for agricultural research in the early phase of the colonial period. The Portuguese explored most of the West African coast from 1415 until the 1600s. Portuguese trade focused on a search for gold, ivory and pepper in West Africa and, gradually, slaves. They noted both urban centres and agricultural systems capable of feeding large populations (Ross, 2002). They exchanged some plants, and pigs, turkeys, ducks, geese and pigeons, were introduced into the African areas where they had a presence. Also new breeds of sheep and cattle were introduced, but they were usually killed by sleeping sickness.

The Madeira archipelago was exploited by the Portuguese for agriculture. Many fishers and peasant farmers left Portugal because farmland on the islands was not controlled by the nobility. There were great possibilities for deep-sea fishing. Forests were cut down for agriculture on both Madeira and the Azores. Timber from the cedar and the yew were shipped to Portugal and Spain.

Madeira's farming capacity was increased by building stone-walled terraces on the mountain slopes and the creation of a system of aqueducts (*levadas*). When wheat production declined in the 1450s it was replaced by sugarcane for export. By the 16th century, that export had grown, requiring more labour, so slaves were imported from West Africa. There was also production of sweet grapes, barley and expensive dyes. Red dye came from the resin of the dragon-tree and blue dye came from woad.

Through the territory of New Spain, which existed from 1521 to 1821, the Spanish claimed territory that extended from Alaska to Florida and the Caribbean in North America, plus Mexico, Central America and South America. With the first Spanish settlement in the Philippines, a new sailing route was open to Mexico. This was an alternative to the old Silk Road for shipping silk, spices, silver, porcelain and gold to the Americas from Asia.

Initially, gold mining was important for the Spaniards but from the mid-16th century ranching and commerce became the important economic activities in New Spain. There were fewer mineral resources in Central America, except for some gold mining in Honduras. Wild indigo, like cacao, was native to this region, and the Spaniards established indigo plantations in Yucatán, El Salvador and Guatemala. Indigo was high in demand in Europe in the 16th century.

In the fertile Bajío region of central Mexico, commercial agriculture was developed to produce food and livestock on large haciendas. The main crops were sugarcane, cotton, cocoa, vanilla and indigo. Since the local population resisted cultivating sugarcane, slave labour became necessary. The Spaniards introduced wheat, grapes and olives. Wheat was grown in Puebla, known as the granary of New Spain, for two centuries. The orange, a native of China, was brought as seed to Honduras in 1493 by Columbus. It became the main crop in

the 16th century in what is nowadays Florida and California. Coffee became important in Colombia and Venezuela. In 1765, a monopoly on tobacco was introduced leading to increased tobacco consumption (Foster, 2000).

All colonial rulers organized agriculture to focus on the establishment of plantations for cash crops. Traditional farming was replaced to maximize profits for Europe. It began with sugarcane in the Caribbean using slave labour. The sweetest cultivated variety of sugarcane was native to South Asia and East Asia. The same profitable model was repeated to produce tea, coffee, cocoa, poppy, rubber, indigo and cotton in the New World (Naithani, 2021).

The colonists also brought new agricultural techniques, such as ploughing, shovelling and crop rotation. Yokes and ploughs of wood or metal were introduced, both for animal and human traction. Some of the new ranching haciendas produced horses, mules and oxen for drafting.

Some developments took place in the initial phase of the English colonial period. Hans Sloane (1660–1753) visited Jamaica in 1687, and other islands, and collected more than 1000 plant specimens, most of which were catalogued as new plant species. Edward Long (1734–1813) was a Jamaican planter and slave owner, who conducted experiments with plough, uncommon in colonial territories. Other English explorers collected plants and seed on Tobago island, where medicinal use of plants was customary practice. A review of the literature revealed a total of 338 different plant species on the island with reputed medicinal properties (Barclay, 2012).

In Africa, a Scottish mission introduced tea plants into Nyasaland in 1878. Coffee was first introduced into Kenya in 1893 by French missionaries. During the British imperial period, several agricultural institutions were established in their colonies, indicating a long-term commitment (Table 8.1).

The French Company of the East Indies (Compagnie Française des Indes Orientales) was founded in 1664 but its attempt to make Madagascar a great centre of trade failed (Wellington, 2006). In Mauritius, the origin of the Botanic Gardens of Pamplemousses (now Sir Seewoosagur Ramgoolam Botanic Garden) can be traced to the first French Governor of

Table 8.1. Some British initiatives in tropical agricultural research between 1880 and 1950 (Simmonds, 1991).

Date	Initiative
1888	Sugarcane breeding was initiated in Barbados and Java
1898	Imperial Department of Agriculture established in Barbados
1905	Indian Agricultural Research Institute opened at Pusa
1919	The Gezira Scheme started in Sudan
1924	Imperial College of Tropical Agriculture founded in Trinidad
1925	Rubber Research Institute of Malaya established
1927	Imperial Agricultural Bureaux founded
1935	Colonial Agricultural Service was named
1949	Colonial Research Service started

Mauritius. It is one of the oldest botanical collections in the tropics, founded in 1768.

In North America, in New France (along the St Lawrence River and around the Great Lakes), the French focused on fishery and also on a profitable fur trade. This started in the 1680s, concentrating mostly on beaver fur. Much later in 1898, in Africa, the French established L'école colonial d'agriculture de Tunis as one of several agricultural institutions. The school offered French students theoretical and practical training to prepare them as agricultural colonists. With the arrival of French farmers, national agricultural production changed and viticulture rapidly became popular. In the 1920s, the French introduced coffee into the Cameroon Grassfields, kept under full control of the colonial administrators, despite the interest of local farmers (Mbapndah, 1994).

Nowadays, Tunisia is much dependent on imports of seed and fertilizers from Russia and Ukraine. Other imports from these countries include rice and sugar, plus wheat (39%) and maize (62%) from Ukraine and sunflower oil (62%) from Russia according to the United Nations Comtrade Database (2023).

Some examples of recent research for agricultural innovations by the private sector

As of today, many actors are involved in developing new technology. They include the private and public sectors, research institutes, universities,

entrepreneurs and others. According to a *Popular Science* (2022) listing of the most important innovations in 2022, the James Webb telescope was top. It can show light that has travelled for 13 billion years. In addition to artificial intelligence (AI), there was a perennial rice developed by Chinese researchers and the discovery of a gigantic bacterium (*Thiomargarita magnifica*), 5000 times larger than an average one.

With reference to the wealth of information it is not easy to highlight major innovations in food and agriculture during since the 2010s. In the following subsections, some examples of innovations from the private sector are listed. The agricultural business transnational corporations (TNCs) play a leading role in the innovation process, usually focusing on consumers in the industrialized countries. At the local level, TNC innovations, may lead to a dehumanizing process of societies since profit is the only objective. This profit is aimed for the TNCs' headquarters, usually in another country. This is in contrast to early industrialization processes when most private entrepreneurs created jobs in their own region and took responsibility for the general well-being of the local population, beyond paying salaries to the employees. This sustained a local market in industry, agriculture and forestry. Nowadays, this only occasionally takes place with certain private firms and entrepreneurs.

The TNCs have focused on export crops and global marketing of various foodstuffs, mostly developed with today's high-tech research. This includes genetic modifications, biotechnology and nanotechnology, and these may not be focused on sustainability. Often, these developments would be of less immediate practical use to most smallholders, especially in sub-Saharan Africa and South Asia. However, there might be a change, as indicated by the TNCs in their Sustainable Markets Initiative (SMI). Its objective is to build a coordinated global effort to enable the private sector to accelerate the transition to a sustainable future (Sustainable Markets Initiative (SMI), 2022) (Chapter 10). If focused on food crops, it may result in useful new innovations and approaches, that can also be used in low-income countries.

Biotechnology

Today, there is a great number of different actors involved in biotechnology research. One example is the American company, Gingko Bioworks Holdings Inc., established in 2008, specializes in synthetic biology, using genetic engineering to produce bacteria that have industrial applications for other biotech companies. In 2023, it entered a partnership with Visolis Inc. to improve one of their existing microbial strains to produce biobased isoprene for commercial synthetic rubber and lower the carbon intensity of sustainable aviation fuel production (Gingko Bioworks, 2023a).

Gingko Bioworks has also entered a partnership with Syngenta Seeds for the next genetic seed technology. Gingko's cell programming and protein engineering capabilities will be used together with its proprietary ultra-high-throughput screening techniques (Gingko Bioworks, 2023b). Another partnership was entered into with Boehringer Ingelheim to offer Gingko's metagenomic database for research to develop new therapeutics for genetically related diseases, which are currently untreatable or difficult to treat (Bioworks, 2023c). As one of the world's largest pharmaceutical companies and the largest private one, Boehringer Ingelheim launched Sound Talks in 2022. This microphonic system continuously analyses the respiratory health status of pigs during rearing and fattening.

Lab-grown food

In the 2010s, cultured (lab-grown) meat was grown from muscle cells but had difficulties in reaching the market. Today, billions of dollars are being globally invested in the cultured products of chicken, beef, eggs, and salmon. According to the Good Food Institute and European Food Safety Authority (EFSA), there were approximately 150 companies working with cultured food products in 2022 (EFSA, 2023).

A kind of lab-grown meat was said to be available in 2023 at a restaurant in Singapore – once a week. In the same year, the Food and Drug Administration (FDA), the United States equivalent of the European EFSA, confirmed that chicken meat from Upside Foods, an American food technology company creating cultured meat, is safe (FDA, 2022). A few months later the FDA confirmed that GOOD Meat's cultured chicken product was safe for consumers (Malleck, 2023). In mid-2023, the United States Department of Agriculture (USDA) has given clearance to both Upside Foods and GOOD Meat to commercially produce and sell lab-grown chicken in the United States (NBC, 2023).

Chains like Burger King are routinely stocking plant-based burgers and the plant-based meat substitute pioneer company, Beyond Meat, is another alternative. The New Age Meats company used stem-cell biology to produce pork sausage in 2018 but has since ceased to operate. The Israeli MeaTech company (now Steakholder Foods) has grown artificial meat in its laboratory, cultured through stem cells. Recently, Israeli authorities gave Aleph Farms approval to sell cultured meat cells from a Black Angus cow called Lucy.

The Jimi Biotech company claims it has developed China's first 100% cultivated chicken. Cells from young roosters were used to grow the meat using the company's technology. The smell and taste are reported to compare well to traditionally slaughtered chicken (Vegeconomist, 2023).

Applications for cultured meat have been filed in the United Kingdom and Switzerland. By using a method of cultivating red algae the Swedish biotechnology company, Volta Greentech, has partnered with Lantmännen to conduct feed trials with cows to create beef and achieve a simultaneous reduction of methane emissions (Askew, 2022). The Danish company, FÆRM, has redesigned plant masses to achieve the textures and flavours of traditional cheese (Vegeconomist, 2024). It licenses a patented method to food manufacturers that mimics dairy fermentation to produce plant-based cheese without coconut oil, starches or additives.

Estimates suggest meat alternatives could account for 10% of the global meat industry by 2029 (Siegner, 2019). A major problem is that they are expensive and are for markets in rich countries. Several European Union (EU) countries are sceptical about meat alternatives since they will influence conventional farming. In Italy, the government wants to ban the production and sale of synthetic food and feed. Moreover, lab-grown beef cannot compete with traditional meat production from pastures in terms of long-term sustainability.

The American company, Shiru Inc., has launched its first product: OleoPro™. It is a plant protein-based fat ingredient as an alternative protein food. It is claimed to be self-standing, browns when cooked and delivers a juicy, fatty mouthfeel in plant-based meat applications. The Israeli PoLoPo company has developed a technology capable of producing egg protein. The potato acts as a protein carrier. It is cheap and hardy and focuses on low-income countries with low-protein diets, where the demand for animal protein is expected to double by 2050 according to the company. This bioengineered product will require approval by both authorities and consumers (PotatoBusiness, 2022).

Genome-edited food developed with the CRISPR/Cas9 technology has reached Japanese consumers via Sanatech Seed, together with its partner for sales, Pioneer EcoScience (Maxwell, 2021). Sanatech Seed planned to introduce Sicilian Rouge High GABA tomatoes through their home gardening channel (Waltz, 2021). Sicilian Rouge tomatoes have been genetically edited to contain considerable amounts of gamma aminobutyric acid (GABA). Oral intake of GABA can help support lower blood pressure and promote relaxation according to the company.

In 2023, investment in agrifoodtech startups reached its lowest point in 6 years according to Global AgriFoodTech Investment Report by AgFunder Inc. (2024). It declined almost 50% from 2022 to 2023 due to fewer and smaller deals. Funding to upstream startups at the farm or in food production had increased in 2023 (62%) compared to both 2022 and 2021 (Marston, 2024).

Agricultural data

In partnership with Microsoft, Bayer launched a cloud-based solution for the agrifood industry in 2023. It is a combination of Bayer's AgPowered Services and Microsoft's Azure Data Manager for Agriculture which provides internal or customer-facing digital solutions. Consumer companies can use the cloud offerings to build solutions to get insight into nutrients, sustainability and production practices. Companies developing on-farm technologies can build digital tools that support agronomic outcomes for the growers (Bayer, 2023).

Agricultural machinery

A recent major advance for precision farming is the self-driving autonomous tractor. John Deere launched its robot tractor, 8R, in 2022 – a fully autonomous driverless machine. It combines the tractor, its tillage technology with a chisel plough, GPS guidance and six pairs of stereo cameras enabling 360-degree obstacle detection and calculation of distance. The Dutch AgXeed company makes autonomous machines and a cloud-based portal of smart algorithms. Its AgBots tractor robots can operate nonstop and unsupervised up to 23 hours. Case IH and Raven Industries have introduced the first robotic system för a fertilizer spreader: the Case IH Trident™ 5550 applicator with Raven Autonomy™ (Raven, 2022).

A new precision fertilizer spreader is the Kverneland Geospread Gyro. It allows adjustment of the spreader angle depending on both the load and the angle vis-á-vis gravity. The Swedish company Väderstad's prototype Proceed high-precision plant seeder can place the crop seed-by-seed in the planting/sowing direction, without granulation or seed sorting. Not only can it be used for wheat but also for maize, beans and sugarbeets. The Norwegian company, Dimensions Agri Technologies, has developed a see-and-spray concept that can be retrofitted to any sprayer. The system reduces the use of chemicals by up to 90% while increasing the yield by up to 10%.

Väderstad has entered into cooperation with the German crop production system Nexat. Its 'all-in, one-system-tractor' system is based on an electrically driven carrier vehicle. Its working width is up to 24 m and handles all steps from cultivation to sowing, and plant protection to harvesting. It is prepared for complete automation (Väderstad, 2021).

The Swiss company, Liebherr, has launched a prototype hydrogen internal combustion engine (H966). The Israeli company, HomeBiogas Ltd., produces anaerobic digesters for home use, allowing customers to convert their own food waste (and human and animal waste) into gas for cooking. The liquid residues can be used as fertilizer.

Electrical drones

The United States company, Zipline, has tested its P1 autonomous electric delivery drones in Rwanda for delivering blood and health supplies to clinics and hospitals. The new P2 model (2023) will make rapid aerial deliveries, carrying up to 3.5 kg of cargo within a ten-mile radius. It can land a package on a table or doorstep, and dock and power up autonomously at a charging station. It may be of effective use to reach isolated livestock farmers.

The public sector and new agricultural technology

International organizations and donor countries

In addition to the three major international organizations focusing on agriculture, namely the FAO, the International Fund for Agricultural Development (IFAD) and the World Bank, there are also numerous international and national organizations giving financial support for agricultural technology development in low-income countries. Specific organizations have been established to support the development of innovations.

The United Nations (UN) started early to promote innovation and entrepreneurship in low-income countries. The United Nations Office for Project Services (UNOPS), with its global headquarters in Copenhagen, was established in 1973 as part of the United Nations Development Programme (UNDP). In 1975, it became independent and self-financing and focused on

implementing infrastructure and procurement projects for different partners, all over the world.

From 2014 to 2021 it expanded its portfolio of projects. Some of its surplus funds were used to start the Sustainable Investments in Infrastructure and Innovation; a first attempt at investment an UN organization. After a few years it was discovered that key investing recommendations were not implemented prior to issuing loans. Some US$ 50 million had been given to companies owned by a British businessperson whose daughter´s company received US$ 3 million as a grant to write a pop song. Due to management failures the investment body was closed in 2023.

Despite such unclarities, Finland got a UNOPS office in 2018 and Sweden in 2019, with government funds (Brattström, 2022). Also, Antigua and Aruba got offices. The United States threatened to stop its financial support and UNOPS was reported to the police in Denmark. Sweden contributed funds in 2021 when the executive director was placed on administrative leave, resigning the next year. Swedish financial contributions have been given since 2010 but without regular reports from UNOPS and no innovation has been presented.

SEED is a global partnership for action on sustainable development and a green economy. It was founded by the United Nations Environment Programme (UNEP), the UNDP, and the International Union for Conservation of Nature (IUCN) at the 2002 World Summit on Sustainable Development in Johannesburg. So far, SEED has supported some 200 innovative small-scale and locally driven entrepreneurships globally, integrating social and environmental benefits into a business model. One example is Ghana in the sectors of waste management, water and sanitation, and agricultural processing.

The European Innovation Partnerships (EIPs) were established by the European Commission in 2021. They bring together private and public partners to help to avoid duplication of investment and contribute to reducing the fragmentation of the research and innovation landscape in the EU (European Commission, 2022).

Development aid started with the Marshall Plan. The United States provided economic and technical assistance to European countries after World War II. Aid can include monetary assistance in form of direct grants, programmes or training to support a developing country's political, social or economic development. Aid is also political, especially so during the Cold War period (1947–1991). The Eastern Bloc countries aided the socialist low-income countries. Sweden gave aid support to both dictatorships, such as Cuba, and for good reasons to the growing freedom movements in southern Africa, such as in Zimbabwe during Robert Mugabe's initial years, when he was fighting white minority rule, although his ZANU party soon approved the one-party state. This move was supported by the Southern Africa Development Community (SADC) since the Zimbabwean opposition was seen as a threat (Mbeki, 2001, 2009). According to Mbeki, Swedish aid to southern Africa had been used to build up totalitarian states in Mozambique, Angola and Namibia.

In addition to many foundations and several international organizations, including the large UN system, all industrialized countries have their own donor agency. The United States Agency for International Development (USAID) was one of the first governmental aid organizations. Created in 1961, it centralized several existing foreign assistance organizations and programmes. Also, Sweden was early in establishing the National Board for International Development Aid (NIB) in 1962, replaced 3 years later by the Swedish International Development Authority (SIDA, and later Sida), followed by other countries.

Some donor governments have set up special organizations to support research in low-income countries. This was a forward-looking step in giving financial support to researchers in low-income countries so that they can take responsibility for the own development of their technology. The first agency was the International Development Research Centre (IDRC) founded by the Canadian Parliament in 1970. Five years later, the Swedish Agency for Research Cooperation with Developing Countries (SAREC) was created. Since 1982, the Australian Centre for International Agricultural Research (ACIAR) has been active.

In 1994, the Swedish government decided that SAREC should cease to operate, and its activities be integrated into the new Swedish International Development Cooperation Agency (Sida), which led to a reduction of Swedish

aid for research in low-income countries. In 2021, the United States announced plans to launch the Agriculture Innovation Mission for Climate (AIM for Climate). The goal is to speed up research and development on agriculture in support of climate action, along with the United Arab Emirates and other partners during 2021–2025.

Over the decades, the focus of international aid has shifted from technical assistance in the 1960s to basic human needs in the 1970s and emphasis on the free market in the 1980s, followed by democracy and sustainability around the year 2000. In general, aid programmes from industrialized countries included significant technology transfer. There was a change during and after a lengthy period of wars in Afghanistan and Iraq. After the 11 September 2001 attacks in the United States, the international military coalition led by the United States invaded Afghanistan, with UN support. The conflict in Afghanistan ended in 2021 when a Taliban offensive overthrew the Islamic Republic, reestablishing the Islamic Emirate. During this period, much development was directed towards rebuilding governments and development aid changed in character.

The current focus of international aid may be described as directed towards climate issues and energy. In 2023, much military and development aid was going to the Ukrainian government for rebuilding after the Russian invasion. As a result of the Israeli war against Hamas in Gaza, there were calls for assistance for the war-struck Palestinian population. This means that there is much less focus on the more conventional type of development aid to low-income countries. The current political superpowers (United States, China and Russia) seem more oriented towards gaining worldwide political and military influence, and far from an approach to food security based on sustainability.

Russian influence in Africa has been through the military support of the Wagner Group. It is reported to be involved in more than ten countries. The group took part when Russia annexed Crimea in 2014 and have also been active in Europe and South America (Venezuela), but the focus is Africa. This challenges the United States whose military bases have recently been questioned, for example, in Niger. Earlier, French military forces were requested to depart from Mali.

China has used another approach, although the political issue with Taiwan is not yet settled. Since 2000, China has offered loans to the value of US$240 billion to 22 low-income countries with weak economic status (Horn *et al.*, 2023). These loans are usually part of China's infrastructure and trade project, the Belt and Road Initiative (BRI), and, officially, are to support good trade relations. Since the loans are expensive (5%) compared to the 2% interest rate of the International Monetary Fund (IMF), they may serve as dept-traps.

The Consultative Group of International Agricultural Research (One CGIAR)

One of the best outcomes of aid donor funds has been the financial support given to the CGIAR with its independent research centres or institutes. Founded in 1971, with only four established centres, the CGIAR became an especially important actor in the Green Revolution. Through new high-yielding varieties of rice (IRRI), and wheat and maize (CIMMYT), agricultural production in low-income countries could increase. Later, other innovations were gradually developed by a growing number of research institutes, established by the CGIAR (Table 8.2). Over time, some institutes have ceased to operate, and some were merged with others. For half a century, the CGIAR has been a productive leader in agricultural science and innovations for agricultural development in low-income countries.

Originally, the individual CGIAR centres operated within a specific mandate of a crop or subject matter of high relevance to agricultural innovation. Each independent centre had a board composed of members with long agricultural experience and professional knowledge in relevant scientific areas. Their decisions were science-oriented and without regard to political considerations; one factor for their progress. Every year at Centre's Week, all the centres had to report progress to their donors. That formed the basis for donors to pledge their funding for the following budget year. There was not only a regular system of reviews to be carried out by individual centres but also System Reviews to measure overall impact, all too rare in development activities. Today, there is a Science Week for highlighting new research findings.

In recent years, there has been a growing concern that the original structure of the CGIAR is no longer relevant for the future. Compared with other development aid, the independent CGIAR centres have been successful, some of them with a very good track record on impact. There has always been a demand that publicly funded agricultural research for development must demonstrate increased impact. Among the reasons for change were escalating climate change, diminishing biodiversity and the need for more multidisciplinary research to better tackle the issue of sustainability. Prior to 2007, developing innovations towards sustainability had been a slow process (Bengtsson, 2007). In addition, funding has not been increasing to meet the CGIAR centres' expectations for steady growth of research budgets.

A process of change began in 2019 under the assumption that the centres could achieve more impact when brought together under fewer institutional boundaries. They were to be supported by empowered management and governance. This would also integrate collaborative partners' perspectives and needs. This led to an ambitious process of change between 2019 and 2023 of both mandate and institutional arrangement with unified governance in an integrated operational structure; One CGIAR.

According to the CGIAR Financial Report Dashboard the 2021 budget amounted to US$815 million for 15 centres (CGIAR, 2022a). It covered 12 research programmes plus excellence in breeding (EIB) (US$17.8 million) and big data (US$5.4 million), e.g. global data in agriculture. The CGIAR system included more than 770,000 germplasm accessions: the largest in the world. In addition, other budget items were the gene bank (US$31.4 million), gender (US$6.2 million) and the nonportfolio (US$114.6 million). The last budget item can be considered administrative costs.

In 2021, CGIAR reported a total of 1152 innovations, identified as significant products or findings (Table 8.3). In recent years, an increasing number of innovations have been directed towards the UN's Sustainable Development Goals (SDGs). The majority of innovations was related to genetic innovation or reducing poverty, and 22 cases were confined to the improvement of natural resources and ecosystem services (CGIAR, 2022b). There were 26 innovations in

Table 8.2. CGIAR funded centres and budgets in selected years (US$ millions).

Centre	Budget (US$ millions)			
	1976	1992	2005	2021
IITA	9.4	25.0	40.2	99.3
IFPRI	–	13.0	39.7	95.6
CIMMYT	8.7	27.1	38.8	94.9
CIAT	6.3	26.5	42.4	81.1
ILRI	8.9[a]	33.7	32.2	74.9
IRRI	9.7	30.6	33.4	48.4
ICRISAT	6.8	27.7	28.4	46.7
ICRAF	–	15.0	30.0	46.1
CIP	4.1	15.2	22.0	41.2
CIFOR	–	–	17.5	37.7
World Fish	–	4.0	15.2	31.5
ICARDA	1.5	21.4	29.1	26.7
IWMI	–	10.0	23.1	24.4
Bioversity	0.9[b]	7.1	34.6	23.4
Africa Rice	0.8[c]	6.4	10.9	19.4
EIB	–	–	–	17.8
Gene Bank	–	–	–	31.4
Gender	–	–	–	6.2
Big Data	–	–	–	5.4
Non-CGIAR Centre	–	–	–	2.1
System Org. Entities	–	–	–	21.4
Non-portfolio	–	–	–	114.6
During early years:				
IFDC	0.8	12.5	–	–
ISNAR	–	7.7	–	–
IBSRAM	–	5.0	–	–
INIBAP	–	3.3	–	–
Non-CGIAR:				
AVDRC	2.0	7.6	10.0[d]	–
ICIMOD	-	2.7	–	–

[a]ILCA and ILRI.
[b]IBPGR/IPGRI.
[c]WARDA.
[d]2003.

improved food and nutrition security for health. The CGIAR reported that 69 innovations were taken up by next users and 203 innovations were available for uptake. In 2021, CGIAR gene banks distributed 96,590 germplasm samples to users: about two-thirds went to recipients in 91 countries. In 2020, a total of 353 PhD students had been incorporated into CGIAR research initiatives (CGIAR, 2022b).

In 2022, the centre boards and leadership developed the Integration Framework Agreement (IFA) to clarify the path to One CGIAR. Twelve centres signed the IFA, while three remained CGIAR centres but operated

Table 8.3. Examples of CGIAR results during 2017–2021 (CGIAR Results Dashboard, 2017–2021; CGIAR, 2022b).

Total SDGs contributions	91
Total innovations	1152
Total policies	95
Total partnerships	261
Total reviewed papers	2539
Total number of trainees	718,136

outside the IFA. In 2019, the Alliance of Bioversity International and CIAT was established. The IFA was approved by all CGIAR centres and their board chairs in early 2023.

The 2022 budget has expanded to about US$900 million, and 17% was spent on general, administrative and system-level costs (CGIAR, 2023). A staff of more than 9000 people worked in 89 countries around the world. During 2022, the CGIAR produced 1837 knowledge products, 476 innovations and 48 policy changes (CGIAR, 2023). According to official CGIAR statistics, 342 innovations contributed to SDG one to reduce poverty and improve livelihoods. In 2023, One CGIAR had produced 2942 research results (CGIAR, 2024).

One CGIAR is to work with a network of partners and provides a participatory mechanism for national governments, multilateral funding, development agencies and leading private foundations to finance important agricultural research. In 2023, there were 1786 partners involved in the CGIAR system (CGIAR, 2024).

The CGIAR-ICARDA's Integrated Desert Farming Innovation Program (IDFIP) of 2023 is one example of the new CGIAR approach. The IDFIP aims to transform desert agriculture across the global drylands into highly productive food systems with the support of the Gulf Cooperation Council countries. The aim is to create a 'public–private–producer' partnerships to build new farming systems leading to sustainable food and nutrition security. It is really a test case, based in an area with declining water supplies, poor soils, expanding desertification, and an increase of pests and plant diseases (ICARDA, 2023).

A collaborator in the network may be the Food Systems Countdown Initiative. It was started in late 2023 with the objective of tracking food systems annually to 2030. This framework and holistic monitoring architecture aims to track food system transformation towards global development, health and sustainability goals, using 50 indicators within five themes (Schneider *et al.*, 2023). The themes were: (i) diets, nutrition and health; (ii) environment, natural resources and production; (iii) livelihoods, poverty and equity; (iv) governance; and (v) resilience.

According to the new governing System Council of the One CGIAR, the new entity was to be 'an institutional environment in which a research programme modality based on integrative thinking on food systems and land and waterscapes can truly thrive'. The System Council is advised on all scientific aspects by an external Independent Science for Development Council (ISDC). In turn, it is supported by the CGIAR Independent Advisory and Evaluation Service. After this change it is critical that the concentration of power into one mechanism of central management instead of independent boards will remain science-based and allows freedom with a minimum of bureaucracy for individual centres.

A Research and Innovation Strategy towards 2030 has been developed and expands the CGIAR's original vision and mandate. The new mission is 'to deliver science and innovation that advance the transformation of food, land and water systems in a climate crisis'. The One CGIAR strategy is to pursue measurable benefits across five impact areas:

- nutrition, health and food security
- poverty reduction, livelihoods and jobs
- gender equality, youth and social inclusion
- climate adaptation and greenhouse gas reduction
- environmental health and biodiversity

The research of One CGIAR is grouped into three action areas which are built up of 33 transdisciplinary research initiatives. The action areas are (i) systems transformation, (ii) resilient agrifood systems and (iii) genetic innovation.

The One CGIAR research portfolio is a work in progress with high aspirations to be the world's largest publicly funded group of agrifood systems research centres. It is too early to tell whether the reform will deliver what has been planned. The new strategy means that innovations are the connector between research outputs and development outcomes (Meinke *et al.*, 2023). It was also noted that insights gained from the CGIAR reform process, and the frameworks, would be useful for other organizations and their collaborations with One CGIAR. They could apply the new framework how to assess the relevance, scientific credibility and effectiveness of research.

The strategic focus on innovation is obvious but it is critical that innovation fits into the social and political environment within which the individual farmer adopts or rejects it to ensure societal development and – sustainability. The scientific advice from ISDC is for global research but one should be aware that most agriculture is regionally and locally specific in implementation – to get impact. This calls for not only qualified scientific advice from ISDC but also a very qualified network of research partners, well informed about the practical aspects of agriculture in their respective regions. It is unclear how farmers will be actively involved in the early problem identification process. It is not very common in most national agricultural research systems.

At COP28, the CGIAR launched a case for US$4 billion in investments towards its 2025–2027 research portfolio. Also, it secured some funds from five of its traditional donors and the United Arab Emirates became a new donor, announcing US$200 million of funding in partnership with the Bill & Melinda Gates Foundation (Accelerate Action on Climate and Strengthen Food Systems Through Investment in Agricultural Innovation). Despite this positive signal it raises a question concerning both the increased immediate influence of rich Middle Eastern oil powers and their potential long-term financial support. At the same COP28, the UN called for a phasing out that kind of energy. This raises a fundamental question about sustainability and One CGIAR. The planned budget increase makes future funding a problem. Moreover, some traditional donors may even prefer to continue to invest their funds in selected, productive CGIAR institutes, based on their own short-term political preferences rather than an administrative mechanism.

Universities

The substantial number of national agricultural universities have a large task in producing innovations, competing with a range of research divisions at TNCs and One CGIAR. Agricultural universities must set their individual strategic research priorities on agricultural problems where practical solutions will give societal impact at the national level to justify public funding.

For many years, internationalization has been important in university education for both students and the institution. This trend has been further emphasized in the recent years and universities like to claim that they are global as a way of attracting foreign students. Alternatively, universities conduct high-quality research, offer a good environment for higher studies, or may have a generous system of supporting PhD students. Rankings of the best global universities may be one way to help applicants around the world to select the best place for higher education. Such rankings have been done by the US News & World Report for many years. Their ranking assesses 13 indicators from academic research and reputation to personal considerations, including location, campus culture and costs (Morse and Wellington, 2024). Among the indicators are global and regional research reputation, international collaboration, books, and the number of publications and highly cited papers among the top 1% of most cited papers in their respective field (Table 8.4).

Academic research has expanded, especially in industrialized countries, both in terms of new research institutions and a growing number of research publications. Looking at many of their titles nowadays, they are dealing with research on sustainability or sustainable development. A basic issue is whether universities contribute to societal development by their research. A small attempt to investigate this was made some years ago with regard to Nordic agrarian universities.

In 2018, the top management of the four Nordic agrarian universities were contacted. They were asked about the three most important innovations, research products and/or research results produced at their respective country's agrarian university during the last decade that had concretely influenced/changed the country's agricultural and forest policy and/or

Table 8.4. Ranking of some selected Best Global Universities (US News and World Report 2022–2023).

University	Ranking
Harvard University, Cambridge, MA	1
Imperial College, London	13
Sorbonne Université, Paris	48
Humboldt University of Berlin	61
University of Helsinki	99
University of Padua, Italy	115
Université de Montpellier, France	194
University of agricultural sciences:	
University of Copenhagen	18
Wageningen University and Research	89
Swedish University of Agricultural Sciences	356
Norwegian University of Life Sciences	736

contributed to solid improvements in practical forestry or agriculture ((Bengtsson and Gäre, 2019).

The responses from the four Nordic agrarian universities did not contain any specific information about research on agricultural sustainability. The Faculty of Agriculture and Forestry of the University of Helsinki mentioned several innovations and concrete research products. Finnish research was reported to have played a key role in achieving the UN SDGs. It was also stressed that research on agriculture and forestry should be given more political attention.

Some specific examples of important research from Finland included: (i) laser scanning technology, describing the properties of the forest with high accuracy at the stand level and individual trees for estimation of all the Finnish forests; (ii) new data on the structures of Fennoscandia forests; (iii) high-quality research on greenhouse gas emissions from mineral soils and peatlands; (iv) increased understanding of viral, bacterial and fungal diseases; and (v) progress in the breeding of broad beans, increased knowledge in molecular plant biology and advances in genomics for animal breeding, nitrogen-fixing plants and agricultural sensor technology.

The Faculty of Science and Biological Sciences (SCIENCE) at the University of Copenhagen declared that there was no register or ranking of great research breakthroughs or any way of systematically searching for their impact. A similar view was held by Copenhagen's Faculty of Health Sciences. Despite that initial declaration, the faculty management at SCIENCE emphasized three particularly productive research areas: (i) synthetic biology with a focus on bioactive natural products and the biosynthesis, transport, storage and degradation of cyanogenic glycosides for plant defence mechanisms; (ii) identification of biosynthetic genes and the transport and regulation of glucosinolates; and (iii) 'reverse photosynthesis'. The latter is the discovery of a natural process, showing that sunlight also contributes to the breakdown of cellulose in plant cell walls and not just to photosynthesis if the enzyme monooxygenases is added. It was predicted that this process could radically change the industry's future production of chemicals.

The Norwegian University of Life Sciences (NMBU) has a vision of global well-being, but its reporting showed its research had also contributed to problem-solving within Norway. The NMBU underlined difficulties in identifying the three most important innovations in a decade. Some examples were highlighted from the Faculty of Life Sciences: (i) a new technique of 'genomic selection' with research underway to apply the technology in animal breeding; (ii) contributions by Norwegian researchers to the work of gene sequencing bread wheat within the International Wheat Genome Sequencing Consortium with emphasis on gene sequencing chromosome 7 B; and (iii) genome sequencing of the Atlantic salmon genome.

The Faculty of Environmental Science and Nature Management at NMBU highlighted airborne laser scanning as an integral part of forest inventories which are used commercially worldwide with improved quality and cost savings of 40–50%. It can also be used in the assessment of forest ecosystem functions, for carbon reporting through forest biomass and for habitat detection for species. Another example was the 'Keep It' sustainability indicator which is an instrument providing ongoing information on a product's sustainability, an improvement of the 'Best Before' label.

The Faculty of Veterinary Medicine at NMBU stressed there was no ranking of research results and innovations. A decade was considered too short to note the full effect of its research. Nonetheless, the Faculty underlined that its research on infectious diseases, including scrapie and foot rot in sheep, and its participation in subsequent control programmes had been of great financial significance. Research had also contributed to improved animal health, higher milk quality in goats and data systematization from milking robots. Pet-related methicillin-resistant *Staphylococcus aureus* (MRSA) was on its way to extinction, in contrast to Denmark.

The response from Swedish University of Agricultural Sciences (SLU) was less specific in nature. It was briefly stated that 'it is the individual researchers' work and publications that report the results of the university's activities' and 'the majority of research activities are carried out as international collaborations'. Therefore, the management of SLU, as 'one of the participating universities in a development process', did not consider itself able to provide specific examples of innovations or research products during the last decade. Instead, reference was given to the university's strategy for 2017–2020. In early 2019, the SLU board decided upon an evaluation of the quality and benefits of research in an international perspective to emphasize both strategic visions and societal benefits.

As a small country, Sweden has never had so many research institutions that all consider themselves academic as it does today: 16 universities and 31 colleges in 2023. These institutions have to compete with each other and also research institutes that are part of RISE Research Institutes of Sweden AB, a Swedish government company. In general, there seems to be little evidence of practical solutions to sustainability for implementation.

One way to further clarify the categorization in Table 8.4 may be to use examples from Sweden to provide more specifics. One major actor is the SLU, and another one is Agtech 2030: a ten-year initiative for developing agricultural technology for tomorrow's agriculture. The focus areas are sensors, digital technology and artificial intelligence. The Agtech centre started in 2018 and is coordinated by Linköping University with the participation of technology and agricultural

companies, agricultural machinery companies, consultants, institutes, the academy and business partners. So far, Agtech claims some 50 innovative projects, some of which have had an international impact, for example, Flexrow, a system for preventing soil compaction, and a technology for determining the sex of chickens at the egg stage (Agtech, 2022).

The overall objective of the SLU strategy for 2021–2025 is to be a leading international university, playing a key role for change towards a sustainable society (SLU, 2020). It has requested a larger mandate from the government to recruit more students, especially from abroad, based on claims of increasing demands, both within Sweden and globally, to work for a sustainable society. Long ago, the Swedish government stated that doctoral theses contribute to global knowledge within the areas of greatest importance. This ought to be reflected in areas such as future food production with sustainability.

SLU Global is a unit in the Vice-Chancellor's Office that supports low-income countries based on the SDGs of Agenda 2030. The Swedish International Agriculture Network Initiative (SIANI), funded by Sida, supports communication, multisector dialogue and actions around sustainable food systems. SIANI is embarking on its fourth period for 2023–2027.

Two MSc programmes in agricultural science and forest science started in 2021, redesigned to comply with the Bologna principle of 3 + 2 years. So did an undergraduate programme in English, forests and landscape. Two completely new programmes were also launched: one undergraduate programme in political science/sustainable development and one master's programme in sustainable food and landscape. There was an early attempt to change the strategy and research direction towards sustainability at the former Department on Plant Husbandry of SLU in the mid-1990s (Strategy Group, The, 1995). It was, however, rejected by the Head of Department, the Dean of the Faculty of Agricultural Sciences and the Vice-Chancellor.

According to the SLU Annual Report (2023a), the university staff published 2200 articles and reviews, a drop of 10% between 2021 and 2022. The share of scientific articles published that 'belong to the ten most cited per cent in their respective subject field amounted to some 16 per cent'. The innovation support by SLU Holding has led to new investments in five companies without mentioning a focus on agricultural sustainability.

SLU is engaged in Sida's bilateral projects for increased national research capacity in Bolivia, Ethiopia, Cambodia, Mozambique, Rwanda, Tanzania and Uganda. SLU supports the doctoral courses and study programmes. Four platforms collaborate with actors outside academia: SLU Future Food, Forest, One Health and Urban Futures.

Since 2019, an average of 110 new doctoral students have been admitted annually according to Swedish University of Agricultural Sciences (SLU) (2023b). The situation over the last five-year period is presented in Table 8.5. Some 40% of the doctoral students have a foreign background and female students are in majority. The number of doctoral degrees in 2021 was considered a reasonable number since an average of 93 doctoral students were annually admitted during the 2016–2021 period. The coronavirus pandemic affected research at SLU during 2020–2021 with travel restrictions making it difficult to recruit students.

The number of active doctoral students has, however, decreased continuously since 2016, following a national Swedish trend. The average number of theses by doctoral students has also decreased since the pandemic (Table 8.6). Relatively few students graduated annually during the five-year period (2017–2022); about 80 students. In 2023, the number of doctoral degrees was estimated at 105, a little more than previously. Doctoral studentship is the most usual form of funding for most of the doctoral students. The number of doctoral students with a research grant has decreased during the last 5 years.

One way to investigate the international approach in research may be a review of doctoral theses and their research orientations (Table 8.7). This categorization is based on the title and abstract of each thesis in selected years. Very few theses had a focus towards the SDGs. During the period, there is a vague orientation towards problems of Swedish agriculture and forestry, and a modest percentage for developing country agriculture. This raises a fundamental question of research being of high relevance and an efficient use of resources.

Table 8.5. The doctoral programme at Swedish University of Agricultural Sciences (SLU) during 2018–2023 (SLU Annual Reports, 2018–2023).

Year	No. of admitted students	Active doctoral students	Students of foreign background (%)	Female students (%)	Number of theses
2023	123	560–580	44	59	105
2022	116	535	43	59	79.5
2021	92	535	44	59	92
2020	107	559	45	56	69[a]
2019	113	528	47	64	91[a]
2018	130	563	46	50	71

[a]Low figure partly explained by fewer admitted students 4–5 years ago.

Table 8.6. Average number of doctoral theses at the Swedish University of Agricultural Sciences (SLU) during five-year periods between 2002 and 2021 (SLU Annual Reports 2003-2022).

Period	Average annual number of theses
2002–2006	98
2007–2011	101
2012–2016	104
2017–2021	82

Table 8.7. Major research focus of theses by
doctoral students at SLU in selected years
Swedish University of Agricultural Sciences (SLU),
2023b.

Research focus	2007	2017	2022
Towards Agenda 2030	–	2	1
High relevance for:			
Swedish agriculture and forestry	32	28	25
Developing country agriculture	20	11	6
New scientific knowledge	28	25	14
Curiosity/others/ unidentified	44	43	32
Total	126	110	77

Internationalization is vital in research, but
it is unclear if there are advantages in doctoral
agricultural students getting their doctor's
degrees abroad but often not returning to their
home country, where they are much closer to
acute agrarian problems. Doctoral students
may decide they prefer to settle in the country
in which they study, where opportunities and
potentials may offer more than their home
country as well as a safer environment. However,
this may deprive their country of origin of
much-needed talent and the skills to tackle their
agricultural challenges. Also, there is a need to
strengthen the national agricultural research
capacity in most low-income countries despite
certain progress. For instance, in Ethiopia, there
are now over 20 government owned and privat-
ized universities with agricultural departments,
although only Haramaya University gives PhD
degrees (G. Tedla, personal communication,
Addis Ababa, Ethiopia, 2024). But there can be
problems, as shown from Sweden, where almost
21,000 people were granted residence permits
for studies in 2023. About 14,000, of them were
permits for higher education studies (Swedish
Migration Agency, 2024). China was the most
common citizenship among those granted
permits in 2023, followed by India, the United

States, Sri Lanka and Singapore. It is rather easy
for international students to get financing in
Sweden and, unlike many other EU countries,
there is no upper limit for how much a foreign
student is allowed to work alongside his or her
studies. This has led to some abuse of residence
permits according to the Swedish Migration
Agency (2022). Focusing on one country, there
were strong indications of "extensive abuse of
resident permits for studies" but the same patterns were believed to apply to other nationali-
ties. One third of M.Sc students had abandoned
their studies for full-time work immediately after
completing 30 credits. Another third deviated
from studies for unknown reasons. It was concluded that the provisions of the Aliens Act and
the Aliens Ordinance created the possibility of
misusing residence permits for studies to work.
In Sweden, unlike in many other EU countries,
there is no upper limit for how much an international student is allowed to work alongside
studies. Moreover, some higher education institutions did not put enough focus on attracting
international students with a sincere intention
to study. Some Swedish universities have also
used recruitment agents to attract foreign
students. The number of doctoral students at
the Swedish University of Agricultural Sciences
have decreased over the last 15 years (Table 8.8).
But the reduction is much less for international
students than from within Sweden.

Table 8.8. Country of origin of doctoral students
at the Swedish University of Agricultural Sciences
(SLU) in selected years between 2007 and 2022
(Swedish University of Agricultural Sciences
(SLU), 2023b).

Origin	2007	2017	2022
Swedish	74	52	39
Developing country	28	29	18
Other countries	21[a]	25[b]	16[c]
Unidentified	3	2	4
Total	126	108	76

[a]China (5), Iran (2), Russia (1), USA (1), Canada (1), Brazil
(1), Ukraine (1) and EU countries.
[b]China (8), Russia (2), Canada (1) and EU countries.
[c]USA (2), China (1), Japan (1), Switzerland (1), Croatia (1)
and EU countries.

References

Agtech (2022) An innovative environment for the agriculture of the future. Linköping University, Sweden. Available at: https://agtechsweden.com/en/ (accessed 16 August 2024).

Alpern, S. (1992) The European introduction of crops into West Africa in precolonial times. *History in Africa* 19, 13–43. Available at: https://doi.org/10.2307/3171994 (accessed 16 August 2024).

Andersen, T.B., Jensen, P.S. and Skovsgaard, C.V. (2016) The heavy plow and the agricultural revolution in Medieval Europe. *Journal of Development Economics* 118, 133–149. Available at: https://doi.org/10.1016/j.jdeveco.2015.08.006 (accessed 16 August 2024).

Askew, K. (2022) We plan to sharply increase production in 2023-24': Methane-reduced beef trial in Sweden 'sold out in less than a week. *Food Navigator*. Available at: https://www.foodnavigator.com/Article/2022/07/11/methane-reduced-beef-trial-in-sweden-sold-out-in-less-than-a-week (accessed 8 September 2024).

Baguilat, C.L. (2011) *The Hungduan, Ifugao Rice Terraces: An Agri-Cultural Heritage*. International Forum on Globally Important Agricultural Heritage Systems, Beijing, China. Available at: https://www.fao.org/fileadmin/templates/giahs/Presentations/beijing/Clarence_Baguilat-Ifugao_Rice_Teraces.pdf (accessed 16 August 2024).

Barclay, G. (2012) Medicinal plants of Trinidad and Tobago. In: Reid, B.A. (ed.) *Caribbean Heritage*. The University of the West Indies Press, Trinidad and Tobago, pp. 221–235.

Barlow, R.S. (2003) *300 Years of Farm Implements and Machinery 1630–1930*. Krause Publications, Iola, WI.

Bayer (2023) Bayer collaborates with Microsoft to unveil new cloud-based enterprise solutions, advancing innovation and transparency in the agri-food industry. Bayer. Available at: https://www.bayer.com/media/en-us/bayer-collaborates-with-microsoft-to-unveil-new-cloud-based-enterprise-solutions-advancing-innovation-and-transparency-in-the-agri-food-industry/ (accessed 13 August 2024).

Bengtsson, B.M.I. (2007) *Agricultural Research at the Crossroads. Revisited Resource-poor Farmers and the Millennium Development Goals*. CRC Press, Enfield, NH.

Bengtsson, B.M.I. and Gäre, S. (2019) *Agriculture, Climate, and the Future* (in Swedish). Ekerlids förlag, Stockholm.

Bennerhag, C., Grundin, L., Hjärtner-Holder, E., Stilborg, O., Söderholm, K. *et al.* (2021) Hunter-gather metallurgy in the early iron age of Northern Fennoscandia. *Antiquity* 95(384), 1511–1526. DOI: 10.15184/aqy.2020.248. (accessed 16 August 2024).

Beraaouz, M., Abioui, M., Hssaisoune, M. and Martínez-Frías, J. (2022) Khettaras in the Tafilalet oasis (Morocco): Contribution to the promotion of tourism and sustainable development. *Built Heritage* 6, 24. Available at: https://doi.org/10.1186/s43238-022-00073-x (accessed 16 August 2024).

Bioworks, G. (2023c) *Ginkgo Bioworks partners with Boehringer Ingelheim to develop breakthrough therapies for hard-to-treat diseases*. PR Newswire.

Brattström, A. (2022) Cultural ideals in the entrepreneurship industry. In: Wennberg, K. and Sandström, C. (eds) *Questioning the Entrepreneurial State. International Studies in Entrepreneurship*, Vol. 53. Springer, Cham, Switzerland.

Campbell, B.M.S. and Overton, M. (1993) A new perspective on medieval and early modern agriculture: Six centuries of norfolk farming, c.1250–c.1850. *Past and Present* 141, 38–105. DOI: 10.1093/past/141.1.38. (accessed 16 August 2024).

Center for International Forestry Research and World Agroforestry (CIFOR-ICRAF) (n.d.) The Future of Food: Serving People and the Planet. CIFOR-ICRAF. Available at: https://www.cifor-icraf.org/agroecology-tpp/ (accessed 6 August 2024).

Consultative Group for International Agricultural Research (CGIAR) (2022a) 2021 CGIAR Financial Report Highlights and Dashboards. CGIAR, Washington DC. Available at: https://www.cgiar.org/food-security-impact/finance-reports/dashboard/overview/ (accessed 8 September 2024).

Consultative Group for International Agricultural Research (CGIAR) (2022b) Harvesting Research and Innovation for Impact. Summarized Annual Performance Report 2021. CGIAR, Washington DC. Available at: https://cgspace.cgiar.org/items/7980a512-6e8b-4687-baac-cf5f4d832dc9 (accessed 16 August 2024).

Consultative Group for International Agricultural Research (CGIAR) (2023) *CGIAR Annual Report 2022: Science to Transform Food, Land, and Water Systems in a Climate Crisis*. CGIAR, Washington. D.C.

Consultative Group for International Agricultural Research (CGIAR) (2024) CGIAR Annual Report, 2023. CGIAR, Washington, DC. Available at: https://www.cgiar.org/food-security-impact/finance-reports/dashboard/overview/ (accessed 8 September 2024).

Crosby, A.W. (2003) *The Columbian Exchange: Biological and Cultural Consequences of 1492, 30th Anniversary Edition*. Greenwood Publishing Group, Santa Barbara, CA.

Earth Observatory (2016) Ancient waterways in Morocco. Earth Observatory. Available at: https://earthobservatory.nasa.gov/images/89133/ancient-waterways-in-morocco (accessed 6 August 2024).

European Commission (2022) *European Partnerships in Horizon Europe*. European Commission. Available at: https://research-and-innovation.ec.europa.eu/funding/funding-opportunities/funding-programmes-and-open-calls/horizon-europe/european-partnerships-horizon-europe_en (accessed 13 August 2024).

European Food Safety Authority (EFSA) (2023) EFSA's Scientific Colloquium 27 "Cell Culture-Derived Foods and Food Ingredients. EFSA, Brussels. Available at: https://www.efsa.europa.eu/en/events/efsas-scientific-colloquium-27-cell-culture-derived-foods-and-food-ingredients (accessed 8 September 2024).

Fishback, P.V. (2022) *Agricultural Crises and Government Responses between the World Wars in the Atlantic Trading Network*. National Bureau of Economic Research. Cambridge, MA. Available at: https://www.nber.org/papers/w30069 (accessed 6 August 2024).

Food and Agriculture Organization (FAO) (n.d.) Agroecology Knowledge Hub. FAO. Available at: https://www.fao.org/agroecology/knowledge/en/ (accessed 6 August 2024).

Food and Drug Administration (FDA) (2022) Subject Cell Culture Consultation (CCC) 000002, Cultured *Gallus gallus* cell material. Memorandum, November 14, 2022, To Administrative File, CCC 000002, Sponsor: UPSIDE Foods, Inc. U.S. Food and Drug Administration, College Park, MD 20470.

Foster, L.V. (2000) *A Brief History of Central America*. Facts on File, New York.

Gingko Bioworks (2023a) Sustainable, Carbon-Negative Materials with Visolis. Available at: https://www.ginkobioworks.com/2023/04/24/sustainable-carbon-negative-materials-with-visolis (accessed 8 September 2024).

Gingko Bioworks (2023b) Next-Gen Seed Technology with Syngenta. Available at: https://www.ginkobioworks.com/2023/04/17/next.gen-technology-with-syngenta (accessed 8 September 2024).

Gollin, D., Hansen, C.W. and Wingender, A.M. (2021) Two blades of grass: The impact of the green revolution. *Journal of Political Economy* 129(8), 2344–2384. Available at: https://doi.org/10.1086/714444 (accessed 8 September 2024).

Greenberger, R. (2006) *The Technology of Ancient China*. The Rosen Publishing Group Inc, Buffalo, NY.

Hammarlund, N. (1956) *A Few Glimpses from a Bygone Era in Northwestern Scania* (in Swedish). AB Skånska Dagbladets Tryckeri, Malmö.

Horn, S., Parks, B.C., Reinhart, C.M. and Trebesch, C. (2023) China as an international lender of last resort. National Bureau of Economic Research, Working Paper 31105, Cambridge, MA. Available at: https://doi.org/10.3386/w31105 (accessed 8 September 2024).

ICARDA (2023) ICARDA-CGIAR desert farming innovation sprint announced at the AIM for Climate summit. ICARDA. Available at: https://www.icarda.org/media/news/icarda-cgiar-desert-farming-innovation-sprint-announced-aim-climate-summit (accessed 13 August 2024).

Investopedia (2024) Great leap forward: What it was, goals and impact. *Investopedia*. Available at: https://www.investopedia.com/terms/g/great-leap-forward.asp (accessed 6 August 2024).

Iordachi, C. and Bauerkamper, A. (eds) (2014) *The Collectivization of Agriculture in Communist Eastern Europe: Comparison and Entanglements*. Central European University Press, Budapest.

Janick, J. (2008) *Islamic Influences on Western Agriculture*. Purdue University, West Lafayette, IN. Available at: https://hort.purdue.edu/newcrop/Hort_306/text/lec22.pdf (accessed 16 August 2024).

Jones, P. (2016) *Agricultural Enlightenment: Knowledge, Technology, and Nature, 1750–1840*. Oxford University Press, Oxford.

Lal, R. (2001) Thematic evolution of ISTRO: Transition in scientific issues and research focus from 1955 to 2000. *Soil and Tillage Research* 61(1–2), 3–12. Available at: https://doi.org/10.1016/S0167-1987(01)00184-2 (accessed 16 August 2024).

Lightfoot, D. (1996) Moroccan khettera: Traditional irrigation and progressive desiccation. *Geoforum* 27(2), 261–273. Available at: https://doi.org/10.1016/0016-7185(96)00008-5 (accessed 8 September 2024).

Malleck, J. (2023) A second lab-grown chicken producer got A step closer to hitting the shelves. *Quartz*. Available at: https://qz.com/good-meat-lab-grown-chicken-fda-clearance-1850251226#:~:text=US%20Food%20and%20Drug%20Administration (accessed 8 September 2024).

Marston, J. (2024) Agrifoodtech startup investment drops 50%, accounts for just 5.5% of global VC dollars. *AgFunderNews*. Available at: https://agfundernews.com/agrifoodtech-startup-investment-drops-50-now-accounts-for-just-5-5-of-global-vc-dollars (accessed 6 August 2024).

Mascarelli, A. (2010) Mayans converted wetlands to farmland. *Nature*. Available at: https://doi.org/10.1038/news.2010.587 (accessed 16 August 2024).

Maxwell, M. (2021) Sanatech seed launches world's first GE tomato. *Eurofruit*. Available at: https://www.fruitnet.com/eurofruit/sanatech-seed-launches-worlds-first-ge-tomato/184662.article (accessed 8 September 2024).

Mbapndah, N.M. (1994) French colonial agricultural policy, african chiefs, and coffee growing in the cameroun grassfields, 1920-1960. *The International Journal of African Historical Studies* 27(1), 41–58. Available at: https://doi.org/10.2307/220969 (accessed 16 August 2024).

Mbeki, M. (2001) *Zimbabwe Before and After the Elections: A Concerned Assessment*. South African Institute of International Affairs, Braamfontein, South Africa.

Mbeki, M. (2009) *Architects of Poverty: Why African Capitalism Needs Changing*. Picador, Johannesburg, South Africa.

Meinke, H., Ash, A., Barrett, C.B., Smith, A.G., Zivin, J.S.G. *et al.* (2023) Evolution of the one CGIAR's research and innovation portfolio to 2030: Approaches, tools, and insights after the reform. *NPJ Sustainable Agriculture* 1, 6. Available at: https://doi.org/10.1038/s44264-023-00005-x (accessed 16 August 2024).

Morse, R. and Wellington, S. (2024) How U.S. News calculated the 2024-2025 Best Global Universities Rankings. US News & World Report. Available at: https://www.usnews.com/education/best-global-universities/articles/methodology (accessed 8 September 2024).

Mullen, S. (2020) Centring transatlantic slavery in scottish historiography. *History Compass* 20(1), e12707. Available at: https://doi.org/10.1111/hic3.12707 (accessed 8 September 2024).

Mumford, J.R. (2012) *Vertical Empire: The General Resettlement of Indians in the Colonial Andes*. Duke University Press, Durham, NC, and London.

Naithani, S. (2021) *History and Science of Cultivated Plants*. Oregon State University, Corvallis, OR.

NBC (2023) U.S. approves chicken made from cultivated cells, the nation'S first "lab-grown" meat. *Associated Press*. Available at: https://nbcnews.com/ (accessed 8 September 2024).

Needham, J. (1986) *Science and Civilization in China. Volume 4, Physics and Physical Technology, Part 2, Mechanical Engineering*. Caves Books, Ltd, Taipei, Taiwan.

Olsson, S.-O. (1978) Agriculture and Swedish food supply during the Second World War. *Scandinavian Economic History Review* 26(2), 191–193. Available at: https://doi.org/10.1080/03585522.1978.10415635 (accessed 6 August 2024).

Peng, X. (1987) Demographic consequences of the great leap forward in China's provinces. *Population and Development Review* 13(4), 639–670.

Phillipson, D. (2005) *African Archaeology*, 3rd edn. Cambridge University Press, Cambridge, UK.

Pingali, P.L. (2012) Green revolution: Impacts, limits, and the path ahead. *Proceedings of the National Academy of Sciences of the United States of America* 109(31), 12302–12308. Available at: https://doi.org/10.1073/pnas.0912953109 (accessed 8 September 2024).

Popular Science (2022) The 100 greatest innovations of 2022. The 35th annual Best of What's New Awards. *Popular Science*. Available at: https://popsci.com/technology/best-of-whats-new-2022/ (accessed 8 September 2024).

PotatoBusiness (2022) Novel technology capable of producing egg protein in potatoes. PotatoBusiness. Available at: https://www.potatobusiness.com/trends-news/novel-technology-capable-of-producing-egg-protein-in-potatoes/ (accessed 6 August 2024).

Raven (2022) Case IH and raven industries collaborate to deliver an advanced, driverless spreading solution. *Raven*. Available at: https://www.ravenind.com/resources/news/industrys-first-autonomous-spreader-debuts-at-farm-progress-show (accessed 13 August 2024).

Ross, E.G. (2002) The Portuguese in Africa, 1415–1600. In: *Heilbrunn Timeline of Art History*. The Metropolitan Museum of Art, New York.

Rundgren, G. (2023, 10 April) Why do we grow food? *Garden Earth*. Available at: https://gardenearth.substack.com/p/why-do-we-grow-food (accessed 6 August 2024).

Schneider, K., Fanzo, J., Haddad, L., Herreros, M., Moncayo, J.R. *et al.* (2023) The state of food systems worldwide: Counting down to 2030. *Arxiv*. Available at: https://doi.org/10.48550/arXiv.2303.13669 (accessed 8 September 2024).

Siegner, C. (2019) Analysts: Cell-cultured and plant-based meat could be 10% of the market by 2029. FoodDive. Available at: https://www.fooddive.com/news/analysts-cell-cultured-and-plant-based-m eat-could-be-10-of-the-market-by/555573/ (accessed 15 August 2024).

Simmonds, N.W. (1991) The earlier British contribution to tropical agricultural research. *TAA Newsletter*. Tropical Agriculture Association.

Strategy Group, The (1995) A roadmap for the year 2005. Future sustainable cropping systems and natural resource management. Scientific tools, methods, and systems for problem-oriented synthesis research. Strategy Group at the Department of Crop Production. SLU, Uppsala, Sweden.

Sustainable Markets Initiative (SMI) (2022) World's leading food and farming businesses launch action plan to scale regenerative farming, warning speed of progress "must triple" to tackle the impact of climate change. SMI. Available at: https://www.sustainable-markets.org/news/world-s-leading-food -farming-businesses-launch-action-plan-to-scale-regenerative-farming-warning-speed-of-progress -must-triple-to-tackle-the-impacts-of-climate-change/ (accessed 6 August 2024).

Swedish Migration Agency (2022) Misuse of a Residence Permit for Studies (In Swedish). Analysrapport från enheten för operativ analys. Europeiska Unionen (Fonden för inre säkerhet), Migrationsverket, EC/004/2022, Diarienummer:1.3.4-2020-1460. Available at: https://www.migrationsverket.se/Engli sh/About-the-Migration-Agency/Current-topics/The-Swedish-Migration-Agency-Answers/Migratio nsverket-svarar/2024-03-20-The-Swedish-Migration-Agency-answers-how-a-residence-permit-for -studies-work.html

Swedish Migration Agency (2024) The Swedish Migration Agency answers: how a residence permit for studies work. Available at: https://www.migrationsverket.se/ommigrationsverket/aktuellt/nyhetsarkiv (accessed 20 March 2024).

Swedish University of Agricultural Sciences (SLU) (2018–2023) *Annual Reports 2018–2023*. SLU, Uppsala, Sweden.

Swedish University of Agricultural Sciences (SLU) (2020) *SLU Strategy 2021–2025*. (in Swedish). SLU, Uppsala, Sweden.

Swedish University of Agricultural Sciences (SLU) (2023a) *SLU Annual Report 2022*. SLU, Uppsala, Sweden.

Swedish University of Agricultural Sciences (SLU) (2023b). SLU Publication Base. SLU, Uppsala, Sweden.

Swedish University of Agricultural Sciences (SLU) (2022) Annual Reports 2003-2022. SLU, Uppsala, Sweden.

Tann, J. (1997) Steam and sugar: The diffusion of the stationary steam engine to the Caribbean sugar industry 1770–1840. *History and Technology* 19(1997), 63–84. Available at: https://doi.org/10.5040/ 9781350018822.0009 (accessed 8 September 2024).

Temple, K.G.R. (1986) *The Genius of China: 3000 Years of Science, Discovery, and Invention*. Simon and Schuster, New York.

US News and World Report (2022) (2022–2023) Best global universities. U.S. News & World Report. Available at: https://www.usnews.com/education/best-global-universities/rankings (accessed 8 September 2024).

Väderstad (2021) Väderstad partner in the DLG gold medal winning Nexat system. Väderstad. Available at: https://www.vaderstad.com/en/about-us/news/news-archive/2021/international/vaderstad-ann ounces-partnership-with-nexat/ (accessed 13 August 2024).

Vegeconomist (2023, 20 March) Jimi Biotech unveils "China's first" 100% cultivated chicken . *Vegeconomist*. Available at: https://vegconomist.com/cultivated-cell-cultured-biotechnology/jimi -biotech-china-100-cells-cultivated-chicken/ (accessed 6 August 2024).

Vegeconomist (2024) Like Oat Drinks Today, We Want High-Quality Plant-Based Cheese to Become Available Everywhere. *Vegeconomist*. Available at: https://vegconomist.com/interviews/faerm-like-oat-drinks-hi gh-quality-plant-based-cheese-available-everywhere/ (accessed 8 September 2024).

Waltz, E. (2021) GABA-enriched tomato is first CRISPR-edited food to enter market. *Nature Biotechnology* 40, 9–11. Available at: https://doi.org/10.1038/d41587-021-00026-2 (accessed 8 September 2024).

Watson, A.M. (1983) *Agricultural Innovation in the Early Islamic World. The Diffusion of Crops and Farming Techniques 700-1100*. Cambridge Studies in Islamic Civilization. Cambridge University Press, Cambridge/London/New York.

Wellington, D.C. (2006) *French East India Companies: A Historical Account and Record of Trade*. Hamilton Books, Lanham, MD.

9 Towards Sustainable Food Production

Abstract

Sustainability is discussed as a conceptual framework in the context of both original and more recent definitions. There is also a brief analysis of the consequences – and costs – of past agricultural research towards sustainability and the role of agricultural sciences. There is a need to turn from the conventional one-factor research to interdisciplinary research to cope with the complex nature of agriculture. No single innovation, without proper validation, will suffice as the ultimate solution. It must also fit into the overall context. Different aspects of agriculture are discussed with a view to possible practical approaches for reaching sustainability in different technical subject matters. This not only calls for changes in plant breeding and genetics, soil fertility and pesticides, but also the introduction of new crops, including underutilized and poorly researched minor crops, diversification of crops and animals, new cropping systems, agroforestry, regenerative agriculture, new industrial farming methods, entomophagy, and the changing of diets and future food consumption, including ultra-processed food.

Sustainability: A Conceptual Framework

The global leaders took an initial step towards sustainability, when they accepted the concept of sustainability in the Brundtland Commission's report at the Earth Summit in 1992. This was based upon important scientific work by the Commission's Advisory Panel on Food Security, Agriculture, Forestry and Environment, which had designed policies for sustainable agriculture.

Over the years, the world economy has been growing, leading to better living conditions for many people globally. The concept of sustainability is used with different meanings and ambiguity as a way of fitting it into many kinds of discussions. The term can be used to signify anything, with nuances and different interpretations (Hedenus *et al.*, 2022). Most governments put priority on the reduction of greenhouse gas emissions. The big issue is to find a fundamental transformation towards a society with a much smaller ecological footprint. Such a transition will come whether people wish it or not – in the end. Political action now, however, will allow populations to influence what a future sustainable society should look like.

It is interesting to note that much of today's thinking on sustainability is similar to some of the measures taken during wartime conditions. During both the World Wars, extra land was cultivated so that agricultural production increased. Food was also grown in green spaces in cities, like today's growth in urban farming. People with their own gardens switched to food plants. In the United Kingdom, it was even forbidden to grow flowers or strawberries so as to make room for food. Rationing concerned both food and commodities, such as gasoline, fertilizers and feed supplements, and goods that were in particularly short supply, such as rubber. Specific food cards were given to the seriously ill, small children and older people.

During World War II, people were forced to change their diets when the authorities told people what they should eat (less meat, more vegetables). Eggs were one of the most rationed goods. The Swedish Food Commission estimated that, due to rationing, consumption was reduced by about two-thirds of the original economic and nutritional value for food products from cows and pigs and by some 90% for chickens and their eggs. No food was wasted. Tops of vegetables were eaten; tea leaves were reused and so was the grease in frying pans. Flour was made from acorns and sausages could be made from badger and fox meat. Less consumption and more daily exercise were health factors since people had to cycle and walk as cars were used less because gasoline was in short supply. Today, the change in food consumption requires people to make individual decisions which is more difficult when voluntary.

Nowadays, environmentalists argue that the planet cannot produce food in a sustainable way if climate change is not tackled. But a decreased carbon dioxide content is only one of the global issues that must be tackled in an integrated approach. This includes questions of where future food should be produced, by whom and at what social, environmental and economic costs? This calls for analysis and the drawing of conclusions from the effects of the past to understand what kind of society we should aim for and whether – and how – the current political/economic systems can provide it.

So far, the Food and Agriculture Organization (FAO), the International Fund for Agricultural Development (IFAD), the United Nations Development Programme (UNDP) and the World Bank have not been at the forefront to show how to transform society towards genuine sustainability by indicating visionary opportunities and proposals for sustainable agriculture and forestry. In contrast, the private sector has been more interested in acting and far ahead of most agricultural universities in researching solutions. For example, in 2010, Unilever introduced its Sustainable Agriculture Code as a guide to sustainable farming for suppliers, smallholders and farmers (Unilever, 2010).

Sustainability is a complex concept requiring interdisciplinary research to get integrated solutions to many of its aspects. Such an interdisciplinary approach, dealing with both the environment and food, towards a sustainable food system is not apparent in the current agri-industrial model with ties to a global food industry. Robért *et al.* (2019) created a handbook for sustainability, outlining a structured, strategic approach for handling the concept in an operational manner. The ideas emanate from an early Framework for Strategic Sustainable Development and are used in some Swedish academic teaching.

Another textbook brings together all the relevant concepts for a serious discussion of food sustainability (Behrens *et al.*, 2020). Using case studies, Hunter *et al.* (2020) presented

cross-sectoral initiatives that have promoted local food biodiversity in Sri Lanka. Brazil, Kenya and Turkey. Lessons were drawn on how to combine concerns for conservation, nutrition and livelihoods that could be utilized at the regional and global levels by policy makers.

Of special importance to policy makers is the synthesis of anthropology, economics, history and politics in a critical analysis of the debates on hunger and food systems in the past two centuries (Stone, 2022). Agribusiness and industrial agriculture were questioned and Western knowledge based on new agricultural technologies was not considered as the solution for feeding the world's population in the 21st century.

The issue of sustainability is not an easy task to tackle as shown in an early study by the Swedish Environmental Protection Agency in 1997. It predicted that agricultural and forestry methods could be adapted to be sustainable by 2021. Biodiversity targets were more difficult to achieve, and the European Union (EU)'s agricultural policy was cited as one of the biggest obstacles towards sustainability. For carbon dioxide, a 20% reduction was estimated, while the targets would be fully met for ground-level ozone. The targets for sulphur and nitrogen dioxides, ammonia and nitrogen would be almost achieved. The findings were political dreaming, which agricultural practitioners also stressed. They argued that the study overestimated consumers' willingness to pay for 'special Swedish qualities' and the willingness of the next generation to become farmers.

Usually, most agricultural research has been one-factor research although the term sustainability is used and featured in most research applications, a prerequisite for funding nowadays. Although interdisciplinary research has increased, leading politicians continue to rely on one-factor research when deciding on political actions in agriculture with great complexity. This is strange since farmers – over centuries – have been forced to function as agroecologists, finding practical solutions within a complex area of biology and economics.

Interactions between the sustainability of food systems and a set of 12 key drivers at the global scale were explored by Béné *et al.* (2020). A set of low-, middle- and high-income countries and their relationships to food security and nutrition, the environment, and social and economic dimensions were analysed. The analysis showed an important data gap that characterizes the statistics of national systems in relation to transformation, transport, retail and distribution. Most of the significant correlations between food system sustainability and the 12 drivers appeared to be negative. Most of them were closely related to the global demographic transition that is currently affecting the world population. This conclusion highlights the magnitude of the challenges ahead.

Although the following discussion deals with the agricultural sector, one should realize that changes in one sector are not a definite solution if the whole society is not integrated into a transformation towards sustainability. Furthermore, ad hoc short-term efforts are not a meaningful way of working with business operations based on biology. Although farmers usually master the biological conditions, they must be supplemented by a well thought out and long-term policy within a clear political and economic framework. This is often lacking at both the international and national level.

Since climates and soils differ worldwide, regionally and even locally, a healthy diet must be based on the use and management of local and regional natural resources rather than a global diet. The Eat-Lancet Commission has proposed healthy diets with regard to climate change (Willett et al., 2019). They focus on energy intake, fat, carbohydrates and protein. These recommended diets would lead to increased transport and its consequent environmental impact. Less attention was given to corporate responsibility and transparency. A healthy diet must not only be sufficient but also balanced in terms of quantity, quality and safety. In high-income countries, it is necessary to reduce the amount of total fat and sugar intake to prevent unhealthy weight gain and obesity in the adult population. The consumption of processed meats, such as bacon, salami and ham should be limited. Also, general food consumption can be reduced in high-income societies, where locally produced food should be given higher priority.

The simplistic solution would be that all human beings should adopt a plant-based diet, even though this is unrealistic. Billions of people can hardly switch rapidly to a fully vegan diet. Dairy and meat are deeply ingrained in most cultures. Some environmentalists have argued that individuals should eat less milk and meat to reduce carbon dioxide and save the tropical forest. One example is Brazil which is using deforested areas to expand beef production for export to high-income countries at a much lower price than their locally produced beef. From a global perspective, reduced meat and milk production is hardly realistic and has been irrelevant in low-income countries for quite some time. Livestock is part of life and integrated with rural and urban life in most low-income countries.

Milk and dairy products are crucial sources of iodine, which is necessary for pregnant women and children worldwide (Tattersall et al., 2023). There is a potential challenge for countries heavily reliant upon dairy products, if there is a switch to dairy substitutes (WHO, 2024a). A diet of nonanimal products will increase the risks of conditions such as goitre and hypothyroidism.

Plant-based, nonanimal products are increasingly popular, but a study in the United Kingdom has shown that most plant-based alternatives are not iodine-fortified (Nicol et al., 2023). It was recommended that those following a nonanimal product diet should seek alternative iodine sources such as seaweed, certain types of fish and iodized salt.

In the past, there has been various views claiming that resilience requires a system of many small players and lots of spare capacity instead of large-scale mass production. A focus on capitalism has brought higher living standards at the cost of deteriorating culture and dehumanizing people. This economic system took command of people's lives. It stripped the satisfaction out of work since workers became anonymous cogs in a large machine. Craft skills were no longer important, nor was the quality of human relationships. But capitalism and economic growth have continued, and digitalization and artificial intelligence (AI) have been added as more effective tools for their pursuit by harnessing data.

The notion that there would be resource constraints on economic development and that human happiness would not be achieved through material wealth might have been idealistic in the 1970s but may regain relevance. Many people, especially the younger generation, view the future with some concern and many feel depressed. For instance, increased mental illness in a materially rich Sweden has led to increased prescription of antidepressants. The general increase is 21% over the past ten years, but among children and young people there is an increase of 204% (Folkhälsomyndigheten, 2024).

Some recent academic studies on sustainability are given as examples. One applied a scientific methodology, using big data from various countries with different environmental conditions (Popkova and Sergi, 2022). It highlighted the environmental footprint of sustainable agriculture in a discussion of the adaptability of circular agriculture in relation to food security, sustainability, food exports and imports. The environmental footprint can be defined as 'the effect that a person, company, activity, etc. has on the environment'. It means how much of the environment's natural resources are being used, the amount and types of harmful gases in the atmosphere, and the chemical products that remain undegraded in nature. The overall objective in all activities is to work towards a zero environmental footprint by conserving, restoring and replacing all natural resources that are used in all work for societal development – not only in agriculture.

By using regression analysis, Litvinova and Zemskova (2022) proved that sustainable agriculture and alternative energy, based on high-tech development of entrepreneurship in the agricultural machinery market, allowed the implementation of Sustainable Development Goals (SDGs) 2 and 7 (see Annex II). This was based on findings from countries leading in exports of agricultural machinery in 2020. There was a 77% increase in the use of AI and big data for the development of high-tech businesses in the agricultural machinery market. This was considered to have increased agricultural sustainability by 9.7%. However, it can hardly apply to low-income countries, where traditional knowledge contributes to the sustainable management of natural resources, as shown by numerous other studies.

Through trial and error, communities dependent on fishing, horticulture or subsistence agriculture accumulate knowledge, promoting biodiversity (Kletskova et al., 2022). This was true in traditional fishing and horticulture through learning by experience. In slash-and-burn agriculture, the accumulation of knowledge took longer. Those farmers that were forced to practice slash-and-burn agriculture learned since they had no other choice in order to grow food for survival. But the process could only be sustainable if they had large areas of land to use. In these circumstances, they could return land back to its natural state after 20 years with new, wild vegetation. Another example is the Inca people, who cultivated plants in such a way that we still find many crop varieties remaining in the Andes. In another study, regression analysis was used to study agricultural practices for sustainable development (Bogoviz et al., 2022). The findings indicated that the impact of responsible agriculture on corporate environmental responsibility and climate change was moderately positive only in low-income countries.

Although people may agree that today's society is unsustainable in the long term, few want to accept that achieving long-term sustainability requires fundamental changes. They go far beyond a 'green technology'. It is not sufficient to simply consider the transformation of resources, reduce carbon dioxide emissions and then rely on the market. Green growth is a myth because current climate measures only treat the symptoms by switching to solar and wind energy, nuclear power, digitalization and new technological innovations (Herrmann, 2022). In high-income countries, most people live in societies of abundance, but there might soon be a shortage of water and increasing prices for water, electricity and food. The average Swede consumes 140l per 24 hours (Swedish Food Agency, 2024a) Elevated temperatures can reduce harvests. For each degree increase, the yield of maize decreases by 7%, wheat by 6% and rice by 3% (Reimer and Staud, 2021; Lauterbach, 2022). Only the state can be responsible for rationing and reasonable distribution in these forthcoming shortages.

In a society that requires eternal growth, capitalism will undermine itself, which is why no politician wants to use the word 'renunciation' (Herrmann, 2022). A necessary change in Germany would require its national product to be reduced by 30%. The British war economy of 1939 is suggested as a possible path for a future circular economy. Society shrank, the use of resources was reduced and the state was forced to make cuts. Similar thoughts were expressed about Japan, where capitalism worked when society was developing but it will not solve the current problems (Saito, 2024). Also, electrification, carbon dioxide sequestration and solar cells will lead to further exploitation of natural resources. Instead, there should be slowdown or degrowth as a model for reorienting society around the maximization of public goods as opposed to the endless pursuit of economic wealth. A model for the future must be chiefly based on biology to harmonize with nature, avoiding systematically introducing high concentrations of chemicals into society and further exploration of natural resources. This also applies to the agricultural sector and its future production towards sustainability.

The first requirement is that biology is more important than conventional economics. We humans are part of a large and complicated biological system. The biosphere refers to the mass of all living organisms within the lithosphere, hydrosphere and atmosphere, but it can also refer to the sum of all of the Earth's ecosystems, which house biological life. Life itself is also of a systemic nature, where all the details of different qualities are in harmony with life at large. No single innovation will be the solution to the problem but can contribute – together with supplementary actions – for change.

When sustainability becomes the political governing principle, agriculture will become a more fundamental aspect in contrast to the neglect it has received since the 1970s. The agricultural industry employs millions of people in all sorts of fields, from machinery to inputs, to researchers, to retailers, to packagers, especially in low-income countries. With sustainability, the availability of safe food of superior quality will be a necessity in times of warfare. Russia's invasion of Ukraine disrupted food supply chains and caused geopolitical upheavals. In times of crisis, countries dependent on food imports face serious problems.

Second, sustainability implies agriculture where no more energy is supplied than produced. Reference can be made to the second law of thermodynamics. The working capacity of energy decreases every time an energy conversion occurs. If energy is increased locally, a decrease occurs elsewhere else in the universe. The energy that does not carry out a job becomes waste or pollution. The overall questions would be 'How much energy should a society consume, for what, and why?'

Third, agriculture has been organic for most of its history, until the arrival of synthetic chemicals, fossil-fuel-driven machinery and genetically modified organisms (GMOs). This implies that sustainable agriculture must not leave dangerous chemical residues that remain in the soil, water and overall environment. As far as organic farming in Sweden goes, there is already a shortage of manure, affecting farms in all production areas that produce organically (Larsson, 2021). More farm manure will require more animals, which conflicts with climate considerations that require fewer animals. The political goal is for 30% of agricultural land to be used organically, which is why organic producers can still get 1-year compensation in 2024 in spite of the current market for ecological products appearing to be saturated. Most commercial fertilizer used in Swedish agriculture is imported, making the country vulnerable to external changes that affect supply chains. Before Russia's invasion of Ukraine in 2021, Russia and Belarus accounted for about 22% of the imports of fertilizer to Sweden and the corresponding EU figure was 40% (Jordbruksverket, 2023). Not all organic farming always meets the requirements of sustainability and some farmers believe that subsidies to, for instance, organic farmers may distort the market. Instead, tougher political governance is required to put greater pressure on both the market and the consumers. If the above-mentioned requirements are met, there will be less need for a special category of organic farming. Today's economic situation seems to show that organic growers in Scandinavia are considering returning to conventional farming. It is difficult to recruit new organic growers.

Fourth, sustainable agriculture must be circular to minimize waste and no food should be allowed to go into waste as is currently the case in most industrialized societies.

Fifth, future measurement of agricultural development cannot be confined to monetary terms and yield in kilograms per hectare. Sustainability must be measured in at least three dimensions: environmental, economic and social. This can be illustrated by interesting experiences with these three dimensions from Indian agriculture. There, starting with the

Green Revolution in the 1960s, agricultural innovations have been unequivocally positive from an economic point of view regarding fertilizers, pesticides, integrated pest management, new heavy machinery and high-price incentives for water-loving plants in semi-arid areas. From a social point of view, they were positive for fertilizers, integrated pest management and water-loving plants. From the ecological point of view, the innovations were only positive for integrated pest management and negative for heavy machinery, leading to soil compaction. Such a matrix for measurement can be of future interest rather than the overall focus on yield in kilograms per hectare.

Finally, there are marked differences in designing agricultural research and development for societal development in emerging and low-income countries compared to industrialized countries. Current focus in research is concentrated on technology, such as AI, drones and sensors, for example, as used in precision agriculture. They are currently relevant to large-scale Western-type agriculture but hardly meet demands for sustainability. They are certainly less relevant to most smallholders in low-income countries.

The Future Role of Agricultural Sciences

Overall perspective

Previous innovations have been mechanical improvements (tools, machinery), more effective fertilizers and genetic advances (better seed). Each of them was a solution to one single aspect of the complex agricultural system. The Green Revolution was a momentous change, where several individual innovations were combined with government supporting services such as credit extension and – to some extent – marketing services. In the future there will be more issues, such as rising temperatures in both industrial and low-income countries and the need for sustainability. As George Bernard Shaw noted long ago 'science never solves a problem without creating ten more'.

A working concept of sustainability will be composed of distinctive features. A few features are already familiar while others are to be added, redesigned or offer new potential. Other aspects include fewer regulations, reduced taxes on fuel, improved marketing for new crops and changed grazing legislation. The focus must shift from individual crops to cropping systems, maintenance of or preferably increased soil fertility, and measures to increase biodiversity, for example, introducing new crop varieties from recent botanic expeditions to find new plants for food and medicine.

In the last few years, large food companies have declared that they will deliver products from green, sustainable farms with innovations and marketing. This is hopeful but may be complicated since innovations cannot be based on technology, only.

Sustainable development requires innovations that meet biological, social and cultural needs, and give economic output in both low-income countries and industrial societies. This means an emphasis on biology without neglecting economics, a basic requirement for the future. Moreover, social structure must also be given special attention.

In the past, the focus on technology has left the people outside the technical implementation process. To most farming people it is a fact that many of them will be left out by new technology, at least initially, and especially if not tested in practical farming. A research process is required that is also influenced by farmers at an early stage and not just developed by scientists. There is, however, a complicated research process, where Campbell's law will still be relevant (Campbell, 1979). It postulated that 'the more any quantitative social indicator is used for social decision-making, the more vulnerable it becomes to corruption pressures and the more prone it becomes to distort and corrupt the social processes it is intended to monitor'. This implies that there is often a tendency for decision-makers to make symbolic decisions or introduce cosmetic measures to convince the public of action, even if the decisions are ineffective. Thus, there is a great need for careful consequential analysis prior to any introduction of an innovation in society on a large scale. Also, there is a need for decision-makers with competence, who dare to take relevant, sometimes courageous, decisions.

A food system in harmony with both humans and their environment was the traditional rice culture on Bali and the Tri Hita Karna philosophy. It meant that life should be lived in harmony with other people, the environment and with some spirituality. That way of life sustained a social system for generations, i.e. it was sustainable. After centuries, it collapsed mainly because of increased tourism, consumerism and the modernity of a Western-type life. Biology lost power to economics. In principle, farm entrepreneurs should be active farmers in a lifestyle, close to nature but not excluding new technical innovations. This would be a second requirement for a future environmentally safe life without hazardous chemicals and in a human – and not an industrial – culture connected to an overall agroecological environment.

All countries of the world have made commitments to reduce greenhouse gas emissions and thousands of private companies are themselves committing to net zero targets. Communities around the world plan to plant seeds and forest plants to restore millions of hectares of land. However, decarbonization by itself is not enough, although it will contribute. The trend to reach net zero will hardly work if nature is destroyed in the process, or if people are left behind. The future focus should be on transforming energy and food systems and involving cities in achieving local impact. For each country it will be different – how to produce food for all its citizens, manage its natural resources, generate sufficient energy for this food production, and manage and control the future urban development as well as food security in times of crisis.

Costs of past research

Past research has contributed to strong modern development, benefitting the industrialized countries. People born in the early 1900s have experienced remarkably positive changes during their lifetime in Western countries right up to 2000. The question is whether new generations can expect a similar modernization process in a globalized world without negative consequences. To what extent are the costs of past agricultural progress summarized, calculated and considered in plans for future agricultural research?

Already in the late 1960s, there were some Swedish reports that environmental problems,

such as acid rain and mercury in freshwater fish, could be universal and interlinked. This led to a breakthrough in environmental awareness among the Swedish public. It also led the ruling Social Democratic Party to initiate the first climate summit under the auspices of the UN in 1972. The Stockholm Declaration had 26 principles on environment and development. The most important one not only recognized the sovereign right of states to formulate their own environmental goals, but also gives states a joint responsibility to ensure that national activities do not have negative effects on the environment of other states. Half a century has passed, and states are still struggling with increasing environmental problems.

In addition to old, remaining environmental problems, there are new ones such as climate change and an increasing number of storms bringing extreme precipitation with risk of damage, and new diseases affecting humans, animals and plants. According to the World Meteorological Congress, extreme weather, climate and water-related events caused 11,778 reported disasters between 1970 and 2021, with just over 2 million deaths and US$4.3 trillion in economic losses (WMO, 2023). Over 90% of reported deaths worldwide occurred in low-income countries.

The environmental problem can be demonstrated by a specific example from agricultural research several decades ago. In the early 1970s, the Dow Chemical Company initiated and partly financed scientific work in Sweden on herbicides. During 1966–1970, the studies comprised a literature survey on the effects of picloram on higher plants, soil microorganisms and higher animals, the persistence of picloram in plant tissue, soil and water, and methods for determining picloram residues. The study revealed there were no existing chemical methods for analysis of picloram residues passing through ecosystems (Ebbersten, 1972). Specific investigations focused on picloram persistence, its movement in the soil, determination of its residues and their biological effects on peas and wheat plants. The applied biological method proved to be sensitive enough to indicate tiny amounts of the chemical. These early findings were, however, criticized not only by the chemical industry but also from within the then Swedish Agricultural College and by some academic staff members of the then Department of Plant Husbandry at the college.

These studies were conducted when a herbicide and defoliant were used by the United States military during the Vietnam War. Based on the early scientific findings, Ebbersten (1983) underlined great caution in the use of pesticides for agriculture in both his teaching and documentation during the late 1970s and the 1980s. Instead of being supported to expand that research, he was opposed. Funding for his research on alternative approaches to agriculture was restricted. The academic management preferred pleasing ruling politicians to secure future funding for business as usual. In retrospect, it would have been constructive for decision-makers to continue the dialogue with the Dow Chemical Company. It could have been an early step towards sustainable agriculture, in cooperation with the private sector. Instead, there was an expanding use of insecticides and herbicides.

Such a dissociation also affected other actors and even Swedish authors in the 1960s, who argued that scientists and scientific research results should be given greater political influence. Critics pointed out that the researchers were elitist, undemocratic and politically naïve. In principle, no research on problem-solving should be banned and results from controversial research should be supported and discussed in academic quarters.

Frankness and open discussions are important in science rather than the current trend setting on research agendas of rejecting opposing views. This was tragically demonstrated years ago in the Soviet Union, when Josef Stalin arrested professor and botanist Nikolai Vavilov and replaced him with Trofim Lysenko, an agronomist and pseudoscientist. Opposing natural selection and modern genetics, Lysenko claimed his agricultural techniques could radically increase the national crop yields (Leone, 1952). They were overstated, the experimental results were fake cited and failures were omitted (Rispoli, 2014).

Despite this, the Lenin Academy of Agricultural Sciences announced in 1948 that Lysenkoism was 'the only correct theory' ten years after his appointment as its director. The Academy, state propaganda and access to the Communist Party provided the platform for Lysenkoism against genetics and science-based

agriculture. Other countries of the Eastern Bloc accepted the concept to varying degrees and so did the People's Republic of China until 1956. This example of fake news, remained a policy of official Soviet genetic science until 1949 (Cohen, 2019).

Swedish ruling politicians have, however, taken decisions that inhibited research on nuclear power. A referendum after the Harrisburg nuclear accident in the United States in 1979 led to a political decision to gradually deescalate nuclear power. After the Chernobyl accident in 1986, the Swedish government introduced what became known as "The Prohibition of Thought". In 1987, a provision was introduced in the Nuclear Activities Act (Swedish Government, 1984:3, Section 6). It stated that 'no one was allowed to draw up design drawings, calculate costs, order equipment or take other such preparatory measures with a view to the construction within the country of a nuclear reactor'. The law was criticised since the question of building nuclear power plants was a political issue; not dependent upon whether research takes place or not. In 2006, the Swedish parliament decided to abolish this section of the law. According to the government bill (Swedish Government, 2005), it had inhibited research for improved nuclear safety in Sweden and was perceived as questionable from a democratic perspective. In 2010, the Swedish parliament accepted a new government proposal that new nuclear power reactors were allowed. As of 2024, nuclear power is part of the government's climate plan.

From one-factor research to interdisciplinary research

There is another problem with today's scientific matters. Most current research deals with details rather than societal problems. Already Horgan (1996) has questioned whether science had been reduced to mere puzzle solving by adding details to existing theories. This calls for caution when a researcher is presenting research results for the solution of a societal, multifaceted problem as in agriculture, mining, etc. It is quite different to an entrepreneur in the private sector. The latter is searching for a new product for sale on the market and so are the transnational corporations (TNCs).

For policy-makers, wishing to follow scientific advice in decision-making, it is crucial that they refrain from trusting only one new research result from one-factor research but consider other relevant factors. The other factors may lead to various consequences for the decision and not entirely support the planned political approach. The current focus on specialization requires a degree of management of ignorance (Burke, 2023). A main concern is that those who have power often lack the skills needed for the best decision, while those who possess these skills often lack power.

Science provides the truth through systematic investigations. In the same way that farmers originally benefitted from trial and error, agricultural science is also influenced by mistakes. They bring the scientist closer to a truth after measuring the outcome of all the elaborations. Good science requires scientific methods, source criticism, representative samples, validation and intersubjectivity as a counterbalance to simple quick information from the Internet, fake news and submissions from lobbyists and different stakeholders.

A general problem in current academic life is that funds for research projects are normally allocated for a few years only, seldom sufficient time for any scientific breakthroughs that serve as a basis for obtaining validated knowledge. Thus, scientific findings are exaggerated rather than confirmed by repeated independent studies by other research institutions. Scientists may claim they can provide the solution to a specific problem and politicians frequently argue that their decisions are based on current research.

A solution to a problem area, such as agriculture usually requires scientific input from other disciplines in an interdisciplinary approach. This insight is a requirement towards achieving sustainability. One, or even two, innovations, although effective on a one-factor basis, will hardly be sufficient to change the whole agroecological system in a controlled manner. The same applies to a single recently published scientific article without validation from other scientific sources confirming that research finding. This problem can be exemplified when a senior Swedish researcher within a research area of 'sustainable food systems from a broad and interdisciplinary systems perspective' argued that milking cows must be forbidden since they produce methane gas. Such a simplification neglected the overall context and various ongoing research to limit the release of methane gas from ruminants. This can be a problem to the media, who are often quick to report one new research result, although it may not have been validated by any other source.

It is important to trust science and current facts but there are difficulties in distinguishing between good and bad science, both for experts and the public. Most people, and the politicians, depend on experts who know better what will facilitate well thought out decisions for improvements in society, as well as for the individual. The decisions must not only be based on facts, values and preferences but also take into account the consequences over time. Nevertheless, experts should not be allowed too much control (Frans, 2023). They do not always agree, which is important. They can even be wrong as in case of COVID-19. The number of deaths was exaggerated in Sweden. Immunity was to be reached earlier than it happened in reality. This was later documented by researchers (COVID-19 Excess Mortality Collaborators, 2022) and by scientists in the World Health Organization, explaining mistakes in high-profile mortality estimates for Germany and Sweden (Van Noorden, 2022). Many international organizations initially had difficulties in advising how best to avoid the negative effects of the new disease.

Future major societal challenges are calling for innovative approaches and solutions in various subject areas, such as health, food and the environment, all with a focus on sustainability. These areas must be prioritized rather than investing public funds in a range of individual research projects in narrow subject areas. Since these subject areas are closely intertwined, this requires interdisciplinary research. Instead of funding individual research projects, there is a need for substantial financial investments to be given to some selected national research centres or groups to tackle specific issues and the most relevant agrarian problems for societal development. This requires connections to an effective international network for problem solving. The exchange of scientists in such an international collaboration will stimulate both creativity and productivity.

Moreover, future innovations must be focused on biology. This will exclude much of the current industrial agriculture with its focus on patented seeds and synthesized agrichemicals. This innovation will be done much better by the TNCs. The major issues may not be confined to creating more innovative technology, they will also focus on social and biological issues for societal development. With an anthropologist's eye, Stone (2022) stressed that small farms change and evolve over time demonstrating the need to consider anthropology, economics, history and politics. A viable alternative to current industrial agriculture must be regenerative and recreated by involving active farmers.

This view is based on long-term experience and was highlighted in an intervention made by representatives of the Local Communities and Indigenous Peoples Platform (LCIPP) at COP28 in Dubai in 2022. In defence of their territories and existence, they stated they have been adapting with extreme resilience to climate change over generations. They hold knowledge on science, practices and innovations that all ecosystems and humanity have benefitted from during the whole period of agricultural development. It would be of critical importance to make beneficial use of this knowledge and learn how societies can be sustainable – in the 2050s.

Studies towards a doctor's degree in higher education and scientific debates in agriculture

Research towards sustainability requires that the whole educational system moves in the same direction, highlighting the concept of sustainability, and being based in a basic knowledge of biology. This is of special importance in higher education for students who are embarking upon a career in this complex scientific field. In Sweden, doctoral education has become expensive, and all doctoral students must succeed because it is too expensive to fail for the institution's future budget. A thesis is no longer a completely independent intellectual achievement, since it is normally approved prior to the public defence of the dissertation.

In 2023, the vice-chancellor of the Stockholm School of Economics (SSE) claimed

that it was not possible to rely on the grades when students are admitted to higher education (Strannegård, 2023). He referred to reports of so-called happiness ratings which he considered to be eroding the social contract, equality, trust and the basis that knowledge should weigh heaviest. Thus, the Stockholm School of Economics was considering abandoning the national admissions system and introducing its own system, with red flags for certain schools. A few days later, the Swedish Minister of Education concluded that grades derived by cheating in Swedish schools are a major problem (Johansson, 2023). National tests may be required and a need for the reinstatement of a written and oral matriculation examination, abolished by the Swedish government in 1968. One of its aims, administered by the authorities, was – and still is in, for instance, Finland and Denmark – to achieve equivalence throughout the whole country. It is critical that all students applying for university education towards a doctorate in agricultural sciences have both a theoretical basic knowledge and a practical experience. The implantation of an innovation must be well-suited to practical farming, be profitable and sustainable. This requires good basic education in the compulsory and upper secondary schools.

The PISA programme measures the ability of 15-year-olds to use their reading, mathematics and science knowledge and skills to meet real-life challenges in 81 countries or regions, including 37 out of the 38 OECD countries. Estonia was the number one country in Europe in this survey (OECD, 2023). Estonia and other countries attach significant importance to mathematics and natural science. Sweden was above the OECD average in all three categories, but the Swedish compulsory school system has declined in quality. According to the Swedish National Agency for Education (2022), one in four Swedish pupils did not pass the requirements in one or more subjects, and one in three pupils left Swedish compulsory schools without a final exam to be able to apply for a university education. More than one in four students leave upper secondary school without a degree.

It is critical that science and work towards future innovations are discussed and debated both within the agricultural scientific community and together with politicians, farmers and the public. There are many issues that merit such a dialogue, but the most important one would be an agenda on sustainable agriculture, steps to achieve it, future benefits and implications for the life of an individual. National agricultural universities should initiate such a dialogue but also involve the young generation. Such a common goal to solve critical problems will bring optimism for the future. Information technology will allow many participants in a countrywide network.

One example can be discussions on the growing wild boar population in Sweden. Activists have argued that the wild boar must remain a component of biodiversity as a wild species. According to the Federation of Swedish Farmers (LRF) the cost of wild boars to agriculture amounts to about US$0.1 billion per year (Nihlén, 2020). The calculation used data from the Swedish University of Agricultural Sciences (SLU) based on the size of the wild boar population in 2015 and the cost of one of the more expensive grocery bags in Sweden in 2019.

A recent study has the most comprehensive details to date on Swedish wild boar, their home areas and the use of resources (Augustsson *et al.*, 2024). It concluded that the exact number of wild boar in Sweden is not known, but the population has increased sharply in recent decades, damaging crops and pastures. One conclusion was that when there are many wild boar in an area, they overuse the fields in their search for food and cause damage. This finding would imply there should be no need to shoot a large number of wild boar to reduce the damage to agriculture.

Swedish policy on wild boar has been inscrutable from a long-term perspective. The issue has received little if any qualified attention by the media or the scientific community, despite its importance for biodiversity. Wild boars came to Sweden durng the last Ice Age (Swedish Hunters Association, 2022). Hunting and domestication led to their extinction about a thousand years ago. In 1723, the Swedish king Fredrik I reintroduced some 50 wild boars as hunting prey on a royal hunting farm (Länstyrelsen, 2020). The farmers protest led to their extermination in the 1770s. Then, they were reintroduced into enclosures but escaped. In 1980, the Swedish parliament decided that wild boar were unwanted animals and should be eradicated. Seven years later, the same parliament ruled that wild boar belonged to Swedish fauna without considering its damage to crops and the environment. The Swedish government decides which species may be hunted, and when, from total of 60 game species. Elk, wild boar and roe deer are the most commonly hunted animals in Sweden. Approximately 80,000 elk, 100,000 wild boar and 200,000 deer are shot annually (Swedish Environmental Protection Agency, 2023).

Before the hunting season of 2016–2017 there was no specific government decision to allow hunting of exactly 100,000 wild boar, although the figure was a guideline. There have been various measures and proposals to deal with the increasing wild boar population and the damage they cause. Discussions included the use of drones to make hunting more efficient, protective hunting to reduce damage to agricultural crops and proposals to drastically reduce the wild boar population The decision to include the so-called wild boar package within the food strategy, will contribute to more wild boar meat reaching the market (Sveriges riksdag, 2021).

In 2015, almost 100,000 wild boar were shot in Sweden, a figure that has significantly increased to 158,809 in the 2020–2021 hunting year (Svensk Jakt, 2021). This figure means twice as many wild boar as elk were shot in Sweden and has been much questioned by nature lovers and environmental activists. At the same time, wild boar meat is imported. There has probably been a rise in the wild boar population in 2024 according to early indications from the Swedish Environmental Protection Agency.

The Swedish approach is very different from that of neighbouring countries. When African swine fever was reported to be getting closer to the country in 2018, the Danish government requested that the Danish Nature Agency exterminate the entire wild boar population (Dahl, 2024). It was considered a major threat to Denmark's important pig industry, even though the number of wild boar in Denmark was only about 100. Today, there are no remaining wild boar in Denmark and there is no African swine fever. Fences were built along the border with Germany to stop the boar. In Poland, hunters are employed by the state to kill wild boar and in Germany hunters are paid for any animal they kill. In Italy, the military is called on for the same job. In Norway, they are classified as unwanted animals. Infected wild boar were detected in Sweden in 2023. The Swedish Farmers' Federation has called for a reduction of the wild boar population by 90%. Otherwise, there is a risk that African swine fever will spread to Swedish pig farms.

Plant Breeding and Genetics

Over the centuries, farmers worldwide have selected new plants and varieties better than the old ones in their climatic region, based on temperature, rainfall, light and soils. This process was adapted to the local agro-ecology. Their relatively slow work and the saving of their own seed was speeded up when farmers themselves began to organize this work.

Around 1890 in Sweden, plant breeders were employed at small firms. They made crossings and tested them in trials across the country to produce new improved seed for farmers to buy. This proved successful and major companies were gradually formed; Weibull AB with some government support and Svalöf AB, jointly owned by Swedish farmers and the government. A changing market and the beginning of privatization trends in the 1980s led to a fusion of the two companies in 1993: Svalöf Weibull AB. The German company, BASF was a joint partner for a decade until Lantmännen, a cooperative of Swedish farmers, took over full ownership in 2010. The name was changed to Lantmännen Seed and it is currently breeding crops for 15 distinct species at two breeding stations, one in Sweden and another in the Netherlands.

Moreover, there have been substantive changes in plant breeding. In the initial traditional breeding work, the DNA of the parents was mixed. This led to desired as well as unwanted genes emerging in the offspring. Later, mutation breeding was introduced by radiation or the use of chemicals, causing random mutations throughout the whole genome. Recently, more advanced breeding technologies, such as genetic modification (GM) have been prominent for more efficient hybrid seed production. By adding to, removing from or altering the DNA sequence in a way that does not occur naturally GMOs are developed. This seed production is controlled by the large global companies because these genetic modifications are proprietary.

Agribusiness TNCs play a key role in controlling most of these new research methodologies. They may take over this research globally, attaining a monopoly. Already in 2008, Chinese scientists had proposed investments in the 'Rice Functional Genomics Project 2020' in collaboration with the International Rice Research Center (IRRI). That resulted in a 'Green Super Rice' for irrigated lowland areas with harvests of up to 15 t/ha without substantial amounts of fertilizer and pesticides. The concept and practices of Green Super Rice mean a shift in goals for crop genetic improvement and models, with functional genomics, of food production for promoting sustainable agriculture (Yu *et al.*, 2022).

In contrast to GMOs, genome editing using the CRISPR/Cas9 technique (the genetic scissor), offers more advanced and better methods. It is targeted at one specific location in the genome for a specific change and takes place without the introduction of non-species DNA. For instance, the reduction of one or two genes by the CRISPR technique can stop wheat from taking up cadmium, which causes cancer, or be a way to find varieties of crops that are better suited to a drier climate by changing the DNA sequence of a plant to cause mutations (CRISPR/Cas9). New varieties from this technique may not yet be patentable but the technique requires significant regulatory changes.

In the future, there is a great need for intensified plant breeding for a warmer and drier climate in the northern countries. Changing temperature, more rainfall or more drought conditions have occurred over the centuries depending upon the region, as exemplified by the Roman Empire. It reached its greatest extent in the 2nd century CE when there were favourable climatic conditions during the so-called Roman warm period. There was a significantly higher cultivation limit (a wider geographical area) than today. Then, solar activity decreased and winters became colder, leading to crop failures, starvation and epidemics. When the climate warmed up again, the empire recovered. With another climate change, high-lying marginal areas began to shrink, leaving people malnourished and starving. The period 350–370 CE has been reported to be the driest in the last 2000 years.

A new period with a warmer climate can not only help European production of fruit and vegetables, but also create better conditions for pests, as well as plant and livestock diseases. This scenario will require the increased use of pesticides in the present approach to agricultural production, which in turn can be detrimental to groundwater, bees and food safety, while disease outbreaks in animal production can result in lost export earnings – not a sustainable approach.

Water shortages may become a serious problem as illustrated by the gravity-measuring satellites of the Gravity Recovery and Climate Experiment (GRACE) in the Middle East. The ancient agricultural area of the Tigris and Euphrates river basins – including parts of Turkey, Syria, Iraq and Iran – lost $144\,km^3$ of fresh water from 2003 to 2009. That amount was equivalent to the volume of the Dead Sea (American Geophysical Union, 2013). It would not be the first instance of water scarcity. About 60% of the loss was attributed to the pumping of groundwater from underground reservoirs.

Another example of change is the world's most famous coffee bean, *Coffea arabica*, grown at higher elevations. The bean is sensitive to changes in both temperature and humidity so more breeding is necessary, or the plant type must be changed. It accounts for about 70% of the coffee beans traded globally, and is regarded as having superior quality and giving the coffee a good taste. An option is the robusta coffee (*C. canephora*), often used in the manufacture of instant coffee and blending. It has less aroma and is more bitter but grows well at lower elevations. Even coffee liberica (*C. liberica*) could attract more interest. Consumers' tastes may have to change, even if liberica coffee now has a bitter flavour, though further breeding might solve that problem. It grows well in hot, wet lowland forests, requiring heavy rainfall and hot temperatures. Consumers may have to adjust to another taste than the one they are used to from arabica varieties.

More perennial crops may become of interest in the future, such as wheat. Through breeding work, crops can be developed to absorb more nitrogen and/or that nitrification is inhibited by so-called nitrification inhibitors in some cereals. Nitrogen losses will be minimized and less water is polluted. Chinese scientists have reported on a new method of making a cheap 'biofertilizer'

originating from rapeseed pollen, which is said to increase yields by 50% at less cost (Cai *et al.*, 2008). A more recent overview of biofertilizers in agriculture focuses on soil microorganisms (Mącik *et al.*, 2020)

Another example is participatory plant breeding. It is a collaborative process used in an old, long-term plant breeding programme for potatoes as practiced in the Andean region. Farmers provide local knowledge of soil and rainfall patterns and even early reports of new insects and diseases which may appear with increased temperatures. Such preparedness will be critical in times of escalating crisis, though so far ruling politicians have neglected to act. Without patenting, this maintains biodiversity in contrast to conventional breeding for monoculture. That is why the monoculture approach, as with the Green Revolution, for instance, led to a decrease in the cultivation of hilly fields in parts of the Andes. There, polyculture has been practiced and has provided long-term sustainability for generations. This is in contrast to global plant breeding for monocultures by various actors, including the TNCs.

The examples given above illustrate the need for certain programmes of national plant breeding and seed production to also consider regional climatic aspects in securing long-term biodiversity. This should include all relevant crops. The experiences from Sweden also illustrate the importance of the integration of government support with initial private initiatives for national seed production. Genuine support for Swedish agriculture disappeared with the trend of privatization and for the pursuit of global ambitions. This historical development is hardly the best model for assisting farmers in low-income countries in achieving 21st century food security and food sovereignty.

At the same time, the TNCs currently focus on a few global crops for large-scale industrial agriculture that are profitable to the shareholders. Their monopoly on breeding and seed production may not be revolutionizing agriculture in low-income countries but it may have a devastating impact on small farmers globally.

Scientists at the Monsanto Company were the first to genetically modify a plant in 1982. They used *Agrobacterium* to introduce a new gene into a petunia plant. Some 5 years later, genetically modified tomatoes, resistant to insects and viruses, were planted in the first outdoor trial. In 1996, the first GMO crops became commercially available, leading to an intensive debate which is still going on. The critics have raised concerns about the risks and long-term uncertainties of genetically modified crops. The main concerns involve allergies, cancer and environmental issues (Konov *et al.*, 2005).

The initial idea of a new type of rice appeared in 1991, and a Golden Rice variety was developed (Potrykus, 2012). In 2017, IRRI and the Philippine Rice Research Institute submitted an application for a biosafety permit to the Philippines Department of Agriculture, Bureau of Plant Industry seeking approval to allow direct use of Golden Rice (GR2E) as food, animal feed, or for processing (Asis, 2017). Formal approval was given in late 2019, when the Bureau of Plant Industry claimed it 'has been found to be as safe as conventional rice'. In 2018, Food Standards Australia New Zealand (FSANZ), Health Canada and the United States Food and Drug Administration (FDA) published positive food safety assessments for Golden Rice. In the autumn of 2022, farmers in the Philippine province of Antique harvested a substantial amount of beta-carotene-enriched Golden Rice for the first time; 67t from 17 fields (Ruegg, 2022).

Representatives of agribusiness, especially Monsanto, have argued for increased cultivation of GMOs to, among other things, meet future global food needs and give producers higher profits. Although other biotech companies have developed GM crop varieties, the Monsanto varieties have been the most successful on the market (Benbrook, 2004). This was mainly due to the company's aggressive acquisition of seed companies during 1996–1998. Monsanto bought or merged with most of the major US seed companies to gain control over seed germplasm. It was not able to buy out Pioneer Hi-Bred but sold Pioneer the rights to use Monsanto genes for Roundup Ready soy and Bt maize traits (Center for Food Safety, 2005).

A second factor in Monsanto's success was that it acquired many patents on both the GM techniques and the GM seed varieties. Around 2005, Monsanto provided the seed technology for at least 90% of the world's genetically engineered crops (Monsanto, 2004). A patent is a government grant of a temporary monopoly over a particular invention, usually for a period of up to 20 years. The patent holder can exclude all others from making, using or selling the invention. When a nonengineered crop becomes contaminated with patented traits, it effectively becomes the property of Monsanto through the patent. This could be done by pollen flow or through seed movement via animals or equipment, a direct economic threat to farmers growing nongenetically engineered seeds.

According to the Center for Food Safety (2005), there was another reason to Monsanto's initial success. It required any farmer purchasing its GM seeds to sign an agreement prohibiting the saving of seed. This forced farmers to repurchase Monsanto's seed annually, being regulated in the Monsanto Technology Use Guide. In 2005, Monsanto had an annual budget of US$10 million and a staff of 75 devoted solely to investigating and prosecuting farmers for not following the signed agreement (Center for Food Safety, 2005).

The Technology Use Guide recognized that genetically engineered crops such as maize, are by nature, transportable from a user's farm onto another farm. But the Guide stated that growers using genetically engineered seeds are under no obligation to prevent the spread of patented genetic traits to other neighbouring farms. The requirement to buy seed annually also violated very old farming practices and traditions worldwide: a farmer had the right to save and replant crop seed. Today, the company offers its expert guidance on the farm through a year-round customized service provided by Channel SeedPro, with eleven additional seed brands (Bayer, 2024).

For several reasons, the rate of increase in GM crops slowed down after 2016, one of which was growing protests against the use of GM technology both by farmers and the public. They considered the new technique to be a risk factor in causing cancer. This was more than two decades after the first commercial planting of Monsanto's herbicide-tolerant Roundup Ready soy in 1996. In Europe, the company consider the market to be too small (Cressey, 2013). There were major delays in approval of crops (GM bananas, cowpea and maize) in five African countries. The United States authorities responsible for the approval of GM crops have also suggested looking back on their approaches

Table 9.1. Genetically modified (GM) crops grown by countries on more than ten million ha in 2019 (Statista, 2024a).

Country	Millions of hectares
United States	71.5
Brazil	52.8
Argentina	24.0
Canada	12.5
India	11.9

in their regulations for GM crops and domestic animals, including a major court case on cancer risks caused by the new technology.

After Bayer's purchase of Monsanto in 2018, Bayer agreed in a settlement in 2020 to pay US$10 billion to settle claims that its weed-killer, Roundup, caused cancer (Farrell, 2003). But in late 2023, juries in four separate cases awarded more than US$2 billion in damages to a handful of the roughly 50,000 claims not covered by the 2020 settlement (Sylla and Wolfe, 2024). Bayer is arguing that Roundup does not cause cancer. The glyphosate, developed by Monsanto, is important for both GM crops and conventional crops. China has become the world's largest producer of glyphosate-based pesticides. In 2023, the Asia-Pacific was the largest region in the glyphosate market, but North America is expected to be the fastest growing region (Research and Markets, 2024).

In 2019, the total area of GM crops worldwide was almost 200 million ha. Five industrialized and 24 developing countries planted biotech crops. The top five countries growing GMOs in terms of crop area were the United States, Brazil, Argentina, Canada and India (Table 9.1). According to the United States Department of Agriculture (USDA), GMO seeds are used for planting more than 90% of all maize, cotton and soybeans grown in the United States (USDA, 2023). Countries with smaller areas of GMO crops include, China (3.2 million ha), South Africa (2.7 million ha) and the Philippines (0.9 million ha). The major biotech crops, planted at more than 1 million ha, are soybeans (96 million ha), maize (59 million ha), cotton (25 million ha), canola (10 million ha), and alfalfa (1 million ha) (ISAAA, 2019a).

More than 6 million farmers have planted 11.9 million ha of Bt cotton (13.5% of all the cotton grown worldwide). It is a genetically modified, pest-resistant cotton plant variety that produces an insecticide to combat bollworm (ISAAA, 2019b). At that time, there was no solid evidence of the early concern that pollen from Bt cotton crops may negatively impact honeybees, but more research has been recommended.

The Brazilian Cerrado, the vast savannah area that covers about one-fifth of the country's surface in the interior of Brazil, has rapidly been converted into arable land for soybeans. It has affected water availability and pushed back endemic plant species. In Brazil, studies have shown that GM soybeans can yield on average up to 26% more per hectare than conventional varieties (ISAAA, 2018). Such a yield increase is mainly attributed to the enhanced resistance to pests and herbicides that the GM soybeans possess. This entails a reduction of the area needed for the same soybean production (Gazzoni et al., 2019). Because of the demand to increase soybean production, there is a difficult choice of whether to either double production in the current soybean areas or to recover degraded pasture areas. Soybean-driven deforestation has been concentrated at nearly half located in the Brazilian Cerrado, and in central Bolivia soybeans are replacing the Chiquitania forests. The authorities claim that tropical rainforest is to be protected in line with stringent environmental law, and the private sector is reported to have made commitments to preserve the Amazon rainforest. The exploitation of these areas for soybean production is neither environmentally nor economically sound (Gazzoni et al., 2019).

Soybean cultivation in Brazil has tripled since 2000 and about three-quarters of production is for export – to China. In fact, soybeans were a recognized staple in Chinese food and were domesticated in China before the existence of written records. They were important as a food crop in north-east China, Korea and Japan. They were taken to the United States in 1804 according to Purseglove (1974). Commercial production did not start, however, until the 20th century, initially as a forage and pasture crop.

The largest sugarcane technology centre in the world, the Centro de Tecnologia Canavieira SA in Brazil, has developed the world's first GM sugarcane, resistant to the sugarcane borer

(*Diatraea saccharalis*). Sugarcane is a major crop and is exported to more than 150 countries. There is the potential to plant GM sugarcane on some 15% of about 10 million ha of sugarcane fields, globally. A Bt toxin has been introduced into the plant, acting as a pesticide against the sugarcane borer, as it causes great losses. Brazil's National Technical Commission of Biosafety has approved the new GM cultivar, but it will take some years before shipments of sugar produced from GM crops can reach export markets in countries without strong regulations and objections for importing GMOs (AgNews, 2017).

In India, the harvests of insect-resistant Bt cotton tripled between 2002 and 2015. Bt cotton had been introduced to India in a joint venture between Monsanto and Mahyco just after 2000. Its commercial use started in the United States in the early 1990s. Five years later, Bt cotton was given formal approval in China. In India, Monsanto has been criticized for putting farmers in a position of dependence and guilt.

Mustard might become India's first food crop that uses GM technology. In October 2022, India's Genetic Engineering Appraisal Committee approved the evaluation, in open fields, of GM mustard (Padma, 2022). Several organizations opposing GM crops contested the approval. In November 2022, the Supreme Court of India ordered a suspension of the GM mustard, DMH 11, from environmental release for a four-year period. GM eggplant or brinjal has already been granted a permit in 2010 which led to significant controversy without a solution.

The first genome-edited food is of special importance. Tomatoes (GABA) with CRISPR/Cas9 technology have been available to Japanese consumers since September 2021 (Waltz, 2021). In Sweden, Lyckeby Starch, together with SLU researchers, has developed a starchy potato with the CRISPR/Cas9 method that holds only amylopectin (and not amylase). It provides better storage properties than potatoes with a mixture of both components. Swedish researchers plan to use the same technology for reducing the content of cadmium in wheat.

A common belief with GM crops has been a reduction of the use of insecticides because they can be integrated into the seed. This may be doubtful since the seed is higher in price, patented and growers seldom wish to take any risks with a new seed. Insecticides are applied on 90% of US maize (Benbrook, 2012) and more than 30% of all soybean fields in Brazil (Stewart and McClure, 2021). These toxic products can remain in the soil for a long time and bind water molecules that can penetrate soil cracks. This would hardly be a sustainable approach.

Advances in biotechnology have much affected the agricultural sector. For example, the American Alltech company was established to supply biotechnologically derived ingredients to the animal and poultry feed industry. In 2015, Alltech bought Ridley Inc., an American company that manufactures and markets a wide range of complete feed rations, nutritional supplements and vitamins and minerals. Alltech's owns Optigen II, which provides a slow release of nonprotein nitrogen to the rumen over time. It concentrates the nitrogen fraction of the diet, creating dry matter space for more fibre and energy. It may be classified as a climate-smart feed (Coffey *et al.*, 2016). Long-term studies in nine countries have shown the ingredient is useful (Salami *et al.*, 2020). It could replace vegetable protein, improve beef cattle performance, reduce their carbon footprint and increase profitability.

The Indiana-based BiomEdit company, launched in 2022 by Ginkgo Bioworks and Elanco, received US$4.5 million research grant from the Bill & Melinda Gates Foundation (Marston, 2023). The objective is to develop microbiome-based solutions, reducing the methane emissions of beef and dairy cattle owned by smallholders and pastoralists in sub-Saharan Africa and South Asia.

In this research, specific microorganisms are to be targeted to reduce methane emissions because ruminants are responsible for about one-third of global methane emissions, cattle accounting for about two-thirds of the latter figure. Methane has a shorter life span than carbon dioxide but traps 84 times more heat than carbon dioxide over the first two decades after it is released into the air. This is an interesting scientific approach but there is a lack of clarity about the long-term risks involved and the consequences of introducing a new strain of bacteria into an unfamiliar environment as a component of globalization.

GM foods are very common in the United States but GMOs are controversial. Four out of ten US consumers believe that GM food is less healthy than non-GM food. Nonetheless, all US foods containing GMO ingredients for consumption must be labelled 'bioengineered food' on the packaging as of 2022. The American Cancer Society has concluded that there is no evidence to link GMO food intake to an increased or decreased risk of cancer, but more long-term research is required.

Most European politicians have mistrusted the original method of GM and GMOs have been banned. Only one GM maize crop has been allowed within the EU and it is grown commercially in the Czech Republic, Spain and Portugal. In 2021, the European Commission opened a dialogue on new genetic methods, based upon on a study on new genomic techniques (NGT). But issues raised in the study did not 'require evidence-based answers about the social, ethical and ecological consequences of the use of NGT'. Because modern technologies have an impact by interacting with both society and the environment experts in the social or ecological sciences should also have been consulted for their views on that study.

The Court of Justice of the European Union has decided that the CRISPR/Cas9 technique is not part of the EU legislation on GM crops. A basic assumption has been that the method does not need to be of relevance to environmental risk assessment that applies to GMOs, for example, in the United States and by the Swedish Board of Agriculture. Data in Höijer *et al.* (2022) showed, however, that DNA had changed after the 'genetic scissor' was used to affect the genome of fertilized eggs. Such unexpected mutations could also be passed on to the next generation, an indication calling for some caution. Mutations can also appear after using the CRISPR/Cas9 technique. Still, this approach might have exciting potential for the future when appropriate regulations are in place and long-term consequences are identified.

According to a study in *Nature*, a research group linked to the Chinese Academy of Sciences has pointed out that the EU may miss a technology with exciting potential to reduce nitrogen use in agriculture (Jia *et al.*, 2021). The European Parliament is to vote whether the CRISPR/Cas9 technique is to be allowed to meet climate challenges. However, the draft law does not include any specific labelling, traceability or monitoring of safety. In early 2024, the

European Parliament adopted by a narrow vote (307 votes in favour, 263 against and 41 abstentions) a report from the European Parliament's Environment, Public Health and Food Safety Committee on the European Commission proposal for a 'new genomic techniques' (NGT) regulation. It would create a new class of GM plants whose genomes have been edited with precisely targeted new laboratory techniques. Such new crops would be treated as broadly equivalent to conventionally bred strains. But patenting of the NGTs is a remaining issue to be resolved by the EU (European Parliament, 2024). In late 2023, the United Kingdom was the first country to approve a medical treatment based on the CRISPR/Cas9 or genetic scissor method. The gene therapy shall apply to patients over the age of 12 who suffer from sickle cell anaemia and thalassemia. The CRISPR/Cas9 technology may be of special interest for medical advancements.

If Chinese developments for GM food crops become hesitant, as in Europe, interest among African politicians and scientists may attract the global agribusiness. Africa has half of the world's untapped cultivated land, a large agrarian population and US$35 billion worth of food imports. In addition, the African population is predicted to double to 2.6 billion by 2050. That is more than half of the expected global increase. So far, only South Africa produces GM crops, while other African countries have approved well-controlled experimental crops (Nigeria, Ghana, Ethiopia, Cameroon, Tanzania).

Plant breeding is a complex activity with patents, cross-breeding agreements, licensing agreements and international treaties taking place in a growing global context. Gene sequencing offers great opportunities to find new gene bank samples with consideration of the replacement rules. The International Treaty on Plant Genetic Resources for Food and Agriculture (ITPGRFA) was finalized in 2001 but needs more clarity to help towards any future and possible increased cooperation with the TNCs. Its purpose is to promote the conservation of plant genetic resources, to protect farmer's rights to access, and have fair and equitable sharing of benefits arising out of their use. The Nagoya Protocol of 2014 functions as a legally binding instrument to set regulations on access and benefit sharing in biological diversity. The aim is to create an equitable distribution of the benefits of utilizing genetic resources (FAO, 2001). So far, there is little evidence that nations rich in biodiversity have received adequate compensation, which was already politically agreed at the Rio de Janeiro Conference in 1992.

Despite global development and technical advances, national states need their own, independent and efficient plant breeding and seed production. It must be adapted to an individual country's different ecological niches, especially in times of crisis when the focus will be on food security and sovereignty. Each country needs a secure supply of seed and access to animal genetic resources and breeding animals for its own breeding of crops and animals, as well as seed production, to supplement and make full use of international breeding efforts. In principle, this requires the involvement of farmers, marketers, processors, consumers and policy makers.

The Introduction of New Crops

When the temperature changes, the whole of nature is influenced in a continuous process. Crops have different temperature requirements, just like different animal species, and insects in particular. The higher the temperature, the more insects from southern latitudes, while those that have been present before may gradually die out. The same applies to fungal diseases and bacteria. Everything is part of a systematic biological cycle, which occurs naturally over a long time. This also influences biodiversity. Rising temperatures and higher rainfall mean, generally, that new crops may be introduced. Likewise, periods of longer drought will facilitate the introduction of drought-tolerant crops such as sorghum and millet. There may be a range of new crops, but also some underutilized crops (Annex V). Tropical regions may face more difficulties.

There is great potential in new crops. Some 30,000 plant species are identified as edible out of the 250,000 species of flora. Only 120 plants are cultivated, and eleven crops satisfy 75% of human nutritional needs. They are wheat, maize, rice, potatoes, barley, sweet potatoes, cassava, soybeans, oats, sorghum and millet. Moreover, wheat, maize and rice make up more

than 50% of the daily calorie intake for the global population (Shelef *et al.*, 2017). This implies that there is ample potential in looking for additional plants that can adapt to changing climatic conditions. So far, investigations of new crops have been marginal, particularly in areas with rich biodiversity where nature has done this task for thousands of years.

Intermediate wheatgrass or perennial wheat (trademarked as Kernza) was developed by the Rodale Land Institute from *Thinopyrum intermedium*, a relative of annual wheat, native to Europe and Western Asia. It has an impressive root system, yielding seeds and forage for both grazing and browsing, and ecosystem services by way of erosion control and adding carbon to the soil organic matter. Average seed yields are low at about 500 kg/ha. So far, hybridization attempts with durum wheat have usually meant the loss of their perennial capacity while other desired characteristics remained in the hybrids. Kernza contains higher values of protein, ash content and dietary fibre content compared to wheat (The Land Institute, 2024).

Industrial hemp, which lacks tetrahydrocannabinol (THC), can become a possibility in regions with higher temperatures. THC and tetrahydrocannabinol are both abbreviations for delta-9-tetra-hydro-cannabinol, which is the main active substance in marijuana. Industrial hemp absorbs more carbon dioxide than other crops and market prospects are good. The same harvest can provide fibre, biofibre and nutritious food for humans and animals. It can replace steel through hemp composite, and cement and concrete via hemp lime and oil. It follows the old plans of Rudolf Diesel and Henry Ford, 'to let automobiles grow out of the earth'. Hemp can replace plastic and it provides three times more fibre than cotton.

When the EU made industrial hemp legal in 2001, the first crop was sown by a Swedish farmer in southern Halland and an agricultural company on the island of Gotland. This gave rise to narcotics offences under the Government's Ordinance (1992:1554) on the control of narcotics. Section 2 of the Ordinance states that substances listed in Appendix 1 of the Ordinance are to be regarded as narcotics under the Narcotic Drugs Punishment Act. It states the definition for cannabis, where certain types of industrial hemp are excluded. Finally, the

European Court of Justice in Brussels gave the farmers the right to grow industrial hemp and ruled 2 years later that industrial hemp is not a narcotic. The Swedish government decided to change the law on the control of narcotics so Swedish farmers received permission after 2003 to commercially grow industrial hemp on Swedish farmland. In 2022, hemp was cultivated on some 220 ha, a modest expansion. In conclusion, industrial hemp are certain specific varieties of cannabis that may be eligible for EU funding and are grown after the application and authorization have been granted. If this is not the case, hemp cultivation will be considered drug production. Reference is made to the EU legal provisions for industrial hemp. Approved hemp varieties in Sweden are listed on the Swedish Board of Agriculture's website. All approved varieties have a tetrahydrocannabinol (THC) content below 0.20%. A level above this is, therefore, not approved and is a criminal offence.

Elsewhere, the legalization of hemp cultivation in various countries has led to the growth of the global industrial hemp market. France had 20,000 ha dedicated to hemp cultivation in 2020, the largest area in Europe, followed by Germany in the second place with 5350 ha (Trenda, 2023). According to a report by Research and Markets (2022), the global industrial hemp market size is estimated to be valued at US$6.8 billion in 2022 and is projected to reach US$18.1 billion by 2027.

Certain observers believe that the market for vegetarian meat alternatives will grow. Health-conscious consumers may prefer other alternatives rather than the health risks associated with excessive consumption of red and processed meats. This will require more plants and plant-based meat substitutes are already a billion-dollar industry. By 2023, the global market had reached US$7.1 billion and is expected to grow by some 19% from 2024 to 2030 according to Grand View Research (2024). In general, the cultivation of vegetables will expand globally. This will require more cultivation of pulses. However, vegetarian meat has two challenges: the taste and the price.

Other new crops in the Nordic countries may be sunflowers and maize, the latter as more than forage. Grapes are already grown in southern Sweden. However, viticulture requires chemicals – the rules stipulate about 60 permitted substances in conventional viticulture but fewer in organic wine production. In recent years there have been no levels of pesticides higher than the permitted maximum level. The problem is greater for imported products, where data are mostly missing.

Underutilized and Poorly Researched Minor Crops

There is a group of small grains and pseudocereals (seeds consumed like grains) of ancient origin. For many years, these crops have been considered as neglected and underutilized crop species. Still, they are dietary staples in many parts of the world, such as China, India, Africa and the Middle East. These minor crops have been re-evaluated because of their high content of protein, fibre, micronutrients and bioactive compounds. They may show potential for the future (Annex V).

Today, new food-processing technologies and products are being developed to encourage companies to process crops native to the Andean region and to increase their consumption. Current research focuses on the reduction of antinutritional factors since they lower the quality of such crops and to improve the baking quality (Pontonio and Rizello, 2019). With rising temperatures and less rainfall, several of these crops can be used as alternatives and they are a relevant step towards sustainability. The Green Revolution certainly delivered new calories globally but led to nutrition problems because of the displacement of traditional crops, for example, in India (Chand *et al.*, 2022).

As long ago as in ancient Rome, scholar Marcus Terentius Varro recommended planting legumes in poor soils, in his book, *Rerum Rusticarum* (37 BCE), as they do not require many nutrients. Such crops offered not only immediate returns in the form of grain, but they also enriched soils for subsequent crops. Also, minor crops can be healthier than conventional ones. Thus, they have gradually become of more interest in Western countries since they are less processed. Some examples are listed in the following paragraphs.

Sorghum (*Sorghum bicolor*) is the fifth most-consumed grain worldwide. The FAO reported it was cultivated on some 40 million ha in 2021. The largest producer in the decade spanning 2012–2022 was the United States, followed by Nigeria. These two countries remained the largest producers worldwide in 2023–2024 when the United States remained the largest producer of sorghum worldwide (Statista, 2024b). Vigorous, drought-resistant sorghum can tolerate a wide range of soil conditions. It is a gluten-free staple in Africa and India, and used as fodder in China. It is rich in nutrients and a useful source of polyphenol plant compounds, which function as antioxidants.

Finger millet (*Eluesina coracana*) has been grown for around 7000 years. In 2019–2020, millet was produced on some 71 million ha according to FAOSTAT (2021). Today, about 20% is produced in India of the Afro-Asiatic type, whereas a highland type is grown in Africa. Millet prefers sandy loams and red laterite loams. Millet can be kept in storage for many years without weevil attacks, thus, making it an important famine crop. It is a nutritious and gluten-free crop used in the diets of about 90 million people in Africa and Asia.

Teff (*Eragrostis teff*) is the world's smallest grain. The gluten-free grains are rich in nutrients, iron and magnesium and have plenty of vitamin C. In Ethiopia, iron deficiency anaemia is rare, due to high consumption of teff grains by Ethiopians.

Fonio or hungry rice is a grain of African origin. It is grown throughout the savannah zone of West Africa, from Senegal to Cameroon. There are two types with small grains: *Digitaria iburua* is grown in Nigeria and *Digitaria exili*. The latter has existed for about 5000 years but has received little scientific research. Fonio competes well with weeds, grows well in poor soils and during drought conditions, and demands little fertilization. Fonio has moderate amounts of fibre and protein and is naturally low in cholesterol, sodium and fat. It provides many B vitamins and minerals such as calcium, iron, copper, zinc and magnesium.

Diversification of Crops and Animals

Crop diversification into high-value crops (HVCs) can be a strategy to improve livelihoods for farmers and their families. Studies have shown that households diversifying towards HVCs are less likely to be poor, the biggest impact being for smallholders. However, such growers need to allocate at least 50% of their growing area to HVCs to escape poverty (Birthal *et al.*, 2015).

Contract farming has the potential for crop diversification beyond its common approach to plantations crops. One example is in fruit and vegetable production for export or sale to domestic supermarket chains, for example, in Madagascar, Kenya and Senegal. This may be more profitable than conventional food crops. This took place with contract farming in the Mekong countries of South-east Asia with crops that include rice, pepper, cashew nuts and fruit. It gave farmers benefits, such as set prices, access to credit and high-quality seeds (Pichdara, 2021). In Ethiopia, a growing demand for beer has led barley farmers to go into a contract agreement with the Assela malt barley factories.

For quite some time, contract poultry production has taken place in Bangladesh, India, Indonesia and Thailand. It is large-scale production of one species only, similar to poultry production by the large company, Tyson Foods Inc., in the United States. Chickens are produced very quickly, usually under poor living conditions and animal care, and often needing additives and antibiotics.

In milk production, a dairy processor is often coordinating the supply of milk to consumers, whereas farmers produce the milk. But various reforms in India have transformed the previous milk sector. Private processors, offering contracts, have grown dramatically relative to the previously dominant dairy cooperatives (Birthal *et al.*, 2008). This transformation process generally means less income for farmers, higher transportation costs, and thus higher prices for most consumers and less sustainability. This trend has continued. Different animal species in combinations may also contribute to better utilization of the natural resources.

Soil Fertility

Farmers are entrepreneurs and easily apply chemicals for the next growing season to ensure good crops and profits. Thus, they see no need to care for long-term soil fertility. Chemical fertilizers are highly effective and nowadays necessary to maintain global food production. The important Haber-Bosch process allowed the synthesis of ammonium nitrate fertilizer on an industrial scale, which increased crop yields. However, as a consequence, it has – over time – also led to a loss of soil fertility and organic substances. The critical issue is whether there are alternatives for the future compared to the current reliance on chemical fertilizers.

In the ancient scriptures of China, Korea and Japan, covering more than 3000 years, reference is made to organic fertilizers. In China, it could be animal/human excreta, urine and cake manure. In fact, chemical fertilizers were only imported and first used at the beginning of the 20th century on tea and cotton on the southern Chinese coast.

Egyptians, Romans, Babylonians and early Germans are all recorded as using minerals and/or manure to enhance the productivity of their farms. The Incas used fertilizers, but they varied within the regions of the empire. In the Cusco valley, human manure was used on the maize fields after being collected, dried and kept as a powder. Animal manure was applied to potatoes in the higher and colder regions. Along the coast, seagull droppings (guano) from the shore were collected to be used. The birds were living on nearby islands and it was forbidden to kill them according to Inca rules. Those who did so during the laying season could be punished by the penalty of death. Also, fish was used as a fertilizer in the early 1600s. The use of wood ash as a field treatment was widespread (Scherer et al., 2000). In the 19th century, this guano was brought in massive quantities from Peru and Chile (and later from Namibia and other areas) to Europe and the United States.

The great loss of grasslands and the permanence of monocultures decreases soil fertility that normally takes a long time to build up. It may take half a century to build up to good soil fertility, depending upon the climatic conditions, type of soil, access to water and the cultivation practices. Several components are involved such as nutrient content, with a balance of essential macro- and microelements in the right quantities. High-quality soil contains ample organic matter (humus), which enhances soil structure, nutrient availability and water retention.

Beneficial microorganisms, such as fungi and bacteria, break down organic matter, releasing nutrients for plants. Another component is the soil pH, a measure of its acidity or alkalinity. In general, most plants prefer a neutral to slightly acidic pH range. Finally, a good soil structure will allow for proper root penetration, water movement and aeration which can prevent compaction and runoff. Altogether, these are the requirements towards sustainability. In the United States, soil is lost 10–100 times as fast as it takes to produce it (Handelsman, 2021). An illustration of the long-time horizon regarding soil fertility are Mollisols. The nutrient-enriched topsoil, with high levels of organic matter, can be between 60–80 cm in depth. This fertile surface horizon is developed by the long-term addition of organic materials derived from plant roots and soft, granular soil structure. It is a real carbon sink.

Mollisols occur in savannahs and mountain valleys in Central Asia and the North American Great Plains. They have been exposed to fire and abundant pedoturbation from ants and earthworms for many years. Mollisols represent 29% of agricultural land and are considered to be one of the most fertile soils in the world. On a global scale, the highest restoring soil carbon concentrations and pools are found in Mollisols from Eastern Europe (including Ukraine and Western Russia) and Asia, while the lowest concentrations are found in Mollisols from South America (Labaz et al., 2024).

These environments, in their natural form, are decreasing and Mollisols have lost about 50% of their antecedent organic carbon pool due to soil erosion, degradation and cultivation activities (Xu et al., 2020) Therefore, restoring soil organic C (SOC) to Mollisols via appropriate management is crucial to sustainable development. The FAO has stressed that such a fertile topsoil could vanish in the next 60 years and 90% of the world's Mollisols will be at risk by 2050 (UN, 2022).

Mollisols are the primary soils in Ukraine and they need to implement conservation tillage and other sustainable land management to reduce soil degradation (Kravchenko et al., 2018), But the war in Ukraine hastens the degradation process since the landscape is destroyed and any upgrading effects are not very likely during warfare.

A long-term perspective is important when considering soil fertility and for national and global food production. It requires an understanding of production biology and relevant conditions from economical, ecological and social perspectives. The focus on quick economic gains means little or no attention is given to efforts to increase the soil humus content for better microbial life, improving soil structures for increased water retention and the general fertility of the cultivated area. Soon, water shortages may become a general concern of significant importance. One fundamental problem is that not enough attention has been given to the need for sufficient drainage, together with a sufficient water supply. Intensified irrigation in the Nile Valley has increased the prevalence of schistosomiasis, or bilharzia. Similar effects are visible in other agricultural operations, for example, in Iraq, Congo, Pakistan, the Mekong Delta and China.

To obtain sustainability, there is a need to rebuild soil fertility. It may require a return to crop rotation, including more grassland with both clover and grass species. Pasture for cattle and/or the use of more meadows is a long-term model, with carbon sinking. It can also justify continued production and consumption of meat since grazing produces fewer greenhouse gas emissions. History tells us that crops and animals have been – and are – complementary. According to some archaeological evidence from both the Middle East and Europe, cropping and herding developed in tandem, and were early demonstrations of the use of organic manure in farming.

Nowadays, the American Farmland Trust uses a metric (PVR) to rate the quality of farmland. PVR stands for productivity, versatility and resiliency. If a rating of the farmland is above 0.65, then it makes great farmland. Currently, the average is 0.43 PVR in the United States (Freedgood, 2020). Nonetheless, the use of resilience in large-scale American agriculture is surprising because resilience requires a system of many small players and spare capacity. Farmland with a high rating is being lost disproportionately quickly and there are already some discussions about US food deserts. This means that suboptimal farmland must be used, requiring more water, energy, fertilizers and pesticides, and more transportation to be productive – and

sustainable. Well-operated farms care for the soil, air and water which produce viable ecosystems.

The other metric of soil fertility is of ancient origin and quite simple. More than a century ago, Charles Darwin was among the first to realize the importance of earthworms. Since 2000, earthworms have decreased (33–44%) in the United Kingdom according to a presentation at the Annual Meeting of the British Ecological Society in early 2023 (British Ecological Society, 2022). As of today, few monoculturing farmers count earthworms in their soils, and they would be surprised to find any.

Biological nitrogen fixation may dispense with some nitrogen fertilizers and the contamination of groundwater with nitrates. No-till farming would conserve the soil that retains water and fixes carbon and may stimulate integrated pest and disease management techniques. But the periods may be too short. This calls for breakthroughs in engineering nitrogen-fixing cereals for nodule organogenesis and infection by nitrogen-fixing bacteria. Although there is a massive body of knowledge, the process of engineering the nitrogen-fixing nodulation trait in non-leguminous crop plants has, so far, given little concrete results (Huisman and Geurts, 2019). Some potentially new biotechnological approaches for incorporating biological nitrogen-fixation capacity into non-leguminous plants have been suggested (Guo et al., 2023).

Chemical fertilizers will also be important in the future but there is a need for more use of crop rotation with legumes and grassland, and the use of animal manure from livestock production. More elaborate systems are needed for nutrient recirculation as manure from cities to the countryside and for better management of the rapid growth of megacities.

The first commercial production of artificial fertilizers in the early 20th century was fossil-free, so this approach might again be possible. This will be necessary because future fossil-free fertilizers will be much more expensive than conventional fertilizers. Recently, researchers at the Federal Institute of Technology (ETH) in Zurich and the Carnegie Institution for Science in Stanford in the United States investigated carbon-neutral production methods for nitrogen fertilizer (Rosa and Gabrielli, 2022). Their findings confirmed that such a transition is possible.

One approach would be using fossil fuels as in business-as-usual but to capture and permanently store the greenhouse gas emissions underground with special infrastructure. It would be efficient but will require more energy and will not reduce dependence on fossil fuels. Another approach would be to electrify fertilizer production by using water electrolysis to produce the hydrogen. This would require much more energy than natural gas. It will demand enormous amounts of electricity from carbon-neutral sources and lead to competition for electricity. A third approach would be to synthesize hydrogen for fertilizer production from biomass. It requires a lot of arable land and water, only available in certain countries. Unless waste biomass can be utilized, this method will compete with food production.

A future approach is likely to be a combination of different methods, depending on the country, the specific local conditions and the available resources. One alternative may be to avoid overfertilization, as is quite common nowadays. In conclusion, all carbon-neutral methods of producing nitrogen fertilizer are more energy intensive than the current use of fossil fuels. This will change farming methods and the relative prices of different agricultural products.

The global trade of food will be reduced, and consumers must accept paying a higher price for food in the future. Price shocks may also have an influence, though more on electricity than on natural gas markets.

In contrast to public research, the private sector has already taken steps forward on the decarbonization operations that are part of the current political process of combatting climate change. Examples of modern technologies for carbon-free production can be found from the German company BASF's work on a gradual conversion to energy from renewable sources (BASF, 2023). Steam reforming is one common way of obtaining hydrogen, but the company is testing an alternative process – methane pyrolysis which is virtually carbon-free – in Ludwigshafen, Germany. Other focus areas include plans to produce a proton exchange membrane water electrolyser in Ludwigshafen for carbon-free hydrogen production and a carbon capture and storage project at the Antwerp site in Belgium.

Leading fertilizer companies and African leaders are searching for practical solutions, including the Sustain Africa Initiative. It aims to support 5 million farmers, hopefully providing food security for 50 million people. It was planned to operate in some ten countries in 2023. The initiative is funded by Rabobank, The Bill & Melinda Gates Foundation, the International Fertilizer Association and the African Fertilizer and Agribusiness Partnership, with the support of the Alliance for a Green Revolution in Africa. The war in Ukraine and its effects on food security makes such commitments even more acute.

Yara International, formerly Norsk Hydro and established in 1905, is a leading global fertilizer producer of nitrate-based mineral fertilizers produced in the European Union and Norway with a carbon footprint claimed to be about 50–60% lower compared with most non-EU fertilizers. This is explained by Yara using the best available technology to develop a catalytic process. Now, Yara has 26 production plants, operating in 60 countries with a portfolio of green ammonia projects in Norway, the Netherlands and Australia (Yara, 2022). Instead of using natural gas to produce ammonia, the company will use renewable energy, such as Norwegian hydropower. It is claimed that such fertilizers will have a 75–90% lower carbon footprint compared to the same fertilizers made with natural gas. Moreover, the carbon footprint of many food products can be reduced by between 10–20% and, in particular, by 12% for bread.

In cooperation with Lantmännen, northern Europe's leading agricultural cooperative, Yara began testing the commercial viability of green fertilizers in 2019 (Yara, 2022). The collaboration led Yara and Lantmännen to sign an agreement 3 years later for the development and commercial launch of fossil-free, mineral fertilizers, produced with renewable energy. It is claimed that they will reduce the climate impact from grain cultivation by 20%. Yara has signed a similar agreement with the German Bindewald & Gutting Milling Group and Harry-Brot, the German market leader in bread and bakery products, to get high-quality flour to consumers by using the new fertilizers. Contract farmers will use the new fertilizers in the 2023–2024 growing season (Yara, n.d.).

Fertiberia is another company with extensive experience in the manufacturing, operation and logistics of mineral fertilizers and ammonia. It is a consortium of 13 leading companies operating throughout Europe including Spain, France, Portugal, the Netherlands, Sweden and Greece. Nowadays, it is a leader in the production of fossil-free, mineral fertilizers using an electrolysis technology based on hydrogen. Its goal is to achieve net-zero emissions by 2035. This is to be achieved by designing biofertilization and biocontrol solutions aiming at precision agriculture and organic farming through agrotechnology, biotechnology and industrial innovations (Fertiberia, 2024).

In 2024, Power2Earth was announced as a new investment initiative by Fertiberia, Lantmännen and Nordion Energi with the objective of establishing Sweden's first fossil-free mineral fertilizer production. It is assumed that by using the new fossil-free fertilizers in wheat cultivation, the impact on the climate will be reduced by some 20% (Lantmännen, 2024).

Power2Earth will be an addition to the industrial landscape in northern Sweden, where hydrogen will be central as a source for energy. Nordion Energi is the transmission grid manager for the Swedish gas transmission system and the infrastructure for biogas and hydrogen needed to secure future energy supply and carbon capture operations. The hydrogen will be derived from water through innovative electrolysis technology (Power2Earth, 2024a).

Power2Earth is also the first step on a 1000 km underground hydrogen pipeline connecting Sweden and Finland. The new hydrogen infrastructure is claimed to be cost-effective, deliver energy security and reduce carbon dioxide emissions. The investment and other industrial activities will depend on secured energy – a huge challenge. The Markbygden Wind Farm is a series of interconnected wind farms in the Markbygden area of northern Sweden. It is the largest onshore wind farm in Europe with over 500 wind turbines (Power2Earth, 2024b).

The investment project for fertilizer production started in 2021 with feasibility studies and the acquisition of land for a factory in Luleå. An environmental permit application is to be submitted in mid-2024. Estimates indicate the potential to reduce emissions of carbon dioxide from Swedish agriculture by about 25% by producing 1 million t annually of fossil-free mineral fertilizers. The renewable energy is to

come from wind and hydroelectric sources in northern Sweden. This will reduce the country's complete dependence on fertilizer imports and strengthen domestic food production. The production is planned to start in 2028 (Bioenergy International, 2024).

Some interesting conclusions can be drawn from these recent developments. Large companies are taking steps towards decarbonization which require vast amounts of energy. This means that countries rich in water and extensive areas of land for solar and wind energy may benefit from decarbonization. In countries with electrification via renewables or the use of biomass dependence on natural gas imports will be reduced.

If fertilizer production is decarbonized, food security may be increased according to Rosa and Gabrielli (2022). Countries dependent on imports of nitrogen or natural gas, such as India, Brazil, China, France and Turkey, are at higher risk. This would contrast with countries producing their nitrogen fertilizers from their own natural gas, exemplified by Russia, Egypt, Qatar and Saudi Arabia. This new situation calls for immediate government action to ensure that that suitable fertilizers are available in time and also affordable to millions of smallholder farmers globally.

Another concluding remark relates to the agreed partnership, in the example, between the private sector (Yara) for research and development and the farmers (Lantmännen) as the ultimate users of the new product from research. This illustrates the need for cooperative arrangements for a transformation of the food system. It is advantageous compared to classic public research that searches for one separate innovation. Yara and Lantmännen are also collaborating on projects on crop nutrition, new farming practices and digital tools in a transition towards decarbonization.

Cropping Systems, Agroforestry and Regenerative Agriculture

Mixed farming

Mixed farming is of ancient origin, being practiced by the Harappa people of the Indus Valley civilization in central Punjab. Nowadays, mixed cropping is common in smallholder production in low-income countries. Five different crop species can tackle different attacks by different parasites and insects in a cropping system which is one component, alongside the rearing of animals. Major food crops in India are wheat, millet, maize, and pulses such as beans and peas. Cash crops include oilseed, sugarcane, coffee, tea, jute, rubber and various horticultural crops. Weeds can be overcome and the need for pesticides can be reduced. Sustainable agriculture requires a transition to cultivation methods with reduced tillage by plough, immediate replanting after the harvest of a main crop, crop rotation systems to bind nutrients into the soil and the return of organic matter to arable land, plus more use of agroforestry and terracing. These possibilities have long existed in developing country agriculture, which is more focused on agroecology than current Western-type food systems.

Recent increased agricultural production in Brazil is associated with new crop management recommendations, the genetic potential of cultivars and the combination of degraded pasture areas into more efficient systems through Integrated Crop-Livestock-Forestry Systems (Gazzoni et al., 2019). Innovative technologies have allowed pasture improvement by inserting agriculture into soil recovery. Cropping systems with new crops offer future potential, even using annual and perennial crops in an integrated system that are capable of demonstrating long-term sustainability.

In general, sustainable cropping systems include animals, where cows are grazing on a minor scale and may not produce maximum amounts of milk but give higher meat quality. Cow dung is used on other land and no extra fodder is bought or needed.

With regard to animals, there are many claims that the solution is to break the habit of consuming animals. This would stem the hunger for more land and cut climate emissions (Rowlands, 2021). About one-third of cropland is used to grow feed for livestock. But several research groups and companies are working on feed compounds that can suppress the formation of methane in cows' stomachs. One example is 3-NOP, from a Dutch company, DSM. In tests,

it reduced methane emissions by 30% (Scott, 2019). Another example is when the United States company, Cargill Inc., and the UK company, Zero Emissions Livestock Project (ZELP), joined forces in 2021 to develop a technology designed to neutralize methane as cows exhale. As a first step, ZELP has created a wearable filter to be worn by cows like a mask (Stark, 2021). It will neutralize the methane gas from belching cows which will decrease the methane emissions by about 50% according to the company.

The European GlasPort Bio biotech company is developing an additive (GasAbate N+) to store cow manure (BioRefine, 2021). This will inhibit the microbial activity and the production of methane. The American company, Windfall Bio, is selling methane-consuming microbes that are grown in the company's fermentation vats. They are delivered to farmers dried and packed. The microbes are released into an environment with a higher methane concentration, such as a dairy barn or a manure storage container. The microbes both consume the methane and capture nitrogen from the air, thus producing an organic fertilizer (Ma, 2024).

The Arla company tested an innovation on 10,000 dairy cows in Sweden, Denmark and Germany in 2022. They were given a methane-reducing feed supplement from the global company Royal DSM NV. It works by controlling the enzyme that triggers methane production in the cow's digestive system. Arla claims that it is scientifically proven that it does not affect the quality of the milk (Arla, 2022). So far, there is little information on the costs for the producer and the consumer.

Bar 20 Dairy is a 7000-cow dairy farm of some 2000 ha in the San Joaquin Valley in California. It uses a biodigester to capture the methane from cows' manure which then generates energy through fuel cell technology. In partnership with BMW, the electricity generated results in annual carbon emissions reductions equivalent to providing power to more than 17,000 electric vehicles. The farm was named 2023 Innovative Dairy Farmer of the Year by the International Dairy Foods Association (Bioenergy International, 2021).

In New Zealand, the previous government had planned to introduce a scheme that by 2025 will require farmers, meeting the threshold for herd size and fertilizer use, to pay for their agricultural greenhouse gas emissions. The plan includes taxing both methane emitted by livestock and nitrous oxide emitted mostly from fertilizer-rich urine – which together contributes to around half of New Zealand's overall emissions output. The price will be set in relation to the national target to cut methane by 10% by 2030, down from 2017 levels (Corlett, 2022). After the general election in late 2023, the new government confirmed its plan for its first 100 days, with 49 items revolving around three focus areas: the economy, law and order, and public services (RNZ, 2023). It is unclear whether the previous government's plan on greenhouse gas emissions will be further elaborated, implemented or stopped.

Agroforestry

Agroforestry is ancient, in principle going back to times of shifting cultivation in Africa, traditional home gardens in Asia and multistorey agriculture in pre-Columbian America. In this land-use system, woody perennials are used on the same land-management units as agricultural crops and/or animals, in spatial arrangement or temporal sequence. It can include trees, shrubs, palms, or bamboo being used as natural resource management systems, crucial to smallholder farmers and other rural people. Different systems provide various economic, sociocultural and environmental benefits. An agrisilvicultural system is a combination of crops and trees (alley cropping or home gardens).

When the rural poor got attention in international dialogue on development, the term agroforestry was coined in the late 1970s, leading to the creation of the International Centre of Research for Agroforestry (ICRAF) in Nairobi, Kenya in 1978, now known as World Agroforestry. It was established to promote agroforestry research in low-income countries and also hold discussions on intercropping and integrated farming. When trees, animals and crops are combined it is known as agrosylvopastoral systems. They combine forestry with domesticated animals that are grazing on pasture, rangeland or farms.

Agroforestry research can also be applied in industrialized countries as shown by Mupepele et al. (2021) from Europe. There was no benefit from agroforestry to biodiversity according to a study between 1991 and 2019. Silvopastoral systems were not more diverse in relation to forest, pasture or abandoned silvopasture. It was probably the case that the history of land-use was more important than the practice of agroforestry. However, the silvoarable systems increased biodiversity compared to cropland by 60%. Specifically, bird and arthropod diversity increased, while bats, plants and fungi did not (Mupepele et al., 2021). The most studied agroforestry practices in high-income countries have been on windbreaks, shelterbelts, hedgerows, riparian buffers and silvopasture systems.

Existing empirical research over three decades in 31 countries has recently been summarized (Castle et al., 2022). The literature was not found to be well documented in supporting or refuting the claims. Some conclusions included that agroforestry could prevent environmental degradation, improve agricultural productivity, increase carbon sequestration, and support healthy soil and ecosystems, and provided incomes to farm families (Brown et al., 2018).

Regenerative agriculture

For centuries, Indigenous cultures have been practising regenerative or organic agricultural techniques. It is a conservation and rehabilitation approach to food and farming systems. It is also a technique that rebuilds the quantity and quality of the topsoil, while restoring local biodiversity (especially native pollinators) and watershed function (Asarum Regenerative Farm, 2021). In the early 1980s, the Rodale Institute began using the term 'regenerative agriculture' but changed it to regenerative organic agriculture and climate change in 2014 to indicate organic agriculture methods.

The term appeared in academic research in the early 2010s. Researchers at Wageningen University in the Netherlands concluded there was no consistent definition of regenerative agriculture – just like sustainability. Instead, research on regenerative agriculture was the authors' attempt at shaping what it meant (Schreefel *et al.*, 2020). The system is based on various agricultural and ecological practices, with an emphasis on composting and minimal soil disturbance. It focuses on topsoil regeneration by recycling farm waste, crop rotation and 'no-till' or 'reduced till' farming as practiced by larger farms. Not all regenerative systems emphasize ruminants, but grassland cultivation and natural grassland can deliver on most of the points of the term regenerative agriculture, a kind of holistic grazing.

In 2018, the General Mills company in the United States planned to convert Gunsmoke Ranch to organic production on some 13,700 ha. It should have been an educational hub to teach other farmers 'how to implement organic and regenerative agriculture practices'. But some neighbours found out the farm was doing more environmental harm (Charles, 2021). The soils are fragile and if disturbed, they turn into fine powder, easily carried away by rain or wind. This is a regular feature of the western part of the state of South Dakota. The cropland is certified as USDA organic land and a soil conservation plan was neglected. Thus, the initiative of General Mills seems to have failed because the Peoples Company responsible for it was seeking lease proposals from 'qualified operators interested in dryland crop production on the Gunsmoke Ranch'. In 2023, only 5560 ha were planted with organic winter wheat.

In early 2021, the US Secretary of Agriculture made reference to regenerative agriculture. This signal led some national and international corporations to initiate or expand activities in this field. PepsiCo had previously worked with US farmers to plant cover crops on more than 34,000 ha, aiming for a reduction of on-farm greenhouse gas emissions. PepsiCo also announced that by 2030 the company would work with farmers to establish regenerative agriculture practices across 2.8 million ha (Peters, 2021). The ambition was to spread regenerative farming practices for more than 25 crops in the supply chain and ingredients from more than 30 countries globally, supporting some 100,000 agricultural jobs. Several other large corporations have also announced regenerative agriculture initiatives to use products from sustainable farms.

Pesticides

For more than a century, the production and use of chemicals, such as fertilizers and pesticides, has had exponential growth, leading to increased productivity and environmental concerns. Ethiopia is one example where pesticides are used intensively, both in large-scale greenhouses and on small and large farms (Negatu *et al.*, 2016). In addition to their acute toxic effects there was illegitimate use of DDT and Endosulfan on food crops and the direct import of pesticides without a formal registration by Ethiopian authorities. A study revealed that most workers did not attain any pesticide-related training, few were aware of modern alternatives to pesticides (Negatu *et al.*, 2016). Only 10% used personal protective equipment when applying pesticides and a minority took a shower or a bath after they had finished their work.

Different kinds of chemicals have eased life for societies in several ways, but their negative effects on the environment have gradually appeared over time. This applies to pesticides, pharmaceuticals for humans, and drugs for animals and pets.

One early agricultural invention was Bordeaux mixture from the late 19th century in the Bordeaux region of France. It is copper and sulphuric acid in a mixture that also includes lime, and was originally used to control fungal infection on grape vines. However, it has also been used to control and prevent potato blight, apple scab and peach leaf curl (Lewis *et al.*, 2016). The copper ions bind to the proteins in the fungus, causing cell damage and killing the fungus.

If the mixture is applied in large quantities annually for many years it will cause heavy metal pollution because copper ions build up in the soil and also accumulate in organisms. Nonetheless, this mixture is advocated by organic gardeners. In fact, all the ingredients of the mixture are common in nature. In late 2018, the European Commission decided that the active substance, copper, would be reapproved as a plant protection agent for a period of 7 years according to the IFOAM Organics Europe (IFOAM-EU, 2018).

Nowadays, the impact of drugs and pharmaceuticals is of growing concern and their impact on the environment may affect future agriculture. Since many of these compounds are produced to function in humans, the presence of these compounds in rivers, streams or in drinking water supplies is alarming (EPA, 2024). All kinds of products can enter the environment through improper disposal, agricultural runoff, and through human and animal waste. Also, disposal of unused or expired medicines via household waste can cause their leakage into the environment (OECD, 2022).

Pharmaceuticals not only affect wildlife and aquatic ecosystems but also remain as residues in water and soil, affecting future agricultural production and, ultimately, food safety. Pharmaceutical residues occur globally in the environment. This is demonstrated in the German database PHARMS-UBA on the Umwelt Bundesamt (UBA) website (UBA, 2024). Residues of pharmaceuticals in the environment have been measured in 89 countries in all UN regions.

An early attempt to highlight the dangers of indiscriminate use of pesticides started in the late 1950s. This was met by fierce opposition from chemical companies but led to initial steps towards reversal of national and international pesticide policy in some countries. In Sweden, certain environmental problems were observed quite early, such as increased car traffic leading to more carbon dioxide in the atmosphere, a growing global population and it was even claimed that environmental problems had their origin in industrial production (Palmstierna and Palmstierna 1968, 1972). The globe is still facing these problems and few effective measures have been taken to stop the increases.

The discussions on environmental and chemical issues were intensified when the US military sprayed defoliants on Vietnamese forests. To achieve long-term persistence, picloram and other herbicides were combined to create Agent White and Agent Orange. The chemicals were produced in the United States from the late 1940s and were originally to be sprayed along railroads and power lines to control undergrowth in forests. The active ingredient was a mixture of 2,4-D and 2,4,5 T with traces of the dioxin, TCDD. According to the Ministry of Foreign Affairs of the Vietnamese government, up to 4 million Vietnamese people were exposed to the defoliant and 3 million people suffered illness because of Agent Orange

(Stocking, 2007). These figures were described as unreliable by the United States government (Tucker, 2011). The US Congressional Research Service (2012) stated there were various estimates that the US military had sprayed approximately 11–12 million gallons of Agent Orange over nearly 10% of what was then South Vietnam between 1961 and 1971. One scientific study estimated that 2.1–4.8 million Vietnamese were directly exposed to Agent Orange. Vietnamese advocacy groups claim that there are over 3 million Vietnamese suffering from health problems caused by exposure to the dioxin in Agent Orange.

But picloram continued to be used on non-crop lands for brush control up to the early 1980s according to the US Environmental protection Agency (EPA). Its Technical Factsheet (EPA, 2023) states that picloram is a systemic pesticide, very persistent and does not adhere to soil so it leaches to groundwater. Picloram is still in use, specifically, Trooper® P+D, containing picloram and 2,4D: an EPA-restricted herbicide. It is designed for use on rangeland, permanent grass pastures, cropland and noncropland (Nufarm, 2024). The University of Minnesota Extension Service (Martinson, 2015) recommended that gardeners using dung as fertilizer confirm that the animal source has not grazed on picloram-treated hay. This dung would maintain the broadleaf-killing potency, problems scientifically indicated half a century ago in early studies of picloram.

In the 1970s, DDT and other pesticides were banned and discussions led to the formation of environmental movements and the creation of national environmental protection agencies. Organic farming was an early response as an alternative to the use of synthetic pesticides.

Today, Bayer AG is the world's leading company in the chemical-technical industry, producing various gardening products, medicines and pesticides. These pesticides are in frequent use in agriculture, forestry and gardening. The products have also become more sophisticated to cope with increasing resistance from agricultural pests and diseases. The global consumption of agricultural pesticides reached 2.7 million t in 2020 (Table 9.2).

One example of current plant protection products of special interest is Gaucho, produced by Bayer AG. It contains neonicotinoids such as

Table 9.2. Agricultural consumption of pesticides worldwide in 2020 (Statista, 2022).

Type	Thousands of metric tonnes
Herbicides	1397.47
Fungicides and Bactericides	605.99
Insecticides	471.24
Others	86.44

clothianidin, imidacloprid and thiamethoxam. They were introduced in the 1980s with the aim of being safer than older plant protection products. They became popular, accounting for about one-quarter of the global market for insecticides. But studies from three countries in 2017 showed that neonicotinoids harmed both drones and queens in bee colonies (Woodcock *et al.*, 2017). These neuroactive insecticides are acting on the nervous systems of bees. This leads to the paralysis and death of honeybees, native bees and bumblebees. Both pollinators and the overall ecosystem health will be affected (Mamy *et al.*, 2023). However, Bayer has introduced a new formulation and packaging for a return to South Africa (Bayer, 2021). Although Gaucho is primarily used for cotton in other territories, the focus in South Africa will be for the control of maize streak virus.

Neonicotinoids are very toxic and can remain in the soil for a long time. Pollen, nectar, plant sap and even dead leaves may hold the toxins. The toxins may reach seas and freshwater since they bind water molecules and penetrate soil cracks. Based upon the updated risk assessment of the European Food Safety Authority (EFSA), a ban on neonicotinoids for use outdoors was approved by all member nations in April to come into force by the end of 2018 (EU, 2018). They were allowed only in closed greenhouses. This was a quicker political response than for glyphosate. It is one of the world's most widely used pesticides, considered for a long time as likely to be carcinogenic according to the World Health Organization (WHO). This view was re-emphasized in 2015 when WHO's International Agency for Research on Cancer announced new findings of studies on five organophosphate insecticides. One of them was glyphosate, again found as 'probably carcinogenic to humans', also with reference to other studies of the chemical

over the last few decades (IARC, 2015). But the use of glyphosate in the EU will continue for 10 more years (Petrequin, 2023). The 27 member countries of the European Commission failed to find a common position for or against a prolongation; not a step towards sustainability.

But coming to one conclusion on scientific matters can be difficult as illustrated from the hiring of a panel of 15 experts for advice on whether glyphosate was carcinogenic (EPA, 2017). They could not reach an agreement. At the same time, the EPA Administrator decided to bar scientists from serving on US boards and scientific panels if they were receiving money through an EPA grant to guarantee future independent advice (Cornwell, 2017).

In California, the authorities decided to put glyphosate on the list of chemicals that cause cancer. This led to a decision in San Francisco's district court that warning labels would be needed from the producer for future sales. It was questioned by Monsanto, who argued that the risks were small when they appealed to the district court. The final decision came in 2023 (Stempel, 2023). In a 2–1 decision, the 9th US Circuit Court of Appeals said it was unconstitutional for California to require Bayer's Monsanto unit, which makes Roundup, and some agricultural producers to provide the warning under a state law known as Proposition 65. After Bayer bought Monsanto for US$63 billion in 2018, the company has faced extensive litigation over whether Roundup causes cancer.

In 2017, the EU approved glyphosate for use within the EU until mid-December 2022 based upon assessments by EFSA (2015) and the European Chemicals Agency (ECHA) (2017), later being extended for another year. Opponents of glyphosate referred to the WHO warning and the fact that Bayer, which bought the manufacturer Monsanto, had settled in court with people who claimed they had got cancer after being exposed to the product. Greenpeace has claimed that the chemical industry influenced an EU committee. The herbicide paraquat was banned in the EU in 2007 and has only been allowed to be used by licensed technicians in the United States. In China, several paraquat formulations have been phased out but the weedkiller is still used in many low-income countries (Prada, 2015). Since, the herbicide kills the weeds on contact it became very popular and has been widely used since the 1950s but also criticized by health experts.

Paraquat is the brand name for paraquat dichloride and glyphosate is the main chemical in the weedkiller Roundup (Clark, 2024). Both chemicals have adverse health effects if inhaled or ingested, including Parkinson's disease and cancers. Paraquat is 28 times more poisonous (Tzvetkova *et al.*, 2019).

ECHA did not share the WHO point of view when, in 2022, the European Parliament discussed a proposal from the EU Commission to halve the amount of pesticides by 2030. The European Parliament rejected it in a 299-to-207 vote on November 22, 2023 – the proposal endorsed by its environment committee (Le Monde, 2023). The decision stopped the EU from taking one step forward towards agricultural sustainability. In addition, the EU was not able to agree on a ban on glyphosate and reached a compromise of approving a new ten-year period for its use within the EU. One argument was that ploughing could be avoided, resulting in reduced greenhouse gas emissions: a victory for agriculture according to the majority on the right.

The occurrence and use of toxic chemicals in societies is of great concern. Elevated levels of hazardous chemicals such as bisphenols, PFAS and phthalates have been measured in the blood and urine of citizens of 22 EU countries, according to 2023 data from the European Human Biomonitoring Initiative (HBM4EU) (Santos, 2023). The impact on health was not measured, but the situation is alarming. Despite known health risks, PFAS, known as 'forever chemicals', are found in several common hygiene products. PFAS stands for per- and polyfluorinated alkyl acids and were initially produced on a large scale in the 1950s. The PFAS molecules have special surface properties, used, for example, to make durable bubbles in firefighting foam and water-repellent impregnation for textiles, as well as use in food packaging and frying pans.

In addition to make-up, PFAS can also be found in shampoos, soaps and lotions. PFAS can be used on the human skin, but they were banned in ski wax in Sweden in 2023. There should be a ban on all other PFAS substances, according to the independent non-profit International Chemicals Secretariat (ChemSec), which advocates banning PFAS for safer alternatives. There are about 10,000 different PFAS

substances, which can be spread via fish from lakes. They affect the skin, kidneys, liver and cholesterol levels.

These few examples show that the time perspective is interesting, but the use of chemicals in food production is of special concern. Long ago, reference was made to FAO data showing the number of pesticide-resistant insect species had doubled from 182 in 1965 to 364 in 1977, a trend that has significantly expanded. In March 2023, the Swedish Chemicals Agency (Kemi) decided on stricter terms of the use for the Galera plant protection product. It contains clopyralid, a selective herbicide used for the control of broadleaved weeds.

Clopyralid is in the picolinic acid family of herbicides, which also includes aminopyralid, picloram, triclopyr and others. A few months later in 2023, Kemi renewed the approval of Ariane S but the conditions for its use were stricter as the product contains clopyralid. Ariane was given renewed product approval, but extension was pending trial, with prohibition of use until April 2025 (Kemi, 2023a). Authorizations for plant protection products are time limited and depend on the approval of the active substance within the EU. Plant material from a treated crop may not be transported from an agricultural field for composting or biogas production. Nor may such plant material be used in greenhouses, mushroom production or as a covering material for plants. After several decades, this regulatory system needs urgent reform if sustainability is to be seriously considered.

This brief overview illustrates the dilemma in finding a consensus view to suit industry and employment, and guarantee human health over a long-term perspective, when ruling politicians must additionally call for sustainability. There is a great need for rapid political action to apply stricter regulations on current pesticides, especially all toxic ones. It will also require more advanced genetic breeding work, and frequent use of nitrogen fixation crops bred with biotechnology qualities for plant protection.

GM cotton is one example which reduces the need for pesticides, although not all cotton pests are affected, so insecticides may still be required. GM cotton is one example where the bacterium, *Bacillus thuringiensis*, is used to create Bt toxins. There are over 200 different Bt toxins that are insecticidal to the larvae of beetles, moths, cotton bollworms and flies. It is not an ultimate solution but may increase biodiversity since the natural enemy populations may grow if pesticides are not sprayed. Resistance is nature's reaction to poison. Bacteria become resistant through supersonic reproduction and the free exchange of genes with each other.

Still, widespread cultivation (more than 95%) of Bt cotton crops may lead to increased resistance as in India. Another effect, in China, is the development of Bt resistant pests, limiting the usefulness of Bt crops. Thus, the US EPA recommends refuge areas with non-Bt crops when growing Bt cotton on a large scale.

The lifelong experience in trying to control the Colorado potato beetle with varying levels of resistance to some 50 different pesticides has been in vain. This crop introduction, centuries ago, and subsequently that of a major pest, illustrates one effect of globalization, that remains today.

Early Spanish colonialists sent the potato (*Solanum tuberosum*) to Spain in the early 16th century. By about 1570, potatoes are considered to have been well established in Spain (Hawkes and Francisco-Ortega, 1992). Then, the crop spread to other parts of Europe, and to the British colonies. In Sweden, in 1655, Olaus Rudbeck was probably the first to introduce the potato at his Botanical Garden in Uppsala. In those early years it was regarded as a botanical curiosity and an ornamental plant called 'Peruvian Night Treasure'. Jonas Alströmer is known as the one first person in Sweden to grow potatoes on his farm outside Alingsås in 1724, probably with potato seed from England (Bodensten, 2020).

When the potato arrived in North America, it was introduced into the western part of the native habitat of the Colorado potato beetle in western United States. The beetles found the potato attractive, and started to cause great damage after several decades. In the mid-1800s the Colorado potato beetle rapidly became the most destructive pest of potato crops. Various control measures were introduced to kill the beetles: copper arsenate paint followed by lead arsenate, Bordeaux mixture, DDT, aldrin, heptachlor and imidacloprid. In the late 1870s, the beetle reached Europe. It can also cause damage to tomato and eggplant crops. The Colorado potato beetle has, however, shown varying levels of resistance to most pesticides.

It became resistant to dieldrin and DDT in the 1950s.

The creation of preparations based on nucleic acids, such as short, single-stranded DNA fragments, may lead to what has been called 'intellectual' insecticides. RNA interference (RNAi) is a valuable technology, which can be applied in both medicine and agriculture. Studies have shown that feeding viruses with double-stranded genomes made of RNA (dsRNA) expressed in bacteria can kill pests, including the beetle (Palli, 2014). The dsRNA is used to silence specific essential genes in the target organism, leading to toxic effects and death. RNAi technology refers to the process where small pieces of RNA can shut down protein translation by binding to the messenger RNAs that code for those proteins. In the past ten years, exploiting the RNAi gene mechanism to silence essential genes in pest insects has offered a promising new control strategy (Guan *et al.*, 2021). This functional genomics through RNAi technology has the potential to contribute towards new pest management methods. But an application of RNAi technology requires more efficient methods for the production and delivery of dsRNA. Large amounts of low-cost double-stranded RNA (dsRNA) are needed. This approach would be like the process when health companies update COVID mRNA vaccines when the virus changes. Here, agricultural companies could deliver an update when the potato beetle overpowers one RNAi. Other control approaches may include antifeedants such as fungicides or products from the neem tree (*Azadirachta indica*). The most important cultural control is conventional crop rotation.

The search for more biological approaches must expand. Long ago, this was highlighted by rice researchers at IRRI. They evaluated various products derived from the neem tree to 'find alternatives to costly and hazardous synthetic pesticides' (Soon and Bottrel, 1994). The neem oil and seed extracts were known to possess germicidal and antibacterial properties without leaving any residue on plants. Research on neem has also been conducted in India, where biopesticides are manufactured and exported. The main ingredient in these biopesticides is azadirachtin. In Kenya, the International Centre of Insect Physiology and Ecology (ICIPE) is at the forefront conducting strategic biological research towards integrated pest management approaches as alternatives to insecticides.

The work towards sustainability will require a gradual trapping down of the use of current chemical pesticides with specific milestones every fifth year up to 2050. Then, the current type of toxic chemicals should not be allowed for use in agriculture and forestry with a few exceptions. This time span would allow for the TNCs and chemical companies to use their great expertise in research and development to design and develop alternative, biologically based products, safe to humans and that do not degrade the environment. It may be difficult, but not unattainable, and there will be a large market for biopesticides.

The use of biological products has shown an increase worldwide (Melo and Swarowsky, 2022). Biostimulants are a diverse group of materials that are used to improve crop vigour, quality and yield, as well as tolerance to abiotic stresses (drought, salinity, heat, etc.). Biofertilizers contain living microorganisms which stimulate plant and root growth. Biocontrol agents composed of microorganisms including bacteria, cyanobacteria microalgae, plant-based compounds and RNAi-based technology have potential as biopesticides (Kumar *et al.*, 2021).

The Belgian company, KitoZyme®, is a leader in manufacturing vegetal chitosan. It is effective against a large spectrum of plant diseases and abiotic stresses by seed coating or foliar applications (Kitozyme, 2024). The products are approved by the US EPA as biopesticides and by the European Commission as basic substances for plant protection for multiple applications and crops. The company has introduced KitoGreen©, the world's first vegetal chitosan to be used in agriculture. It is an activator of natural defence mechanisms against fungal threats such as *Botrytis* and mildew. The product is claimed to be a sustainable solution for farmers to grow more resistant and healthier plants, while reducing the need for chemical pesticides. In 2024, DPH Biologicals, a biotechnology company in the state of Illinois in the United States, partnered with KitoZyme to develop and commercialize the first sustainably sourced chitosan seed treatment technology for US growers. The product, ChitoNox FC, is reported to effectively protect seedlings against nematodes and soil-borne diseases. This product is going to be commercially available for the 2025 crop season.

Finally, old methods of weed control must be reconsidered, such as more mechanical weeding, interplanting, and mixed cropping, and using new versions of sea salt and sulphates, and environmentally friendly nitrates of copper and iron.

New Industrial Farming Methods

New farming methods could help the agriculture industry reduce its environmental impact while still increasing productivity. One new significant transformation is being driven by digital tools, in particular AI, but this would primarily benefit the industrialized countries. These technologies usually require energy, water, nutrients and sometimes plant protection inputs.

Automation is increasing in Western-type agriculture, for example, the use of robots, drones and autonomous tractors to make farming more efficient. Another aspect is precision farming using irrigation, fertilizers and pesticides at variable rates. They relate to the needs of different crops, rather than uniform application at set times, quantities and frequencies. One example is a monitoring system developed by the Israeli company, Phytech, and the Swiss agrichemical company, Syngenta. It includes sensors for plant growth and soil moisture so that farmers can continually monitor crop growth and soil health.

Vertical farming

Vertical farming (growing crops in vertical layers) saves both water and soil. An increasing share of vertical cultivation is expected to reach a global value of almost US$20 billion by 2029, according to Fortune Business Insights (2022). The creators, AeroFarms, of the world's largest vertical farm in Newark, New Jersey have claimed that vertical farming is 390 times more productive per square foot than a field farm (DWIH New York, n.d.).

The German company, Infarm, started vertical farming in 2013 and has evaluated a hydroponic method of farming resembling a giant glass-door fridge for herb growing or sale, saving land and reducing imports of vegetables from foreign (tropical) countries. The farming is done around the year in supermarkets, restaurants, bars and warehouses which lowers the carbon footprint and the costs of transportation. The company has deployed operational farms in Copenhagen, London and Seattle (Epp, 2020). Recently, technology has been developed for indoor wheat. It is very efficient but requires a lot of energy, water and large spaces. These requirements also apply to crops grown in Sweden by companies such as Urban Oasis, Eco Blom and Ljusgårda. So far, the crops are mostly lettuce, spices and tomatoes. They are more expensive than those conventionally grown.

Another example is Pure Harvest Smart Farms outside Abu Dhabi in the United Arab Emirates. The proprietary Controlled-Environment Agriculture (CEA) System is a combination of high-tech greenhouses and vertical farms for tomatoes in a stable year-round climate. In 2022, the total growing area was 16 ha, producing 14 types of leafy greens and 30 tomato varieties (Chilton, 2022). The vegetables are 60% cheaper than air-freighted imports of comparable quality. In addition, Pure Harvest runs a 6 ha farm in Saudia Arabi and another one may be initiated in Kuwait.

In 2022, the world's largest vertical farm opened in Dubai, also in the United Arab Emirates. The ECO1 farm is a collaboration between Emirates Flight Catering and Crop One Holding to produce leafy greens, such as arugula and spinach, in a process that uses 95% less water than crops grown in the fields.

Hydroponics

Hydroponics is growing plants in nutrient-rich water and aeroponic cultivation is growing plants in water, air or fog environments completely without soil. Some observers believe that the future of Chinese agriculture can be conducted without using land through hydroponics, aeroponic cultivation and drip irrigation. LED lights are not only energy efficient but also have the right light frequency for plants.

Aquaponics

Aquaponics is a food production system that combines conventional aquaculture (cultivation of aquatic animals in basins) with hydroponics (cultivation of plants in water) in a symbiotic environment. In traditional aquaculture, the faeces accumulate in the water, which therefore eventually becomes toxic to the aquatic animals. But in the hydroponic system, it is broken down by nitrogen-fixing bacteria. The water is filtered by the plants, which can then absorb the nutrients. The purified water is then returned to the fish. The fresh vegetables are sold to retailers, while the fish are delivered to local restaurants. Aquaponics is an environmentally friendly system that produces little waste, creates new jobs and can lead to better land use in urban environments provided that cheap land and markets can be found.

Urban farming

Urban farming is food production within the city limits of highly populated areas. It is different from community gardening, subsistence farming or homesteading. This latter type of food production in urban areas was once common when cities were established and particularly in times of warfare. Now, there are various techniques, such as hydroponic systems, and vertical or rooftop gardens, that fit into any free urban space that is available. Vertical farming and aquaponics can make tiny spaces, like shipping containers or rooftops, into full-scale operations for food production. In the future, urban farming will be even more important and profitable. There are now a special types of urban farming business, for example, Green City Acres with urban farms around Kelowna in British Columbia, Canada (Urban Vine, 2024). Nature's Always Right is an urban market garden in San Diego, California, featuring regenerative farming practices. FARM is located in London, UK, with a rooftop chicken coop, aquaponic fish farming and small-scale vegetable farming in a polytunnel, among other enterprises.

Urban farming also benefits the communities where the food is produced, helping boost economic growth and provide jobs. It increases food security and involves the local community, bringing the residents of an area together to work towards a common goal. Urban people will understand how food is produced and know where it comes from. It creates green space and food can be harvested just for a few days, creating less waste than supermarkets, which must stock more fresh food than they can sell. Microgreens and edible mushrooms can be profitable; they take little space to grow and can easily integrate with local restaurants or cafes. Vegetables, herbs and fruit from urban farms are more likely to be perfectly ripe, more nutritious, taste better and be produced in season, although there are restrictions in regions with a winter climate. Fresh, nutritional food produce will be closer to where it is to be consumed. Less food must be transported long distances which may cut down on carbon emissions. But urban farming may not always be better for the climate than conventional agriculture over the life cycle. In a study of several agricultural sites, it was found that, on average, the urban agriculture was six times more carbon intensive per serving of fruit or vegetables than conventional farming (Hawes *et al.*, 2024).

Entomophagy

Entomophagy, the human use of insects as food, is everyday practice in many parts of the world. Estimates indicate there are over 2000 types of edible insects but bugs have never been a part of the diet in Western countries (Kourimska and Adamova, 2016). According to the FAO around 2 billion people already include insects in their diets (Food and Agriculture Organization (FAO), 2019). Using insects would be more accessible, less environmentally problematic than animal protein and maybe cheaper. Nonetheless, the most important first step must be to change peoples' perceptions which may take quite some time.

Climate activists and the media have championed the benefits of people starting to eat insects. They are a high source of protein, essential fats, vitamins, fibre and minerals, such as iron and calcium. There have been arguments that they may be one solution for providing adequate nutrition in densely populated urban areas. Large-scale insect consumption will hardly solve the issue of global food security, but it may be a contribution to food in low-income countries in crisis situations and in periods of food shortages. After 2000, the FAO began to consider topics on edible insects worldwide and later outlined the food safety aspects of edible insects (FAO, 2021).

Researchers at ICIPE have used insect nutrients to transform African porridge from a basic, often low-nutrient meal, into a superfood that exceeds micronutrient requirements for people. Finger millet was fortified with high-quality nutrients from the edible African cricket (*Scapsipedus icipe*) and grains of amaranth. The insect was identified by ICIPE researchers a few years ago. The new flour has twice the amount of protein, iron and zinc, 3–4 times more fat and a higher content of vitamins. Some novel porridge products have been blended with edible crickets (Maiyo *et al.*, 2022).

A company in Guatemala has turned bugs into flour and baked bread and cookies. Indonesian entrepreneurs are trying to develop a new kind of oil extracted from the larva of the giant mealy worm (*Zophobas morio*) with plans to replace palm oil (Herranz, 2018). The company, Biteback, produces different kinds of oils and fats for cooking and baking, and also for cosmetic and personal care products where palm oil is often used. The larvae are rich in iron, calcium and magnesium and reproduce so rapidly that they outdo oil palm in yield by at least 37 times on the same area of land (AFN, 2021). The fatty acids they produce have the same properties as those from palm oil, such as palmitic, oleic and linoleic acids, and they have high levels of unsaturated fats and omega-3. Palm oil is the most used vegetable oil on the planet, accounting for 65% of all vegetable oil. But its cultivation causes environmental damage and it is not good for human health. Its saturated fatty acids increase the amount of harmful cholesterol in the blood, consequently increasing the risk of cardiovascular disease, diabetes and stroke.

Another example of an insect factory is Biocycle. Since 2018, this Indonesian company has used the dried larvae of the black soldier fly (*Hermetia illucens*) for products in aquaculture, agriculture, animal husbandry and for pets. Its larvae effectively break down the organic substrates of agricultural waste, but it is also a rich source of protein. At the global level, the multilateral Protix company, based in the Netherlands, is the world's largest operating insect factory farm. In 2023, the US company, Tyson Foods, took a minority stake in the company but is also planning to build a US factory (Wiener-Bronner, 2023). The deal is not to impact on its products for human consumption.

In 2021, following a review of the scientific basis by EFSA, the EU-260 Commission approved mealworms and migratory grasshoppers as food within the EU (EFSA, 2021). Other insect species are also being considered and evaluated, such as the house cricket, tropical house cricket, honeybee, drone pupa and black soldier fly. The Swedish Axfoundation is evaluating the potential of the larvae of the American gunfly to eat food waste and then become feed for chickens and pigs to avoid carbon dioxide emissions. Furthermore, insects are being tested as replacement for soy to produce animal and fish feed. This will call for changes of the regulatory system.

Food and Food Consumption

The concept of agricultural sustainability covers not only all aspects of food production but also all processes thereafter, right up to the ultimate consumption at the table. Over the last century, consumption patterns have changed dramatically in high-income countries but lately also in low-income countries.

In an agrarian society, raw materials were fresh or had been treated solely to extend their durability (drying, smoking, fermentation, freezing, vacuum packaging or pasteurization). Then, the food industry produced processed ingredients to be used in cooking (butter, honey and sugar, vegetable oils, potato and maize flour, salt). Such processed food had clearly changed from the original product. The next step was creating processed foods, produced by combining these two processes. The third step was the development of ultra-processed foods. So far, there is no agreed definition, but ultra-processed foods are foods that have been processed to the greatest degree.

One may conclude that ultra-processed foods contain one or more of the raw materials, such as maize, wheat and sugar, that have been broken down and then reassembled; ingredients that are used almost exclusively by the food industry, and a combination of fat, salt and sugar is usually added (Swedish Food Agency, 2024b).

Although ultra-processed foods may have less calories, some may not be of top quality when the body needs to assimilate available nutrients. This makes it hard for food companies to claim they are healthy (Yeo, 2021). Chips, ice cream, candy, French fries and sausages are often referred to as ultra-processed foods. Several of these foods, and ultra-processed meat, are not healthy, which is why it is relevant to ask whether these processes have led towards sustainability.

Changes in society and lifestyles have given rise to new eating habits. One reason is convenience, people prefer precooked food or quick cooking. Ultra-processed foods are given a long shelf life through the addition of preservatives. This reduces food waste and these products are cheaper to produce and buy. They are often designed to be very flavourful, containing high levels of sugar, salt and fat, making them addictive. Finally, there is effective marketing of ultra-processed products. High consumption of ultra-processed foods can have negative health effects, such as an increased risk of obesity, heart disease and other diseases (Silva et al., 2018). It is important to conclude that not all ultra-processed food is bad for health, and a report published by the FAO examined the peer-reviewed literature on the effects of ultra-processed foods on diet quality and on health (Monteiro et al., 2019)

Food safety is determined, in part, by the presence of chemicals, some of which may not be permitted in food. Current problems in the industrialized world are that we eat the wrong food and that food is not distributed fairly. This was highlighted in a study conducted by the Institute of Tropical Medicine Antwerp, in Belgium, based on data from EFSA (Mertens et al., 2022). The researchers compared the eating habits of the populations in 23 European countries. The Swedes get 40.6% of their daily food intake from things like sausages, cakes, ready meals, sweets and soft drinks. The corresponding figure in the United Kingdom is 39.7%, followed by Germany with 38%, People in Italy consumed the lowest percentage of junk food (13%). But ultra-processed food is much more common in the United States and Canada, and has also reached supermarkets in large cities in low-income countries as an effect of globalization.

Dietary nutritional supplements, like multivitamins and minerals, have become very popular, although they are needed only for those who have a confirmed deficiency of a particular substance. According to a consumer survey, Americans seem to have a positive view of dietary supplements. Some 74% of US adults take vitamins, prebiotics and other supplements, and more than 90% of the users claimed that dietary supplements are essential to maintaining their health (CRN, 2023). They trust the dietary supplement industry but may be unaware of the lack of science behind the product claims (Zhang et al., 2020). For most people, supplements may offer questionable benefit for most healthy people with even a remotely balanced diet. In a study analyzing data from 400,000 healthy US adults, followed for more than 20 years, there was no association between regular multivitamin use and lower risk of death (Loftfield et al., 2024). Fruit and vegetables are a better source for vitamins.

Margarine became an alternative to butter when Hippolyte Mége-Mouriés received a patent for it in 1869. Palm oil is semisolid at room temperature, unlike coconut oil. Therefore, it became a common ingredient in a range of products during the 1960s and 1970s. With the increased use of the cholesterols, rapeseed oil was preferred for use in margarine. However, palm oil is the cheapest choice with a long shelf life although it can affect our health, even globally (Zuckerman, 2021). One example was when Swedish the table margarine Lätta was selected as 'The Food Scam of the Year' in 2019 by the consumer association Äkta. The package shows a nice picture of rapeseed flowers, but the small print has a declaration that the product has a higher proportion of palm oil (21%) than rapeseed oil (18%). In 2023, the producer claimed the content was without palm oil (Askew, 2019).

Reliable information is critical to consumers. In 2023, Oatly was accused of selling fake vanilla sauce in Sweden, since it lacked the expensive vanilla bean. The same company sells its drink of only oats and water but it is twice as expensive as the nutritious milk. Therefore, Oatly ended up in third place in the Food Scam of the Year 2018 competition. Plant-based milk may be allowed to be called milk in the United States, but only with explanations of the nutritional differences. In the United Kingdom, products by Oatly have been criticized by the Advertising Standards Authority (ASA) for not being climate friendly (Miller, 2021).

Globally, obesity has nearly tripled since 1975, adult obesity worldwide has more than doubled since 1990, and adolescent obesity has quadrupled (World Health Organization (WHO), 2024b). In 2022, one in eight people in the world were living with obesity. About 2.5 billion adults (18 years and older) were overweight, which is a condition of excessive fat deposits. Of these people 890 million were living with obesity. Obesity is a chronic complex disease defined by excessive fat deposits that can impair health. Obesity can lead to increased risk of type two diabetes and heart disease.

In 2022, some 43% of adults aged 18 years and over were overweight and 16% were living with obesity (WHO, 2024b). But also, the young generation is affected. In 2022, over 390 million children and adolescents aged 5–19 years were overweight, including 160 million who were living with obesity. Some 37 million children under the age of 5 were overweight.

Most of the world's population live in countries where being overweight and obesity kills more people than being underweight. In the early 2000s, more people died of obesity than of starvation. The Global Burden of Disease Study, led by the Institute for Health Metrics and Evaluation in Seattle, was published in 2010 and pointed to obesity as a more widespread health problem than world hunger. It stated that about 30% of the global population was overweight or obese and that the latter condition caused approximately 5% of all deaths. All the findings of this comprehensive epidemiological study on changing health challenges worldwide were later published with a comprehensive analysis in *The Lancet* (Lim et al., 2012). In the past, poverty meant a lack of food and wealth an abundance of food. Now, abundance is not as obviously associated with wealth because people on low incomes in high-income countries cannot afford to eat healthy food and they may also lack sufficient knowledge of how to avoid consumption of empty calories. Obesity has become a sign of poverty.

The major cause of obesity is an increased intake of energy-dense foods, high in fat and sugars. In addition, many people are less active due to the sedentary nature of many forms of work, changing modes of transportation and increasing urbanization. Nonetheless, obesity is preventable.

Studies show that food allergies are increasing, especially in the Western world (De Martinis *et al.*, 2020). In 2008, there were 725 products available for consumers on the Swedish market classified as allergic. Twelve years later that number was 4539 according to the Swedish Chemicals Agency (Kemi, 2023b). Some products not previously classified have been added in addition to new products in paints, toys, clothes, adhesives and cleaners. The increase highlights the importance of proper labelling and safety measures to protect individuals with allergies.

In Europe, 2% of adults are reported to be affected and food allergies are especially increasing among children. Studies show that the intake of antibiotics at an early age increases the risk of food allergy later in life (Li *et al.*, 2019). At the same time, the use of antibiotics in health care is globally increasing.

In mid-2023, the new Nordic Nutrition Recommendations (NNR) were presented; an update from 2012 (Nordic Council of Ministers, 2023). They form the scientific basis for Swedish dietary guidelines and are now focused both on human health and what is good for the planet. One-third of the greenhouse gas emissions in Sweden can be linked to food. But scientists were protesting about NNR recommendations on ultra-processed foods. They claimed there is evidence to limit the intake of ultra-processed foods, but it is not stated by the NNR despite there being strong links to mortality, obesity and diabetes. The reason is uncertainty about the negative effects according to NNR. Certain additives are, however, not nutritious, such as emulsifiers and stabilizers. The industrial process can even reduce the fibre content, which is important for the breakdown of food.

The NNR recommends lower meat consumption, especially of processed meats such as ham and salami. Instead, consumption of fish, vegetables and legumes should increase. As far as alcohol is concerned, NNR has no limit but 'alcohol should be avoided'. Despite objections from researchers, a doubling of the sugar content from 5% to 10% of energy intake is proposed (Nordic Council of Ministers, 2023). The WHO recommends 5% to minimize the risk of type two diabetes and obesity.

Reducing meat consumption is currently high on the political agenda but this type of protein food is necessary. Although current trends that exercise and healthy eating habits lead to a longer and healthier life one should realize that a long life also depends on genetics, chance and the consumption of a variety of foods rather than dieting, using vegetables and adopting trends in media. Food producers argue that the media is the driver of food system transformation (Phillipov and Kirkwood, 2018).

It would be rather unrealistic for an individual to save the climate by consuming less meat and not drinking coffee. Other aspects are more critical. Humans need food and some of it is produced in low-income countries, benefitting their populations. Socrates coined the phrase 'others live to eat, I eat to live'. At the same time, major changes are taking place in the corporate sector to meet an expected increase in the global livestock production, which has already begun in China where the demand for meat is high. Companies are merging and becoming larger and new global companies are included in the livestock industry.

About one-third of food is lost between the farm and the dining table, and this is especially true for fruit and vegetables. Therefore, shorter transportation times and methods that slow down the maturation of products are required. In addition, we throw away too much food at home. For a long time, even grocery stores have thrown away groceries that have been slightly damaged or are not of the highest quality. For those that remember food shortages, it is a familiar lesson that food or groceries must not be thrown away. It sits in the spinal cord of everyone who grew up in the 1940s and 1950s. Regrettably, modern society has been spoiled and many people find it obsolete to respect and care for food. However, it is a necessity for most people in low-income countries and would contribute to sustainability.

The prices of food and nonalcoholic beverages in Sweden have increased by over 20% during 2023 according to Statistics Sweden (SCB, 2024). Such an increase is exceptional and has not occurred since 1951. It is exemplified by leeks (79.5%), cauliflower (72.2%), peppers (54.1%) and honeydew melons (51.4%). In the past, Swedish policy has been characterized by a steady hunt for cheap food. In addition, families spend less money of food nowadays. Food accounts for some 13% of the budget compared to one-third in 1950. Thus, there are margins for good lives even after priorities! Cheaper food is an illusion in the long-term during insecure times and with steadily increasing costs for large-scale electrification, including a build-up of more nuclear power in many countries, together with a transformation to a sustainability society..

Solutions include the avoidance of imported food produced in greenhouses with high transportation costs. Food should not be wasted as has been the case for spoiled population in many Western countries. Food consumption can easily be based on availability in the respective season, vegetables in summer, root crops in winter but few strawberries in northern Europe since it has a different climate compared to the tropics.

References

AFN (2021) Meet Biteback and Cellular Agriculture: The GROW Impact Accelerator's food ingredient futurists. *AgFunderNews*. Available at: https://agfundernews.com/biteback-cell-ag-grow-impact-ac celerators-food-ingredient-futurists (accessed 16 September 2024).

AgNews (2017) *Agricultural News* (Brazil approves world's first commercial GM sugarcane). Available at: https://news.agropages.com/News/NewsDetail---22616.htm (accessed 16 September 2024).

American Geophysical Union (2013) Groundwater levels drop at "alarming" rate in large swath of Middle East. American Geophysical Union. Available at: https://news.agu.org/press-release/groundwater-le vels-drop-at-alarming-rate-in-large-swath-of-middle-east/ (accessed 15 September 2024).

Arla (2022) Arla Foods and DSM start large-scale on-farm pilot programme to reduce methane emissions from dairy cows by 30 per cent. Arla, Viby, Denmark. Available at: https://www.arla.com/company/n ews-and-press/2022/pressrelease/arla-foods-and-dsm-start-large-scale-on-farm-pilot-programme -to-reduce-greenhouse-gas-emissions-from-dairy-cows-by-30-per-cent/ (accessed 16 September 2024).

Asarum Regenerative Farm (2021) About our farm. Asarum Regenerative Farm. Available at: https://rege nerativefarm.ca/about/ (accessed 13 August 2024).

Asis, M. (2017) Golden Rice moving forward in Philippines. Alliance for Science, Boyce Thompson Institute, Ithaca, New York. Available at: https://allianceforscience.org/blog/2017/08/golden-rice-mo ving-forward-in-philippines/ (accessed 15 September 2024).

Askew, K. (2019) Food bluff of the year' anti-prize puts Lätta in spotlight over palm oil content. *Food Navigator*. Crawley, United Kingdom. Available at: https://www.foodnavigator.com/Article/2019/01 /31/Food-bluff-of-the-year-anti-prize-puts-Laetta-in-spotlight-over-palm-oil-content (accessed 16 September 2024).

Augustsson, E., Kim, H., Andrén, H., Graf, L., Kjellander, P. *et al*. (2024) Density-dependent dinner: Wild boar overuse agricultural land at high densities. *European Journal of Wildlife Research* 70, 15. Available at: https://doi.org/10.1007/s10344-024-01766-7 (accessed 15 September 2024).

BASF (2023) BASF Report 2022. Integrated corporate report on economic, environmental, and social performance. Ludwigshafen, Germany.

Bayer (2021) Gaucho®: Back and better than ever. Available at: https://www.seedgrowth.bayer.com/en-u s/news-stories/gaucho-back-and-better.html (accessed 16 September 2024).

Bayer (2024) A partner in your fields and in your success. Bayer Crop Science, United States.

Behrens, P., Bosker, T. and Erhardt, D. (eds) (2020) *Food and Sustainability*. Oxford University Press, Oxford, UK.

Benbrook, C. (2004) Genetically engineered crops and pesticide use in the United States: The first nine years. *BioTech InfoNet* Technical Paper Number 7. Available at: https://www.organic-center.org/rep ortfiles/Full_first_nine.pdf (accessed 15 September 2024).

Benbrook, C.M. (2012) Impacts of genetically engineered crops on pesticide use in the U.S.: The first sixteen years. *Environmental Sciences Europe* 24, 24. Available at: https://doi.org/10.1186/2190-47 15-24-24 (accessed 16 September 2024).

Béné, C., Fanzo, J., Prager, S.D., Achicanoy, H.A., Mapes, B.R. *et al.* (2020) Global drivers of food system (un)sustainability: A multi-country correlation analysis. *PLOS One* 15(4), e0231071. Available at: http s://doi.org/10.1371/journal.pone.0231071 (accessed 19 August 2024).

Bioenergy International (2021) Bloom Energy Deploys Bar 20 Dairy Farms Biogas power project. Bioenergy International, Stockholm. Available at: https://bioenergyinternational.com/bloom-energy-deploys-ba r-20-dairy-farms-biogas-power-project/ (accessed 16 September 2024).

Bioenergy International (2024) Power2Earth consortium to develop sweden's first fossil-free mineral fertilizer facility. *Bioenergy International.* Available at: https://bioenergyinternational.com/power2ea rth-consortium-to-develop-swedens-first-fossil-free-mineral-fertilizer-facility/ (accessed 13 August 2024).

BioRefine (2021) Irish biotechnology company awarded the coveted Sustainable Energy Authority of Ireland (SEAI) award for Excellence in Energy Research and Innovation. Ghent University, Belgium. Available at: https://www.biorefine.eu/news/irish-biotechnology-company-awarded-the-coveted-su stainable-energy-authority-of-ireland-seai-award-for-excellence-in-energy-research-and-innovation / (accessed 16 September 2024).

Birthal, P.S., Kumar, A. and Datta, T.N. (2008) Trading in livestock and livestock products. *Indian Journal of Agricultural Economics* 63(1), 58–63.

Birthal, P.S., Roy, D. and Negi, D.S. (2015) Assessing the impact of crop diversification on farm poverty in India. *World Development* 72, 70–92. Available at: https://doi.org/10.1016/j.worlddev.2015.02. 015 (accessed 19 August 2024).

Bodensten, E. (2020) A societal history of potato knowledge in Sweden c. 1650–1800. *Scandinavian Journal of History* 46(1), 42–62. Available at: https://doi.org/10.1080/03468755.2020.1752301 (accessed 16 September 2024).

Bogoviz, A.V., Postnikova, L.V., Postnikova, D.D., Lobova, S.V. and Alekseev, A.N. (2022) Successful global practices in responsible agriculture for sustainable development. In: Popkova, E.G. and Sergi, B.S. (eds) *Geo-Economy of the Future: Sustainable Agriculture and Alternative Energy.* Springer, Cham, Switzerland, pp. 365–371.

British Ecological Society (2022) New research uncovers hidden long-term declines in UK earthworms. British Ecological Society, London. Available at: https://www.britishecologicalsociety.org/new-resea rch-uncovers-hidden-long-term-declines-in-uk-earthworms/ (accessed 16 September 2024).

Brown, S.E., Miller, D.C., Ordonez, P.J. and Baylis, K. (2018) Evidence for the impacts of agroforestry on agricultural productivity, ecosystem services, and human well-being in high-income countries: A systematic map protocol. *Environmental Evidence* 7, 24.

Burke, P. (2023) Ignorance: A global history. Yale University Press, New Haven, CN.

Cai, L., Zhou, B., Guo, X., Dong, C., Hu, X. *et al.* (2008) Pollen-mediated gene flow in Chinese commercial fields of glufosinate-resistant canola (*Brassica napus*). *Chinese Science Bulletin* 53, 2333–2341. Available at: https://doi.org/10.1007/s11434-008-0305-6 (accessed 15 September 2024).

Campbell, D.T. (1979) Assessing the impact of planned social change. *Evaluation and Program Planning* 2(1), 67–90. Available at: https://doi.org/10.1016/0149-7189(79)90048-X (accessed 19 August 2024).

Castle, S.E., Miller, D.C., Merten, N., Ordonez, P.J. and Baylis, K. (2022) Evidence for the impacts of agroforestry on ecosystem services and human well-being in high-income countries: A systematic map. *Environmental Evidence* 11, 10. Available at: https://doi.org/10.1186/s13750-022-00260-4 (accessed 19 August 2024).

Center for Food Safety (2005) Monsanto vs. U.S. Farmers. Center for Food Safety, Washington, DC. Available at: https://www.centerforfoodsafety.org/files/cfsmonsantovsfarmerreport11305.pdf (accessed 15 September 2024).

Chand, R., Joshi, P. and Khadka, S. (eds) (2022) *Indian Agriculture Towards 2030: Pathways for Enhancing Farmers´ Income, Nutritional Security and Sustainable Food and Farm Systems.* Springer, Singapore.

Charles, D. (2021) A giant organic farm faces criticism that it's harming the environment. NPR. Available at: https://www.npr.org/2021/05/03/989984124/a-giant-organic-farm-faces-criticism-that-its-harmi ng-the-environment (accessed 19 August 2024).

Chilton, N. (2022) Farming the unfarmable. *Time.* Available at: https://www.everand.com/article/5834534 54/Farming-The-Unfarmable (accessed 19 August 2024).

Clark, C. (2024) Paraquat versus glyphosate: How each lawsuit differs. Schmidt and Clark LLP. Available at: https://www.schmidtandclark.com/paraquat-vs-glyphosate (accessed 16 September 2024).

Coffey, D., Dawson, K., Ferket, P. and Connolly, A. (2016) Review of the feed industry from a historical perspective and implications for its future. *Journal of Applied Animal Nutrition* 4, e3. Available at: https://doi.org/10.1017/jan.2015.11 (accessed 16 September 2024).

Cohen, J. (2019) The rise, fall and resurrection of Russian seed bank pioneer Nikolai Vavilov. *Genetic Literacy Project.*

Congressional Research Service (CRS) (2012) Vietnamese Victims of Agent Orange and U.S.-Vietnam Relations. CRS Report RL34761, Washington DC. Available at: https://crsreports.congress.gov/pro duct/details?prodcode=RL34761 (accessed 16 September 2024).

Corlett, E. (2022) New Zealand farmers may pay for greenhouse gas emissions under world-first plans. The Guardian. Available at: https://www.theguardian.com/world/2022/oct/11/new-zealand-farmers -may-pay-for-greenhouse-gas-emissions-under-world-first-plans (accessed 13 August 2024).

Cornwell, W. (2017) Trump's EPA has blocked agency grantees to serve of science advisory panels. Here is what it means. *Science Insider.* Available at: https://www.science.org/content/article/trump-s-epa-has-blocked-agency-grantees-serving-science-advisory-panels-here-what-it (accessed 16 September 2024).

COVID-19 Excess Mortality Collaborators (2022) Estimating excess mortality due to the COVID-19 pandemic: A systematic analysis of COVID-19-related mortality, 2020-21. *The Lancet* 399, 1513–1536. Available at: https://doi.org/10.1016/S0140-6736(21)02796-3 (accessed 15 September 2024).

Cressey, D. (2013) Monsanto drops GM in europe. *Nature* 499, 387. Available at: https://doi.org/10.1038/499387a (accessed 15 September 2024).

CRN (2023) 2023 Consumer Survey on Dietary Supplements. Council for Responsible Nutrition (CRN), Washington, DC. Available at: https://www.crnusa.org/2023survey (accessed 16 September 2024).

Dahl, J. (2024) Risk assessments and risk mitigation to prevent the introduction of African swine fever into the Danish pig population. *Animals* 14(17), 2491. Available at: https://doi.org/10.3390/ani14172491 (accessed 15 September 2024).

De Martinis, M., Sirufo, M.M., Suppa, M. and Ginaldi, L. (2020) New perspectives in food allergy. *International Journal of Molecular Science* 21(4), 1474. Available at: https://doi.org/10.3390/ijms210 41474 (accessed 19 August 2024).

DWIH New York (n.d.) The world's largest indoor, vertical farm. DWIH New York. Available at: https://www.dwih-newyork.org/en/2020/10/01/the-worlds-largest-indoor-vertical-farm/ (accessed 13 August 2024).

Ebbersten, S. (1972) Studies of persistence in soil and plants and methodological studies on biological determination of picloram residues (in Swedish). Department of Crop Production, Swedish University of Agricultural Sciences, Upsala, Sweden.

Ebbersten, S. (1983) Residues of chemical pesticides with special regard to the preparation lontrel kombi: some results and a discussion of principles (in Swedish). Department of Crop Production, Swedish University of Agricultural Sciences, Uppsala, Sweden.

Epp, M. (2020) Infarm, the new generation of farm. *Direct Industry/Emag.* Available at: https://emag.directindustry.com/2020/08/11/infarm-the-new-generation-of-farm-urban-farming (accessed 16 September 2024).

European Chemicals Agency (ECHA) (2017) Glyphosate not classified as a carcinogen by ECHA. ECHA. Available at: https://echa.europa.eu/-/glyphosate-not-classified-as-a-carcinogen-by-echa (accessed 19 August 2024).

European Food Safety Authority (EFSA) (2015) Glyphosate: EFSA updates toxicological profile. EFSA. Available at: https://www.efsa.europa.eu/en/press/news/151112 (accessed 19 August 2024).

European Food Safety Authority (EFSA) (2021) Edible insects: The science of novel food evaluations. EFSA, Parma, Italy. Available at: https://www.efsa.europa.eu/en/news/edible-insects-science-novel -food-evaluations (accessed 16 September 2024).

European Parliament (2024) Plants produced by certain new genomic techniques. In A European Green Deal. Legislative Train Schedule, Strasbourg. Members' Research Service.

European Union (EU) (2018) Current status of the neonicotinoids in the EU. European Commission Directorate-General for Health and Food Safety, Brussels. Available at: https://food.ec.europa.eu/plants/pesticides/approval-active-substances-safeners-and-synergists/renewal-approval/neonicoti noids_en (accessed 16 September 2024).

FAOSTAT (2021). Food and Agriculture Organization of the United Nations (FAO), Rome. Available at: http s://www.fao.org/faostat/en/#data (accessed 16 September 2024).

Farrell, M. (2003) Years after Monsanto deal, Bayer's roundup bills keep piling up. *The New York Times*. Available at: https://www.nytimes.com/2023/12/06/business/monsanto-bayer-roundup-lawsuit-sett lements.html (accessed 15 September 2024).

Fertiberia (2024) Thirteen leading companies with a common purpose. Available at: https://www.fertiberia. com/en (accessed 16 September 2024).

Folkhälsomyndigheten (2024) Statistik om psykisk hälsa i Sverige. Available at: https://www.folkhälsom yndigheten.se/livsvillkor-levnadsvanor/psykisk-halsa-och-suicidprevention/statistik-psykisk-halsa/ (accessed 15 September 2024).

Food and Agriculture Organization (FAO) (2001). International Treaty on Plant Genetic Resources for Food and Agriculture (ITPGRFA). FAO, Rome. Available at: https://www.fao.org/in-action/right-to- food-gl obal/global-level/itpgrfa/en (accessed 16 September 2024).

Food and Agriculture Organization (FAO) (2019). The contribution of insects to food security, livelihoods and the environment. FAO, Rome. Available at: https://www.fao.org/fsnforum/resources/reports-a nd-briefs/contribution-insects-food-security-livelihoods-and-environment (accessed 16 September 2024).

Food and Agriculture Organization (FAO) (2021) Looking at edible insects from a food safety perspective. Challenges and opportunities for the sector. FAO, Rome. Available at: https://openknowledge.fao. org/handle/20.500.14283/cb4094en (accessed 16 September 2024).

Fortune Business Insights (2022) With 25.9% CAGR, vertical farming market worth USD20.91 billion by 2029. *Yahoo! Finance*. Available at: https://finance.yahoo.com/news/25-9-cagr-vertical-farming-132 000749.html (accessed 16 September 2024).

Frans, E. (2023) *The Expert Paradox: Should I Always Listen to the Scientists* (in Swedish). Bonnier Fakta, Stockholm.

Freedgood (2020) Farms Under Threat: State of the States project, version 2.0. American Farmland Trust, Northampton, MA. Available at: https://farmlandinfo.org/wp-content/uploads/sites/2/2020/09/AFT_ FUT_StateoftheStates_rev.pdf (accessed 16 September 2024).

Gazzoni, D.L., Cattelan, A.J. and Nogueira, M.A. (2019) *Does the Brazilian Soybean Production Increase Pose a Threat on the Amazon Rainforest?*Embrapa Soja, Londrina, Brazil.

Grand View Research (2024) Plant-based Meat Market Size, Share & Growth Report, 2030. Grand View Research, San Francisco, CA. Available at: https://www.grandviewresearch.com/industry-analysis/ plant-based-meat-market (accessed 16 September 2024).

Guan, R., Chou, D., Han, X., Miao, X. and Li, H. (2021) Advances in the development of microbial double-stranded RNA production systems for application of RNA interference in agricultural pest control. *Frontiers in Bioengineering and Biotechnology Sec. Industrial Biotechnology* 9, 21. Available at: https://doi.org/10.3389/fbioe.2021.753790 (accessed 16 September 2024).

Guo, K., Yang, J., Yu, N., Luo, L. and Wang, E. (2023) Biological nitrogen fixation in cereal crops. Progress, strategies, and perspective 2(*Plant Communication*), 100499. Available at: https://doi.org/10.1016/j. xplc.2022.100499 (accessed 19 August 2024).

Handelsman, J. (2021) *A World Without Soil: The Past, Present, and Precarious Future of the Earth Beneath Our Feet*. Yale University Press, New Haven, CN.

Hawes, J., Goldstein, B. and Newell, J. (2024) Urban agriculture isn't as climate friendly as it seems, but these best practices can transform gardens and city farms. *The Conversation*. University of Michigan, Ann Arbor, MI. Available at: https://theconversation.com/urban-agriculture-isnt-as-clim ate-friendly-as-it-seems-but-these-best-practices-can-transform-gardens-and-city-farms-221537 (accessed 16 September 2024).

Hawkes, J.G. and Francisco-Ortega, J. (1992) The potato in Spain during the late 16th century. *Economic Botany* 46(1), 86–97. Available at: http://www.jstor.org/stable/4255411 (accessed 16 September 2024).

Hedenus, F., Persson, M. and Sprei, F. (2022) *Sustainable Development: Nuances and Perspectives*. Studentlitteratur AB, Lund, Sweden.

Herranz, A.G. (2018) The Indonesian palm oil alternative that's powered by insects. *RESET Digital for Good*. Available at: https://en.reset.org/switching-our-diets-vegetable-oils-bug-one-05062018/ (accessed 16 September 2024).

Herrmann, U. (2022) *The End of Kapitalism* (in German). Verlag Kiepenheuer & Witsch GmbH & Co. KG, Cologne, Germany.

Höijer, I., Emmanouilidou, A., Östlund, R., van Schendel, R., Bozorgpana, S. *et al.* (2022) CRISPR-Cas9 induces large structural variants at on-target and off-target sites *in vivo* that segregate across

generations. *Nature Communications* 13, 627. Available at: https://doi.org/10.1038/s41467-022-282
44-5 (accessed 16 September 2024).

Horgan, J. (1996) *The End of Science: Facing the Limits of Knowledge in the Twilight of the Scientific Age.*
Addison-Wesley, US.

Huisman, R. and Geurts, R.A. (2019) A roadmap toward engineered nitrogen-fixing nodule symbiosis. *Plant
Communications* 1(1), 100019. Available at: https://doi.org/10.1016/j.xplc.2019.100019 (accessed
19 August 2019).

Hunter, D., Borelli, T. and Gee, E. (eds) (2020) *Biodiversity, Food and Nutrition: A New Agenda for
Sustainable Food Systems.* Routledge, New York.

IFOAM-EU (2018) News of December 12, 2018. Political Hotspot.

International Agency for Research on Cancer (IARC) (2015) IARC monographs volume 112: Evaluation
of five organophosphate insecticides and herbicides. IARC, Lyon, France. Available at: https://ww
w.iarc.who.int/wp-content/uploads/2018/07/MonographVolume112-1.pdf (accessed 16 September
2024).

International Service for the Acquisition of Agri-biotech Applications (ISAAA) (2018) Study: 20 Years of GM
Adoption in Brazil Increased Farmers' Profits, Boosted Economy, and Preserved the Environment.
Biotech Updates, ISAAA, Metro Manila, Philippines. Available at: https://www.isaaa.org/kc/cropbiot
echupdate/article/default.asp?ID=16796 (accessed 15 September 2024).

International Service for the Acquisition of Agri-biotech Applications (ISAAA) (2019a) Global Status of
Commercialized Biotech/GM Crops: 2019. ISAAA, Metro Manila, Philippines. Available at: https://w
ww.isaaa.org/resources/publications/briefs/55 (accessed 15 September 2024).

International Service for the Acquisition of Agri-biotech Applications (ISAAA) (2019b) Biotech Crops Drive
Socio-Economic Development and Sustainable Environment in the New Frontier. ISAAA Brief 55,
Executive Summary. ISAAA, Metro Manila, Philippines. Available at: https://www.isaaa.org/resources
/publications/briefs/55/executivesummary/default.asp (accessed 15 September 2024).

Jia, H., Powell, S. and Schienmakers, K. (2021) China's leading researchers set their sights on new frontiers. *Nature* 593, S24–S27. Available at: https://doi.org/10.1038/d41586-021-01405-0 (accessed
16 September 2024).

Johansson, M. (2023) Minister: Cheating grades a big problem (in Swedish). Svenska Dagbladet.
Stockholm.

Jordbruksverket (2023) *Gödselmedelsproduktion i Sverige: Aktuella initiativ, tekniker och förutsättningar.*
Bilaga till Missiv 2023-03-31 Diarienummer 3.1.17-17378/2022, Jönköping, Sweden. Available at:
https://www2.jordbruksverket.se/download/18.7dcc8c181886a656b837d93/1709117085578/ra23
_9.pdf (accessed 15 September 2024).

Kemi (2023a) Product Details (in Swedish). Kemikalieinspektionen (Kemi), Sundbyberg, Sweden.

Kemi (2023b) Products Register (in Swedish). Kemikalieinspektionen (Kemi), Sundbyberg, Sweden.

Kitozyme (2024) Natural solutions for healthier and greener planet. Available at: https://www.kitozyme.
com/en/ (accessed 16 September 2024).

Kletskova, E.V., Klimova, O.V. and Zhidkikh, A.A. (2022) Developing a sustainable system of natural
resource management. In: Popkova, E.G. and Sergi, S.G. (eds) *Geo-Economy of the Future:
Sustainable Agriculture and Alternative Energy.* Springer, Cham, Switzerland, pp. 111–118.

Konov, A., Velchev, M. and Parcel, D. (2005) Plant genetic engineering in monsanto company: From the
first laboratory experiments to worldwide practical use. *TSitologiia i genetika* 39, 3–12. Available at:
https://www.researchgate.net/publication/7516871_Plant_genetic_engineering_in_Monsanto_com
pany_From_the_first_laboratory_experiments_to_worldwide_practical_use (accessed 15 September
2024).

Kourimska, L. and Adamova, A. (2016) Nutritional and sensory quality of edible insects. *NFS Journal* 4,
22–26. Available at: https://doi.org/10.1016/j.nfs.2016.07.001 (accessed 19 August 2024).

Kravchenko, Y., Balayev, A., Mazăre, V., Zhang, X., Liu, X. *et al.* (2018) Conservation practices on Ukrainian
Mollisols: A mini review. *International Journal of Energy and Environment* 12, 1–9.

Kumar, J., Ramlal, A., Mallick, D. and Mishra, V. (2021) An overview of some biopesticides and their
importance in plant protection for commercial acceptance. *Plants* 10(6), 1185. Available at: https://
doi.org/10.3390/plants10061185 (accessed 19 August 2024).

Labaz, B., Hartemink, A.E., Zhang, Y., Stevenson, A. and Kabała, C. (2024) Organic carbon in Mollisols of
the world: A review. *Geoderma* 447, 116937. Available at: https://doi.org/10.1016/j.geoderma.2024.
116937 (accessed 16 September 2024).

Länstyrelsen (2020) Rälla-Ekerum (in Swedish) Länstyrelsen i Kalmar Län, Kalmar. Available at: https://lan styrelsen.kalmar.se/kalmar/besoksmal/naturreservat/ralla (accessed 20 October 2024).

Lantmännen (2024) Fertilizer. Lantmännen, Stockholm. Available at: https://www.lantmannen.com/sustai nable-development/important-issues/plant-nutrients-fertiliser (accessed 16 September 2024).

Larsson, L. (2021) Brist på stallgödsel begränsar ekologisk odling. AgriFood Policy Brief. Lunds universitet, Lund, Sweden. Available at: https://www.lu.se/artikel/brist-pa-stallgodsel-begransar-ekologisk-odling (accessed 15 September 2024).

Lauterbach, K. (2022) *Bevor es zu spät ist: was uns droht, wenn die Politik nicht mit der Wissenschaft Schritt halt. s 164*. Rowohlt, Berlin.

Le Monde (2023) EU parliament scraps proposal to halve pesticide use. *Le Monde*. Available at: https://www.lemonde.fr/en/international/article/2023/11/22/eu-parliament-scraps-proposal-to-halve-pesti cide-use_6278915_4.html (accessed 16 September 2024).

Leone, C.A. (1952) Genetics: Lysenko versus Mendel. *Transactions of the Kansas Academy of Science* 55(4), 369–380. Available at: https://doi.org/10.2307/3625986 (accessed 19 August 2024).

Lewis, K.A., Tzilivakis, J., Warner, D. and Green, A. (2016) An international database for pesticide risk assessments and management. *Human and Ecological Risk Assessment: An International Journal* 22(4), 1050–1064. Available at: https://doi.org/10.1080/10807039.2015.1133242 (accessed 16 September 2024).

Li, M., Lu, Z.K., Amrol, D.J., Mann, J.R., Hardin, J.W. *et al.* (2019) Antibiotic exposure and the risk of food allergy: Evidence in the US medicaid paediatric population. *The Journal of Allergy and Clinical Immunology: In Practice* 7(2), 492–499. Available at: https://doi.org/10.1016/j.jaip.2018.09.036 (accessed 20 August 2024).

Lim, S.S., Vos, T., Flaxman, A.D., Danaei, G., Shibuya, K. *et al.* (2012) A comparative risk assessment of burden of disease and injury attributable to 67 risk factors and risk factor clusters in 21 regions, 1990–2010: A systematic analysis for the global burden of disease study 2010. *The Lancet* 380(9859), 2224–2260. Available at: https://www.thelancet.com/journals/lancet/article/PIIS0140-6736(12)61 766-8/abstract (accessed 16 September 2024).

Litvinova, T.N. and Zemskova, O.M. (2022) Sustainable agriculture and alternative energy based on high-tech entrepreneurship development in the agricultural machinery market. In: Popkova, E.G. and Sergi, S.G. (eds) *Geo-Economy of the Future: Sustainable Agriculture and Alternative Energy*. Springer, Cham, Switzerland, pp. 871–877.

Loftfield, E., O'Connell, C.P., Abnet, C.C., Graubard, B.I., Liao, L.M. *et al.* (2024) Multivitamin use and mortality risk in 3 prospective US cohorts. *JAMA Network Open* 7(6), e2418729. Available at: https://doi.org/10.1001/jamanetworkopen.2024.18729 (accessed 16 September 2024).

Ma, M. (2024) Methane-eating microbes show promise for wiping out planet-warming emissions. *Bloomberg*. Available at: https://www.bloomberg.com/news/articles/2024-04-08/methane-eating-microbes-promise-to-wipe-out-planet-warming-emissions (accessed 16 September 2024).

Mącik, M., Gryta, A. and Frąc, M. (2020) Chapter two – biofertilizers in agriculture: An overview on concepts, strategies and effects on soil microorganisms. *Advances in Agronomy* 162, 31–87. Available at: https://doi.org/10.1016/bs.agron.2020.02.001 (accessed 15 September 2024).

Maiyo, N.C., Khamis, F.M., Okoth, M.W., Abong, G.O., Subramanian, S. *et al.* (2022) Nutritional quality of four novel porridge products blended with edible cricket (*scapsipedus icipe*) meal for food. *Foods* 11(7), 1047. Available at: https://doi.org/10.3390/foods11071047 (accessed 20 August 2024).

Mamy, L., Pesce, S., Sanchez, W., Aviron, S., Bedos, C. *et al.* (2023) Impacts of neonicotinoids on biodiversity: A critical review. *Environmental Science and Pollution Research*. Available at: https://doi.org/10.1007/s11356-023-31032-3 (accessed 16 September 2024).

Marston, J. (2023) BiomEdit lands $4.5m Bill & Melinda gates grant to reduce cattle emissions in South Asia and Africa. *AgFunder News*. Available at: https://agfundernews.com/biomedit-lands-4-5m-bill-melinda-gates-grant-to-reduce-cattle-emissions-in-south-asia-and-africa (accessed 16 September 2024).

Martinson, K. (2015) Beware of Feeding Herbicide Treated Hay. University of Minnesota Extension, St. Paul, MN. Available at: https://onpasture.com/20150615/beware-of-feeding.herbicide.treated-hay (accessed 16 September 2024).

Melo, A.A. and Swarowsky, A. (2022) Application technology of biopesticides. In: Rakshit, A., Meena, V.S., Abhilash, P.C., Sarma, B.K. and Singh, H.B. (eds) *Biopesticides, Volume 2: Advances in Bio-Inoculants*. Elsevier, Amsterdam, pp. 31–36.

Mertens, E., Colizzi, C. and Peñalvo, J.L. (2022) Ultra-processed food consumption in adults across Europe. *European Journal of Nutrition* 61, 1521–1539. Available at: https://doi.org/10.1007/s00394-021-02733-7 (accessed 16 September 2024).

Miller, J. (2021) Is Oatly really eco-friendly?. *Slate*. Available at: https://slate.com/business/2021/07/oatly-eco-friendly-oat-milk-investors-controversy.html (accessed 16 September 2024).

Monsanto (2004) *Annual Report*. Monsanto Company. Available at: https://www.annualreports.com/HostedData/AnnualReportArchive/m/NYSE_MON_2004.pdf (accessed 15 September 2024).

Monteiro, C., Cannon, G., Lawrence, M., Louzada, M.L. and Machado, P. (2019) Ultra-processed foods, diet quality, and health using the NOVA classification system. Food and Agriculture Organization of the United Nations (FAO), Rome. Available at: https://openknowledge.fao.org/server/api/core/bitstreams/5277b379-0acb-4d97-a6a3-602774104629/content (accessed 16 September 2024).

Mupepele, A.C., Keller, M. and Dormann, C.F. (2021) European agroforestry has no unequivocal effect on biodiversity: A time-cumulative meta-analysis. *BMC Ecology and Evolution* 21(1), 193. Available at: https://doi.org/10.1186/s12862-021-01911-9 (accessed 20 August 2024).

Negatu, B., Kromhout, H., Mekonnen, Y. and Vermuelen, R. (2016) Use of chemical pesticides in Ethiopia: A cross-sectional comparative study on knowledge, attitude and practice of farmers and farm workers in three farming systems. *The Annals of Occupational Hygiene* 60(5), 551–566. Available at: https://doi.org/10.1093/annhyg/mew004 (accessed 20 August 2024).

Nicol, K., Thomas, E.-L., Nugent, A.P., Woodside, J.V., Hart, K.H. *et al.* (2023) Iodine fortification of plant-based dairy and fish alternatives: the effect of substitution on iodine intake based on a market survey in the UK. *British Journal of Nutrition* 129(5), 832–842. Available at: https://doi.org/10.1017/S0007114522001052 (accessed 15 September 2024).

Nihlén, A. (2020) Wild boar damage in agriculture is massive food waste (in Swedish). Lantbruksnytt. September 27, 2020, Agrar, Kristianstad, Sweden.

Nordic Council of Ministers (2023) *Nordic nutrition recommendations 2023*. Nordic Council of Ministers. Available at: https://www.norden.org/en/publication/nordic-nutrition-recommendations-2023 (accessed 19 August 2024).

Nufarm (2024) Selective herbicide. Alsip, IL. Available at: https://nufarm.com/uscrop/product/trooper-pd/ (accessed 16 September 2024).

Organisation for Economic Co-operation and Development (OECD) (2023) Programme for International Student Assessment (PISA). Directorate for Education and Skills, OECD, Paris. Available at: https://www.oecd.org/en/about/programmes/pisa.html (accessed 15 September 2024).

Organisation of Economic Co-operation and Development (OECD) (2022) Management of Pharmaceutical Household Waste: Limiting Environmental Impacts of Unused or Expired Medicine, OECD Publishing, Paris. Available at: https://doi.org/10.1787/3854026c-en (accessed 16 September 2024).

Padma, T.V. (2022) When will india approve its first GM food crop? *Nature* 611, 648–649. Available at: https://doi.org/10.1038/d41586-022-03738-w (accessed 20 August 2024).

Palli, S.R. (2014) RNA interference in colorado potato beetle: steps toward development of dsRNA as a commercial insecticide. *Current Opinion in Insect Science* 6, 1–8. Available at: https://doi.org/10.1016/j.cois.2014.09.011 (accessed 20 August 2024).

Palmstierna, H.A.K. and Palmstierna, L. (1968) *Looting, Starvation, Poisoning* (in Swedish). Rabén & Sjögren, Stockholm.

Palmstierna, H.A.K. and Palmstierna, L. (1972) *Upon Reflection* (in Swedish). Rabén & Sjögren, Stockholm.

Peters, A. (2021) PepsiCo is scaling up regenerative agriculture on seven million acres of land. *Fast Company*. Available at: https://www.fastcompany.com/90626895/pepsico-is-scaling-up-regenerative-agriculture-on-7-million-acres-of-land (accessed 20 August 2024).

Petrequin, A. (2023) EU commission to extend use of glyphosate for 10 more years after member countries fail to agree. *AP News*. Available at: https://apnews.com/article/eu-pesticides-glyphosate-2fe352fa4426dd90bc32acf1d3304b99 (accessed 16 September 2024).

Phillipov, M. and Kirkwood, K. (eds) (2018) *Alternative Food Politics: From the Margins to the Mainstream*. Routledge, New York.

Pichdara, L. (2021) *Contract Farming in Mekong Countries: Best Practices and Lessons Learned*. Cambodia Development Resource Institute (CDRI), Phnom Penh, Cambodia.

Pontonio, E. and Rizello, C.G. (2019) Minor and ancient cereals: Exploitation of the nutritional potential through the use of selected starters and sourdough fermentation. In: Preedy, V.R. and Watson, R.R. (eds) *Flour and Breads and Their Fortification in Health and Disease Prevention*. Elsevier, Amsterdam, pp. 443–452.

Popkova, E.G. and Sergi, B.S. (eds) (2022) *Geo-Economy of the Future: Sustainable Agriculture and Alternative Energy*. Springer, Cham, Switzerland.

Potrykus, I. (2012) Golden rice", a GMO-product for public good, and the consequences of GE-regulation. *Journal of Plant Biochemistry and Biotechnology* 21(1), 68–75. Available at: https://doi.org/10.1007/s13562-012-0130-5 (accessed 15 September 2024).

Power2Earth (2024a) Reduced climate impact and strengthened food supply: Fertiberia, Lantmännen, and Nordion Energi are investing to develop Sweden's first fossil-free mineral fertiliser factory. Power2Earth. Available at: https://www.power2earth.se/news/reduced-climate-impact-and-strengthened-food-supply (accessed 16 September 2024).

Power2Earth (2024b) Power2Earth and the Hydrogen Plant. Power2Earth. Available at: https://www.power2earth.se/our-initiative (accessed 16 September 2024).

Prada, P. (2015) Paraquat: A controversial chemical's second act. *Reuters*. Available at: https://www.reuters.com/article/markets/commodities/paraquat-a-controversial-chemicals-second-act-idUSL2N0WY2V7/ (accessed 16 September 2024).

Purseglove, J.W. (1974) *Tropical Crops: Dicotyledons*, Vol. 1 and 2. Longman, London.

Reimer, N. and Staud, T. (2021) *Deutschland 2050. Wie der Klimawandel unser Leben verändern wird, s 255*. Kiepenheuer & Witsch, Cologne, Germany.

Research and Markets (2022) Global Industrial Hemp Market Report 2022: Increasing Legalization of Industrial Hemp Cultivation Drives Growth. GlobeNewswire. Available at: https://www.globenewswire.com/en/news-release/2022/12/06/2568001/28124/en/Global-Industrial-Hemp-Market-Report-2022-Increasing-Legalization-of-Industrial-Hemp-Cultivation-Drives-Growth.html (accessed 20 August 2024).

Research and Markets (2024) *Glyphosate Global Market Report 2024*. Available at: https://www.researchandmarkets.com/reports/5741480/glyphosate-global-market-report?srsltid=AfmBOor5J9_T5Pd9Es8GIxLANgNMsvn_QAS6xHV4NqG7NFMB36wyeC8l (accessed 15 September 2024).

Rispoli, G. (2014) The role of Isaak Prezent in the rise and fall of Lysenkoism. *Ludus Vitalis* 22, 42.

RNZ (2023) Government confirms its 100-day plan. *Radio New Zealand*. Available at: https://www.rnz.co.nz/news/political/503534/government-confirms-its-100-day-plan (accessed 16 September 2024).

Robért, K.-H., Broman, G., Waldron, D., Ny, H., Hallstedt, S. *et al.* (2019) *Sustainability Handbook*. Studentlitteratur AB, Lund, Sweden.

Rosa, L. and Gabrielli, P. (2022) Energy and food security implications of transitioning synthetic nitrogen fertilizers to net-zero emissions. *Environmental Research Letters* 18(1), 014008. Available at: https://doi.org/10.1088/1748-9326/aca815 (accessed 20 August 2024).

Rowlands, M. (2021) *World on Fire: Humans, Animals, and the Future of the Planet*. Oxford University Press, New York.

Ruegg, P. (2022) For the first time, farmers in the Philippines cultivated golden rice on a larger scale and harvested almost 70 tons. ETZ Zurich. Available at: https://phys.org/news/2022-11-farmers-philippines-cultivated-golden-rice.html#google_vignette (accessed 15 September 2024).

Saito, K. (2024) *Slow Down: The Degrowth Manifesto*. Astra House, Westminster, MD.

Salami, S.A., Moran, C.A., Warren, H.E. and Taylor-Pickard, J. (2020) A meta-analysis of the effects of slow-release of urea supplementation on the performance of beef cattle. *Animals* 10, 657. Available at: https://www.mdpi.com/2076-2615/10/4/657#:~:text=There%20was%20no%20effect%20of,)%20and%20FE%20(18%20vs

Santos, T. (2023) Largest ever public screening finds "alarmingly high" chemical exposure. *News, European Environmental Bureau (EBB), Brussels*. Available at: https://eeb.org/european-citizens-alarmingly-high-chemical-exposure/ (accessed 16 September 2024).

SCB (2024) Prices in Sweden - Consumer Price Index (KPI) (in Swedish). Statistics Sweden (SCB) Solna, Sweden. Available at: https://www.scb.se/hitta-statistik/sverige-i-siffror/samhallets-ekonomi/kpi (accessed 16 September 2024).

Scherer, H.W., Mengel, K., Kluge, G. and Severin, K. (2000) Fertilizers. In: *Ullmann's Encyclopaedia of Industrial Chemistry*. Wiley-VCH, Weinheim, Germany.

Schreefel, L., Schulte, R.P.O., de Boer, I.J.M., Pas Schrijver, A. and van Zanten, H.H.E. (2020) Regenerative agriculture: The soil is the base. *Global Food Security* 26, 100404. Available at: https://doi.org/10.1016/j.gfs.2020.100404 (accessed 20 August 2024).

Scott, A. (2019) DSM seeks approval of additive that minimizes methane from cattle. American Chemical Society, *Chemical and Engineering News* 97(30). Available at: https://cen.acs.org/buiness/food-ingredients/DSM-seeks-approval-additive-minimizing/97/i30 (accessed 16 September 2024).

Shelef, O., Weisberg, P.J. and Provenza, F.D. (2017) The value of native plants and local production in an era of global agriculture. *Frontiers in Plant Science* 8, 2069. Available at: https://doi.org/10.3389/fpls.2017.02069 (accessed 16 September 2024).

Silva, F.M., Giatti, L., Figueiredo, R.C. de, Molina, M. del C.B., Cardoso, L. de O. *et al.* (2018) Consumption of ultra-processed food and obesity: Cross sectional results 34 from the brazilian longitudinal study of adult health (ELSA-Brasil) cohort (2008–2010). *Public Health Nutrition* 21(12), 2271–2279. Available at: https://doi.org/10.1017/S1368980018000861 (accessed 16 September 2024).

Soon, L.G. and Bottrel, D.G. (1994) *Neem Pesticides in Rice: Potential and Limitations*. International Rice Research Institute (IRRI), Manila, Philippines.

Stark, A. (2021) This burp-catching mask for cows could slow down climate change. WIRED. Available at: https://www.wired.com/story/cows-climate-change-methane-stop (accessed 16 September 2024).

Statista (2022) Agricultural consumption of pesticides worldwide from 1990 to 2021. Statista, July, 2022. Available at: https://www.statista.com/statistics/1263077/global-pesticide-agricultural-use/ (accessed 19 August 2024).

Statista (2024a) Acreage of genetically modified crops worldwide 2003-2019. Statista. Available at: https://www.statista.com/statistics/263292/acreage-of-genetically-modified-crops-worldwide/ (accessed 15 September 2024).

Statista (2024b) Sorghum production by leading country worldwide 2023/2024. Statista. Available at: https:// www.statista.com/statistics/1134651/global-sorghum-production-by-country/ (accessed 16 September 2024).

Stempel, J. (2023) Appeals court blocks california warning requirement for glyphosate. *Reuters*. Available at: https://www.reuters.com/legal/appeals-court-blocks-california-warning-requirement-glyphosate-2023-11-07/ (accessed 16 September 2024).

Stewart, S. and McClure, A. (2021) Insect Control Recommendations for Field Crops Cotton, Soybean, Field Corn, Sorghum, Wheat and Pasture. UTIA Institute of Agriculture, The University of Tennessee Institute of Agriculture, Knoxville, TN. Available at: https://utcrops.com/wp-content/uploads/2021/03/PB_1768_2021_Insect_Control_Recommendations_for_Field_Crops.pdf   (accessed   16 September 2024).

Stocking, B. (2007) Agent orange still haunts Vietnam, US. *The Washington Post*. Available at: https:// kval.com/news/nation-world/agent-orange-still-haunts-vietnam-us (accessed 16 September 2024).

Stone, G.D. (2022) *The Agricultural Dilemma: How Not to Feed the World*. Routledge, New York.

Strannegård, L. (2023) Happiness grades can force us to introduce exams (in Swedish). DN Debatt, Dagens Nyheter, Stockholm.

Svensk Jakt (2021) Record shooting: Wild boar is the most common hunting prey today (in Swedish). *Svensk Jakt*, Nyköping, Sweden. Available at: https://svenskjakt.se/start/nyhet/rekordavskjutning-vildsvin-ar-det-vanligaste-jaktbytet-i-dag/ (accessed 15 September 2024).

Sveriges riksdag (2021) Hunting wild boar (in Swedish). Motion till riksdagen 2021/22:677. Sveriges riksdag, Stockholm. Available at: https://www.riksdagen.se/sv/dokument-och-lagar/dokument/motion/jakt-pa-vildsvin_h902677/ (accessed 15 September 2024).

Swedish Environmental Protection Agency (2023) Frequently asked questions about hunting (in Swedish). Swedish Environmental Protection Agency, Stockholm.

Swedish Food Agency (2024a) Drinking water (in Swedish), Livsmedelsverket 2024/02/14, Uppsala. Available at: https://www.livmedelsverket.se>livsm (accessed 22 October 2024).

Swedish Food Agency (2024b) Ultra-processed food (in Swedish). Available at: https://www.livsmedelsverket.se/livsmedel-och-innehall/tillagning-forvaring-hallbarhet/tillagning/ultraprocessad-mat (accessed 16 September 2024).

Swedish Government (1984) Lag (1984:3) om kärnteknisk verksamhet. Klimat- och näringslivsdepartementet, Regeringskansliets rättsdatabaser. Available at: https://www.vertic.org/media/National%20 Legislation/Sweden/SE_Act_on_Nuclear_Activities_1984.pdf

Swedish Government (2005) Regeringens proposition 2005/06:76 Kärnsäkerhet och strålskydd. Regeringskansliets rättsdatabaser, SFS-nummer 1984:3, Ändring SFS 2006:339. Available at: https://www.regeringen.se/rattsliga-dokument/proposition/2006/02/prop.-20050676

Swedish Hunters Association (2022) Historik (in Swedish). Available at: https://jagarforbundet.se/..../dag gdjur/vildsvin/vildsvin-historik (accessed 20 October 2024).

Swedish National Agency for Education (2022) Grades and study results in upper secondary school (in Swedish). Solna, Sweden, No. 2022:2882. Available at: https://www.skolverket.se/publikationer?id= 10522 (accessed 15 September 2024).

Sylla, Z. and Wolfe, E. (2024) Bayer ordered to pay $2.25 billion after jury concludes Roundup weed killer caused a man's cancer, attorneys say. *CNN*. Available at: https://edition.cnn.com/2024/01/29/us /roundup-cancer-verdict-philadelphia-bayer-monsanto/index.html (accessed 15 September 2024).

Tattersall, J., Arai, M., Rayman, M.P., Stergiadis, S. and Bath, S.C. (2023) A systematic review of the iodine concentration in milk from around the world. *Proceedings of the Nutrition Society* 82(OCE5), E357. Available at: https://doi.org/10.1017/S0029665123004561 (accessed 15 September 2024).

The Land Institute (2024) Transforming Agriculture, Perennially: Kernza® Grain. The Land Institute, Salina, KS. Available at: https://landinstitute.org/our-work/perennial-crops/kernza (accessed 16 September 2024).

Trenda, E. (2023) Agricultural land devoted to the culture of hemp in Europe in 2020, by country. Statista. Available at: https://www.statista.com/statistics/1204146/area-for-hemp-cultivation-by-country-eur ope/ (accessed 16 September 2024).

Tucker, S.C. (ed.) (2011) *The Encyclopedia of the Vietnam War: A Political, Social, and Military History*, 2nd edn. ABC-CLIO, Santa Barbara, CA.

Tzvetkova, P., Lyubenova, M., Boteva, S., Todorovska, E., Tsonev, S. *et al.* (2019) Effect of herbicides paraquat and glyphosate on the early development of two tested plants. *IOP Conference Series: Earth and Environmental Science* 221, 012137. Available at: https://iopscience.iop.org/article/10.10 88/1755-1315/221/1/012137/pdf (accessed 16 September 2024).

Umwelt Bundesamt (UBA) (2024) PHARMS-UBA. Available at: https://www.umweltbundesamt.de/en/dat abase-pharmaceuticals-in-the-environment-0 (accessed 16 September 2024).

Unilever (2010) Unilever Sustainable Agriculture Code, Version 1. Unilever, London. Available at: https:// www.unilever.com/files/origin/97da207c4d3a07bfa611e72454c8ab3637fe77b4.pdf/ul-sac-v1-marc h-2010-spread.pdf (accessed 14 September 2024).

United Nations (UN) (2022) FAO warns 90 per cent of Earth's topsoil at risk by 2050. *UN News*. Available at: https://news.un.org/en/story/2022/07/1123462 (accessed 16 September 2024).

United States Department of Agriculture (USDA) (2023) Adoption of Genetically Engineered Crops in the U.S. Economic Research Service, U.S. Department of Agriculture, Washington DC. Available at: https://www.ers.usda.gov/data-products/adoption-of-genetically-engineered-crops-in-the-u-s (accessed 15 September 2024).

United States Environmental Protection Agency (EPA) (2017) Revised Glyphosate Issue Paper: Evaluation of Carcinogenic Potential. EPA, Washington DC. Available at: https://www.epa.gov/sites/default/ files/2016-09/documents/glyphosate_issue_paper_evaluation_of_carcincogenic_potential.pdf (accessed 16 September 2024).

United States Environmental Protection Agency (EPA) (2023) Technical Factsheet on: PICLORAM. EPA, Washington, DC. Available at: https://archive.epa.gov/water/archive/web/pdf/archived-technical-fact-sheet-on-picloram.pdf (accessed 16 September 2024).

United States Environmental Protection Agency (EPA) (2024) The Impact of Pharmaceuticals Released to the Environment: Pharmaceuticals in the Environment Shown to Negatively Affect Ecosystems. EPA, Washington DC. Available at: https://www.epa.gov/household-medication-disposal/impact-pharma ceuticals-released-environment (accessed 16 September 2024).

Urban Vine (2024) Global Indoor Farming Companies: The Top 750 Report, 2023. Urban Vine. Available at: https://www.urbanvine.co/guide-2023 (accessed 16 September 2024).

Van Noorden, R. (2022) COVID death tolls: Scientists acknowledge errors in WHO estimates. *Nature* 606, 242–244. Available at: https://doi.org/10.1038/d41586-022-01526-0 (accessed 15 September 2024).

Waltz, E. (2021) GABA-enriched tomato is first CRISPR-edited food to enter market. *Nature Biotechnology* 40, 9–11. Available at: https://doi.org/10.1038/d41587-021-00026-2 (accessed 8 September 2024).

Wiener-Bronner, D. (2023) Tyson Foods, one of the biggest meat producers, is investing in insect protein. *CNN Business*. Available at: https://www.cnn.com/2023/10/20/business/tyson-insect-ingredients (accessed 16 September 2024).

Willett, W., Rockström, J., Loken, B., Springmann, M., Lang, T. *et al.* (2019) Food in the Anthropocene: The EAT-lancet commission on healthy diets from sustainable food systems. *The Lancet* 393(10170), 447–492. Available at: https://doi.org/10.1016/S0140-6736(18)31788-4 (accessed 14 September 2024).

Woodcock, B.A., Bullock, J.M., Shore, R.F., Heard, M.S., Pereira, M.G. *et al.* (2017) Country-specific effects of neonicotinoid pesticides on honeybees and wild bees. *Science* 356(6345), 1393–1395. Available at: https://doi.org/10.1126/science.aaa1190 (accessed 16 September 2024).

World Health Organization (WHO) (2024a) Prevention and control of iodine deficiency in the WHO European region: Adapting to changes in diet and lifestyle. WHO. Regional Office for Europe, Geneva, Switzerland.

World Health Organization (WHO) (2024b) Obesity and overweight. WHO, Geneva, Switzerland. Available at: https://www.who.int/news-room/fact-sheets/detail/obesity-and-overweight (accessed 16 September 2024).

World Meteorological Organization (WMO) (2023) World Meteorological Congress. WMO, Geneva, Switzerland. Available at: https://wmo.int/media/news/economic-costs-of-weather-related-disasters-soars-early-warnings-save-lives (accessed 15 September 2024).

Xu, X., Pei, J., Xu, Y. and Wang, J. (2020) Soil organic carbon depletion in global Mollisols regions and restoration by management practices: A review. *Journal of Soils and Sediments* 20, 1173–1181. Available at: https://doi.org/10.1007/s11368-019-02557-3 (accessed 16 September 2024).

Yara (2022) On course to a native-positive food future. Yara Integrated Report 2022, Oslo, Norway.

Yara (n.d.) Low-carbon footprint fertilizers: A game changer. Yara, Oslo, Norway. Available at: https://www.yara.com/sustainability/transforming-food-system/low-carbon-footprint-fertilizers/ (accessed 13 August 2024).

Yeo, G. (2021) *Why Calories Don't Count: How We got the Science of Weight Loss Wrong*. Orion, London.

Yu, S., Ali, J., Zhou, S., Ren, G., Xie, H. *et al*. (2022) From green super rice to green agriculture: Reaping the promise of functional genomics research. *Molecular Plant* 15(1), 9–26. Available at: https://doi.org/10.1016/j.molp.2021.12.001 (accessed 20 August 2024).

Zhang, F.F., Barr, S.I., McNulty, H., Li, D. and Blumberg, J. (2020) Health effects of vitamin and mineral supplements. *The BMJ* 369, m2511. Available at: https://doi.org/10.1136/bmj.m2511 (accessed 16 September 2024).

Zuckerman, J.C. (2021) *Planet Palm: How Palm Oil Ended up in Everything—and Endangered the World*. The New Press, New York.

10 Major Actors Who Could Implement the Concept of Sustainability in Agriculture with Societal Development

Abstract

The conventional agricultural actors, such as the national governments and aid donors, are meeting increased competition from the agribusiness transnational corporations (TNCs), who are also operating in many low-income countries. They may even be even assisted by the TNCs, if food production becomes more attractive to these companies, having advanced their research and expertise in certain areas. This will call for specific political requirements covering the involvement of agribusiness TNCs and for more attention to be paid to sustainability issues. For many years, the United States has exerted its global leadership in agriculture and been copied by other countries, but a transformation towards sustainable agriculture for 2050 requires significant changes. Other actors may become more influential in societies where ecological footprints are less significant due to their current systems of smaller scale farming and agriculture.

Governments and Aid Donors

National governments have responsibility for future food security with sustainability for their populations. They have all agreed to this through their international commitments in accepting Agenda 2030, the Sustainable Development Goals (SDGs) and statements at COP28 in Dubai. So far, there are few, if any concrete results showing that governments in the industrialized countries have taken any major steps to change their agriculture and their methods to achieve sustainability. The overall theme is business as usual, often relying on food imports to obtain the cheapest food. This means a postponement of their commitments to Agenda 2030 and the SDGs.

Most low-income countries are still agrarian communities with a sizeable proportion of the population living in rural areas. It may be easier for the politicians in these countries to find a balance between the priorities of all sectors of the economy in the interests of long-term sustainability, than in high-income countries. They are totally dependent on economic growth, fossil fuel energy and have populations whose consumption is steadily increasing. The problem is, however, that low-income countries follow the conventional modernization process, pursuing Western-type development. That model has several weaknesses in terms of long-term development for a sustainable Earth, and the United Nations (UN) has agreed with this view.

In the late 1950s, international development assistance was a response to decolonization by the major economic powers. It was seen as an approach for them to maintain economic and political relations with the 'Third World'. To the United States it was also a way to fight communism and expand capitalism. Since the mid-1960s, donor agencies from most industrialized countries have spent millions of US dollars on thousands of rural and agricultural projects in low-income countries. So far, there have been certain positive changes in low-income countries. Some have made more progress than others, but the overall effects seem to have been less than expected after more than half a century of large financial investments by donor agencies working together with the governments of low-income countries. There is an increasing trend of people who wish to migrate to find jobs and better living conditions in other countries, mainly in the industrialized world. Poverty has not been eradicated and new issues have emerged, some of which emanate

© Bo Malte Ingvar Bengtsson 2025. *Agricultural Innovation for Societal Change: Towards Sustainability* (B.M.I. Bengtsson)
DOI: 10.1079/9781800627802.0010

from the technological solutions introduced by aid donors.

There have been some critical views in the effectiveness of foreign aid and how most donors and practitioners have thought about development. Most aid donors have identified problems in low-income countries as purely technical, and solutions being easy to implement with the help of dictatorial regimes, while disregarding the rights of ordinary people (Easterly, 2016). Many of these technical solutions applied by high-income countries in low-income countries have not transformed them successfully.

There is a different story with South Korea, which was a poor country after the Korean war in the early 1950s. In 1962, representatives of the country and the International Development Association (IDA) of the World Bank signed a credit agreement for a railroad project (Ferro and Nishio, 2021). New loans were taken for irrigation, schools and universities which led to industrial growth. After 1980, it received no more development aid. Instead, it started to export in the early 1980s and joined the Organization for Economic Co-operation and Development (OECD) in the following decade. In a little more than half a century, South Korea became the first former aid recipient to join the OECD's Development Assistance Committee in 2009; a major forum for providers of development aid.

A common approach by most donors has been to design development projects of 3–5 years which may have to be prolonged for one or even two consecutive periods. Projects can be shorter (Narayan et al., 2022) but some donors have also had long-term country programmes. In general, donors have focused on short-term results, rather than building up national institutions, except in education, that the recipient government has sufficient administrative capacity to take over when a donor turns to finance a new project. There has often been an urge by donors to maximize the appearance of short-term success for their projects, maybe at the expense of public sector capacity in the recipient country (Knack and Rahman, 2007). Sustainability was a nonexistent concept until after the millennium.

Donors seem to have acted, principally, as former colonial powers with too little long-term institution building and have not taken into full account how to best use natural resources to the benefit of the recipient countries towards long-term sustainability, together with necessary poverty reduction and economic development. Major reasons for the slow progress may have been conflicting opinions on the priority given to the offered foreign investment, lack of government policy and qualified staff, or national funds competing with national activities and other donor projects. One lesson from the Swedish development cooperation with Tanzania over half a century was that only formal institutions matter in a development process but the full involvement of the informal ones is important (McGillivray et al., 2016).

During colonial times, the joint-stock companies initially ruled the colonies. To them, their institution building efforts were confined to well-functioning operations that could provide the companies with their desired produce at minimum cost to yield maximum profit. This is still a major policy for most transnational corporations. French trading companies paid African producers prices that were only a small fraction of competitive prices, implying an annual loss of almost 2% of gross domestic product (GDP) during colonial rule (Tadei, 2020).

In the smaller colonies, indirect rule was introduced. Local rulers were allowed to rule their own people on behalf of the colonial power. This arrangement usually required the stationing of some military and civil service personnel for administration and control, but existing plantations continued to operate under company control.

One example illustrating how the rule of a colony is indirectly influenced by private companies is the sugarcane industry in Trinidad. The tropical island became a British colony in 1802, followed by Tobago (1804), but sugarcane had been introduced much earlier by the Spanish in about 1542 and continued to be produced under Spanish rule (Besson, 2018). Some two hundred years later, a Frenchman introduced the Otaheite variety which could flourish in the Port of Spain area. The first sugar mill was established in 1787 and about 100 sugar mills were operating by 1882. This expansion was due to the import of indentured workers from India brought by private ships to work for 5 years. They were later followed by more workers from India, and some from China, Madeira and Portugal to work on sugarcane.

In 1882, eight men accepted an offer by the Director of the Colonial Company Ltd to use parcels of abandoned land on the sugar estates to grow cane. The company had also invested in Trinidad's sugar factory, Usine Ste Madeleine. In 1924, caterpillar tractors were imported by Charles Massy for ploughing and grading, replacing oxen, which led to the repatriation of more than 9000 Indian workers. More than 130,000 animals were working in sugar cultivation 1955. In 1937, the UK company, Tate & Lyle, purchased several small estates in central Trinidad and soon absorbed most of the smaller sugar factories. During the 1950s, the company bought the Usine Ste Madeleine factory, ultimately making Tate & Lyle the nation's largest producer of sugar, molasses, rum and bagasse.

After independence, rising nationalism led the government to purchase Tate & Lyle's Caroni Limited holdings under the name Caroni Limited in 1975. The Caroni Distillery, established in 1918, became part of Caroni Limited's rum division, called Rum Distillers Limited. In 2001, the government sold its 49% holding to Angostura. A year later, the Caroni Distillery lost its access to local molasses and was closed. Today, Angostura remains the only distillery in the country and must import its molasses for rum production. The same applies to sugar for consumption by Trinidadians. In 2003, Caroni Limited closed and about 20,000 workers lost their jobs (World Bank Group, 2017). As recently as 2013, Trinidadian rum producers had to grow sugarcane for their products in Barbados.

The French takeover and colonization of the Kingdom of Dahomey began in 1872 with intensive fighting. It became part of French West Africa from 1894 to 1958 (Chafer, 2002). It gained full independence in 1960, and was renamed Benin in 1975. The colonial power constructed a port and railways were built. Roman Catholics missions expanded the schools. Manning (1982) argued that only 40% of Dahomey government revenue was spent in the colony in 1910, while the rest was sent to Dakar and Paris. During the colonial period, the educational system expanded, and a Dahomeyan elite was formed (Ronen, 1971).

When the more important colonies grew larger the companies lost power, as illustrated in India, where local people also resisted colonial rule. The Indian Rebellion of 1857 almost threw the East India Company (EIC) out of India (Brown, 2005). Company rule was replaced with the British Raj; direct rule by the British government. This called for institution building in developing a huge civil service and military bureaucracy to administer the colony, in addition to the few institutions that could the deliver profitable products, such as tea, coffee and sugar, to the British rulers. Another example can be taken from the large British involvement in mining in Africa. From its copper mining in Northern Rhodesia during 1924–1964, Britain had kept £2,400,000 in taxes for itself from the Copperbelt, while Northern Rhodesia received £136,000 in grants for development from Britain (Roberts, 1982).

Strong and effective institutions are required for the achievement of the 2030 Agenda for all the Sustainable Development Goals (SDGs) adopted by all UN members in 2015. This means that governance and institution building is a priority, and this has been investigated by the UN (2021). It has looked in detail at the changes in national institutional arrangements for implementing the SDGs since 2015. The UN has assessed the development, performance, strengths and weaknesses of follow-up and review systems for the SDGs. There has also been an examination of efforts made by governments and other stakeholders to implement the SDGs, drawing attention to the institutional dimension of SDG implementation and highlighting lessons for national policy makers on this aspect. Finally, the report has concluded that most countries have adjusted their institutional frameworks to implementing the 2030 Agenda.

Although the UN report covers the subject matter in a general sense, it may be useful to illustrate the time dimension for a process of national capacity building in research, and when the national government managed to take over the responsibility for the continuous operation in a time prior to the SDGs.

In the early 1980s, a national committee brainstormed the idea of establishing a national rice research centre in the Philippines. It was strongly supported by the Director General of the International Rice Research Institute (IRRI). IRRI was already operating after a Memorandum of Understanding had been signed in 1959 by the Philippine Government and the Ford and Rockefeller Foundations (IRRI, 1959). In 1985, an Executive Order (EO1061) for Phil Rice was signed and it started operations 2 years later, asking for infrastructure and technical assistance from the Japan International Cooperation Agency (JICA). In 1990, the Los Baños branch within the University of the Philippines Los Baños campus, became the principal office of Phil Rice and a strong national rice research institute, actively collaborating with IRRI on several aspects of rice research (PhilRice, 2024).

In 1978, more than 90% of the teaching staff were expatriates at the Eduardo Mondlane University (UEM) in Maputo, Mozambique, when the Swedish Agency for Research Cooperation with Developing Countries (SAREC) began providing financial support to build up research capacity. It was later broadened to institutional capacity by further support from SAREC and other Sida departments. With reference to an evaluation in 2017, Openaid (2024) stated that the support had resulted in a significantly increased number of the Mozambican academic staff, as well as teaching for doctoral degrees and equipment for research laboratories and libraries. It has strengthened national research capacity building and the model for PhD training of academic staff of UEM has been built on cooperation between UEM, and South African and Swedish universities. Nevertheless, more financial resources are required after 2024 to reach a stable research infrastructure and adequate management capacity at faculty and university levels. Current Sida support is directed to nine faculties, concentrating on the faculties of science, engineering, arts and veterinary sciences.

Each individual donor has had its own approach, rules and priorities for financial support, not forgetting their ideological aspect. Each donor makes priorities for its financial support, which are also formulated based on political self-interest. Long ago, the United States put a priority on trade to expand its own economic growth. In Sweden, the priorities of current Swedish donors are support for human rights, culture, LGBTQ movements and the media. They are important aspects of development but foreign aid through charity organizations has not been able to act as a catalyst for

development in Africa (Easterly, 2016). Most African states rely on the centuries-old tradition of exploitation of their populations by local elites and their foreign partners. The rights, opinions and insights of poor people have been ignored or neglected.

Another example is the Ujamaa policy in Tanzania, where President Nyerere opposed the parliamentary democracy of Sweden, which was a major provider of foreign aid to the country. President Nyerere forced displacements of 11 million people between 1973 and 1977: Operation Vijiji (Reimers and Luth, 2018). This took place in a country with a total population of about 16 million people and led to a total collapse of Tanzanian agriculture. Development assistance focused on 'raising the living standards of poor people' became a purely political matter during the 1970s. The attempt at rural modernization failed miserably in many respects and affected the local people and the ecology (Lawi, 2007).

Self-interest was a driving force in discussions during the early years of Swedish development assistance according to Lundberg *et al.* (2021). In addition to humanitarian reasons, the fear of a population explosion in low-income countries was identified as the strongest motivation for Swedish development assistance in the 1950s. Therefore, well-organized development in the Third World was required. This led to the creation of the Central Committee for Swedish Technical Assistance in 1952. Its foundation was the Swedes high willingness to give aid via the tax bill, later implemented in a Swedish aid policy in 1962 with the volume target of 1% of the gross national income. It was accepted by all the political parties and other actors. They could all find resonance with their own issues. The business community saw a chance for increased trade and exports. The religious communities and civil society desired to continue their mission and education, while the trade unions wanted to manifest solidarity. As for foreign policy, aid offered an opportunity to strengthen Sweden's international prestige. Initially, farmers were probably less interested.

The first Swedish development projects were directed towards family planning, health and agriculture. With a permanently growing aid budget, all the different motives could continue to coexist. In 2020, Sweden was one of the world's largest donors of state aid calculated per capita; 4% of the state budget (Lundberg *et al.*, 2021). Now, some people began to argue for removing the 1% target and the global issues were different. India and China had become major economic and technological powers. The remittances, sent by private individuals, primarily to their relatives in home countries, were estimated to be higher than global aid.

When the Swedish Minister of Development Cooperation presented a record high development assistance framework in 2021, he underlined its importance. It was to give Sweden weight in foreign policy and could make life better for millions in the world (Göteborgs-Posten, 2021). Globally, only Norway and Luxembourg had a 1% target for development assistance. The minister also underlined that Sweden's economy was export-dependent and it needed other countries to trade with. It was an echo of the early ideas in aid discussions of the 1950s. Self-interest remained and aid should be beneficial to all.

In low-income countries, there are many issues that may be considered more or less important depending on the perspective taken but it is their economic value that may affect whether they are regarded as the most critical issues for immediate – or future – action towards sustainability. This is of concern since it emphasizes that economic growth is the priority, just like the industrialized world. Moreover, development projects have frequently been technology transfers from industrialized countries. They have been assumed to be relevant for rapid development in new environments, although they were originally designed for industrial Western societies – and are far from the concept of sustainability. Very often, new aid funds have been easily available for spending and no low-income country would easily reject new funds from development assistance agencies and foreign governments.

Certain long-term development assistance has sometimes been directed at individual countries by some donors, based on a dialogue about the conditions and direction of development. If such aid is concentrated on important sectors of society, continues for a long time and is well-integrated with both the country's own financing and that of other like-minded donors, there might be potential for impact.

Such long-term assistance at the country level has been a feature of Swedish development assistance to, among others, Tanzania. It began in 1962, focusing on poverty reduction for 85% of its population. Swedish aid to Tanzania (and Tanganyika) was 8.4% of Sweden's total bilateral aid between 1962 and 2013 (McGillivray *et al.*, 2016). The country became the largest recipient of Swedish bilateral aid (Frühling, 1986). Swedish financial support to a number of development projects in Tanzania has usually been higher than the average for other donors. Another issue relates to donor support to countries with authoritarian governments or where a clan society has great power. For example, Afghanistan has been Sweden's largest aid recipient since 2013. The new Taliban emirate meant that two decades of aid disappeared meaning no further poverty reduction. The donor community seems to have neglected the country's internal conditions, particularly the strong power of the clan society with long-term internal conflicts. The largest ethnic group in Afghanistan is the Sunni Muslim Pashtun, with a long history of challenging state authority (Walsh, 2021). The return of the Taliban to Kabul in 2021 means a return to a rural Pashtun order.

Prior to Russia's invasion of Ukraine, Swedish aid was provided during 2020–2024 through Sida and the Swedish Institute to civilian Russian organizations working on democracy, gender equality and human rights, or the environment and climate. Russian law requires organizations that receive foreign support to classify themselves as foreign agents. Another example is the significant financial Sida support to Islamic Relief in Sweden (IRS) during the last ten years.

Many projects imply fragmentation when 40 different donors simultaneously provide aid with good intentions to a country. This has generally been considered as a problem. Fragmentation undermines budgets and can lead to inefficiencies, increased transaction costs and a danger that donors spend their funds in an uncontrolled manner (Knack and Rahman, 2007). Experience shows a negative effect on growth when a leading donor is lacking, but donor fragmentation appears to be beneficial in primary education (Gehring *et al.*, 2017).

Fragmentation occurs when, for instance, 30–40 donors provide development assistance to a country without adequate coordination due to the insufficient administrative capacity of the recipient country. In 2014, all of Tanzania's donors financed more than 3008 activities, twice as many as in 2000 (McGillivray *et al.*, 2016). For a long time, China has provided both significant economic assistance to Tanzania, and been directly involved in executing development projects like railways, roads, ports and airports. In recent years, it has reduced its support to Tanzania since there are challenges for some of the Tanzanian leaders responsible for these projects (Makwega, 2024). There is also a discussion about the potential for a debt trap, potentially leading to China's increased influence being contested. Instead, China is developing its influence elsewhere in East Africa, where it was the second largest loan provider during 2015–2020 (Owino, 2022). The focus has been on loan commitments to Kenya and Uganda in transport and power. This illustrates the great demand placed on the recipient country to have qualified staff to manage all these different actors and adjust projects to fit smoothly into internal political framework and national policy. This dilemma has seldom been highlighted in discussions about aid.

Although there have been evaluations of individual development projects, there have been few overall studies of the impact of the activities of all involved aid donors, or by UN specialized agencies, with high relevance to agriculture, on large programme areas, sectors and at the country level. The Food and Agriculture Organization (FAO), the International Fund for Agricultural Development (IFAD) and the World Bank have been conducting their own project assessments to ensure their projects and programmes are effective and having an impact, using their own detailed guidelines and manuals. Recently, IFAD (2022) revised the 2015 edition of its manual into a living electronic document that will be adapted over time. The new manual can provide a comprehensive institution-wide approach through which self- and independent evaluations are planned, conducted and used.

The World Bank started to assess its overall aid to agricultural projects in the 1990s. The central finding was that aid works best when it supports local governments that practice 'good management' of their social, political and economic institutions (Dollar and Pritchett,

1998). All aid was aimed at promoting growth and reducing poverty. It was also concluded that projects that involve local ownership and social investments are most effective. In addition, it was noted that aid can work in poor policy environments if donors are willing to share knowledge and technical capacity rather than focus on budgetary matters. Donors were recommended to be less proactive in financial aid disbursements to recipient countries with distorted policy environments.

The evaluation of development assistance within the UN has evolved more recently. The United Nations Development Assistance Framework (UNDAF) is a tool for the monitoring and evaluation of the effectiveness of development aid (UN, 2017). The framework outlines, as a guideline, learning from the experience of development projects and improving future projects. The UN has also set up comprehensive monitoring and evaluation (M&E) guidelines (United Nations (UN), 2017). They provide recommendations on how to set up technical working groups, multiyear M&E plans and periodic reviews in the context of Agenda 2030.

As a large donor, the United States Agency for International Development (USAID) is supporting agricultural capacity development on farms and, in the private sector, producer organizations, research and extension institutions, government agencies, rural financial institutions, and civil society. The programmes can be capacity-building agricultural research, policy analysis and the essential skills needed to create and run farmer associations and agribusinesses (USAID, 2024). After the launch of the Feed the Future's Borlaug 21st Century Leadership Program in 2011, it has been a major platform for capacity development investments. Among its results are the estimate that some 23 million more people now live above the poverty line, more than 1000 innovations have been developed and deployed, and almost US$14 billion was generated in new agricultural sales between 2011–2018.

In 2013, a Swedish government committee was established with a mandate to independently evaluate and analyze Sweden's international development assistance. One of the Expert Group for Aid Studies' (EBA) first commissioned reports was on Swedish development assistance to Tanzania during 1962–2013 (McGillivray *et al.*, 2016). Swedish aid had focused on poverty reduction for about 85% of the population and this is still a great challenge in the country's rural areas. One conclusion from the report was that Swedish aid to the country over half a century had not contributed to the expected poverty reduction. Especially in the 1970s and 1980s, the Tanzanian government had been unable to implement the aid. During the next two decades, there was too little awareness of local capacity constraints. Another finding was that the spread and fragmentation of aid efforts had been serious. The evaluators emphasized there was a contradiction between short-term results and long-term effects. The donor's desire or requirement to justify aid spending often led to a focus on short-term results, neglecting sustainable long-term effects. Another conclusion was that, in addition to politics, policy and institutional capacity play a major role in the effectiveness of development assistance.

Despite it not being an evaluation, a study has examined the effects of the five major donors in Ethiopia between 1991–2014 (Abegaz, 2015). It investigated the effectiveness, besides long-term growth and poverty reduction, of development assistance in fostering state and business institutions. It is a comparative analysis of the record of donor–recipient relations on what the donors identified as improved service delivery, empowerment and accountability. The five donors that were contributing about two-thirds of the official aid to Ethiopia were the World Bank, the African Development Bank, the European Union (EU), the Department for International Development (DFID) of the United Kingdom and USAID.

The study's major conclusion was that self-interested and pro-poor aid to Ethiopia had enhanced the technocratic capacity of public institutions. Little was achieved to bolster political legitimacy by widening the space for nonstate stakeholders (Abegaz, 2015). Decentralization should ultimately be about the distribution of power to local actors and cannot be confined to administrative control or expediency. In conflict-prone countries, the usual donor focus is only on technocratic approaches and it is not effective in institution building or beneficial to local actors. To minimize institutional harm in such regions, donors must couple poverty alleviation with inclusiveness of Ethiopian institutions.

To get continued public support for development activities in times of emerging crisis, both the donor community and the UN system must provide convincing and transparent facts about the results and progress of development assistance. This can be illustrated by other studies commissioned by the Swedish EBA.

One study concerns agricultural development with strong poverty reduction leverage in low-income countries, which used to be a large sector in Sida's initial years. Official statistics, covering the period 2005–2020, showed a relatively stable agricultural share of around 3% of the aid budget, and some 5.5% of funds distributed by Sida. Recent mapping, commissioned by the EBA indicates underreporting (Virgin *et al.*, 2022). Swedish international development and aid to agriculture had increasingly been directed to sub-Saharan Africa. It was, however, significantly lower than comparable shares from other donor countries, such as Finland, Denmark, Spain, Switzerland, Canada and Belgium. In official statistics, agricultural aid was distributed under other labels, such as climate adaptation or gender equality. These findings led to the question of whether underreporting is common in the donor community. The study concluded that the real share of agricultural aid in Swedish development cooperation was probably 5–6% and between 8–9% of the total aid distributed by Sida.

The second example concerns Sida's efforts to increase employment in sub-Saharan Africa. It is an important sector in Swedish development cooperation and employment and decent work can be one of the different ways out of poverty for individuals and groups in low- and middle-income countries. In a recent study, it was investigated whether the quantitative results on employment presented in Sida's annual reports for 2021 and 2022 are accurate, reliable and communicated in a factual and transparent manner. The agency uses so-called performance examples by selecting results to provide an understanding of the results to which development assistance has contributed. An effect means, by definition, that the aid has caused the results, but Sida's efforts cannot be seen in isolation.

The study was limited to the quantified performance statements on employment. Burman (2024) concluded there was often inconsistent correspondence between the original source and the result reported. In the two annual reports, Sida had reported very good employment results, especially for women and young people. No weak, unexpected or delayed results were reported. The results demonstrated impact or good examples from an unknown total volume of inputs since their representativeness was unknown. According to data in the annual reports, Sida's assistance in Zambia, Ethiopia and Zimbabwe had led to the creation of 124,000 new jobs. But 67,500 jobs in the 'green sector' in Zimbabwe related to the objective. This was also the case in Tanzania. In Ethiopia, the problem was an error in reporting the number of jobs created.

The question is 'How can lessons be learnt to improve future development assistance?' With new issues on the global agenda, the critical question remains as to whether the current approach to development aid can be sufficiently effective, provide impact and reach the political and technical objectives of all concerned and lead to agricultural and societal sustainability. Donors' past dependency on new technical solutions must be supplemented by biological and social components. What development aid can then be most effective in low-income countries, except for emergency operations and humanitarian aid? Other questions would be 'To what extent have recipient governments appreciated past aid?', 'Has past aid contributed to societal development?' and 'What changes would they request if they were asked without any risk of reduced funding?'

The current situation calls for international long-term support so that the skills and knowledge of low-income countries can be utilized, and their own institutions be strengthened according to their own aspirations and plans. Their staff must carry out research of their own and develop innovations, sometimes in research cooperation with advanced institutions and global networks. Innovative ideas and approaches, particularly individual innovations, are needed to tackle the issue of sustainability rather than more funds. So far, aid donors seem to have less to offer in this area compared to the private sector and transnational corporations (TNCs). This applies especially to agriculture, forestry and biological aspects. Agriculture requires long-term planning and must be shaped by a policy that will be sufficient in both normal

times and manage a crisis. The major responsibility lies with the national government. A first step would be to invite all donors and interested TNCs and attempt to agree with them on one common agenda for implementing the national government's agricultural policy towards sustainability. Then, it would be possible to identify what should be done, by whom and over what timeframe, looking at a planning period of 10–15 years, at least. The preferences of individual aid donors will be of less importance.

Each individual donor ought to focus on a smaller number of low-income countries in frank and open dialogue, including with other donors. Recipient countries should decide for themselves where – and what – they want in long-term development assistance towards their own sustainability and minimize fragmentation. Future aid providing societal development should be limited only to the UN's agencies. Across both aid donors and recipients this would require professional staff for technical matters rather than general bureaucrats. Annual reporting should be mandatory, supplemented with five-year reports on impact. A substantial proportion of aid must continue to remain humanitarian in case of natural disasters or disasters caused by human action. Russia's invasion of Ukraine shows such a need and so do future global geopolitical concerns.

There is a new political landscape with new powerful states beyond the United States, Japan and other Western powers. China and India are growing in economic and political power. Authoritarian regimes are emerging and there is a notable process of deglobalization in trade, despite which the TNCs will still attempt to maintain or even expand their power and influence.

The problem is that the UN has little or no power over individual countries. Its role as a global power is marginal in a world divided into just over 200 sovereign states. The principle of sovereignty trumps international solidarity. Each state must design its own military budget. Since the TNCs may be expected to continue to grow in numbers and power, they will further challenge the governments of individual low-income countries. The focus of the TNCs is, so far, mainly on short-term profits to their global shareholders. Their central global power is not easy to control by governments in low-income countries aiming for sustainable agriculture, and involving local institutions. It requires a long-term perspective and is also an issue that must be agreed upon at the agenda setting of any possible cooperation. On the other hand, some agribusiness TNCs have shown more interest in attempting to change their operations towards sustainability. The results of these operations would be useful as a basis for further deliberations. This may be a step forward in influencing future food production and food security.

China is responsible for 15% of all world exports (Seong, 2024). Its export products have changed from labour-intensive to knowledge-intensive. The share of China's total exports to low-income countries is expected to increase from 42% in 2017 to more than 50% in 2023. China is also the world's second largest economy, the third largest global investor, the largest trader of goods and the third largest trader of services. BRIC was a political initiative by China for countries at a similar stage of newly advanced economic development. It was formalized as a group in 2009 and included China, Brazil, Russia and India. South Africa was added in 2010 and the abbreviation was changed to BRICS. These countries have 42% of the globe's population and one-fifth of global GDP. There are plans for a new BRICS currency to switch away from the US dollar, and also to create a new world order, though this might take quite some time (Dasgupta, 2024).

In 2023, some 40 countries had formally asked to join this group. At the 2023 Summit, BRICS invited Argentina, Saudi Arabia, the United Arab Emirates, Egypt and Ethiopia to become members in 2024. Also, Venezuela has proposed its membership with reference to its huge oil reserves. These developments may further challenge and weaken the UN system and current international trade agreements. They will also influence development assistance activities.

The Increased Role of Agribusiness Transnational Corporations in Low-income Countries

For most low-income countries, all the previous challenges of food security, agricultural development and how to harness technology in agriculture to attain societal development based on sustainability, still remain. In addition, they face the effects of climate change in a changing geopolitical landscape. Bearing in mind the marginal effects of past development assistance, this raises the question of whether other actors than conventional aid donors may play a more effective role in the future.

There was an interesting development emanating from actions by the CEOs from seven Task Force members prior to the opening of the 2022 United Nations Climate Change Conference. They announced a joint action by the private sector to achieve a near-term greenhouse gas emissions reduction towards sustainability. This Sustainable Markets Initiative (Rushe, 2022) was in line with the Terra Carta, a 17-page climate recovery plan for large corporations seeking a sustainable transition. This plan had been launched by Prince Charles in 2021 at the One Planet Summit, with guiding principles and actions for large corporations by 2030. It stated that the 'fundamental rights and values of nature need to be our prime focus for the sake of our economy'.

The mission is to build a coordinated global effort by the private sector to accelerate the transition to a sustainable future. Task forces of CEOs will set priorities within sectors and individual countries will be engaged, through linkages with expertise and investments, to support national transition plans. The SMI will encourage three major market transformations: (i) a shift in corporate strategies and operations; (ii) a reformed global financial system; and (iii) an enabling environment that attracts investment and incentivizes action.

It is unclear to what extent food production is included at this stage. According to the report, regenerative agriculture has already been implemented on 15% of all cropland but the rate of transition must be tripled by 2030. Agricultural practices must change to avoid damage (Rushe, 2020). By the end of 2021, regenerative agriculture programmes amounted to more than 200,000 ha in the United States. To accelerate a transition, certain measures were proposed, such as creating metrics for measuring how much farming is sustainable, and paying farmers who will change their farming practices to more sustainable ones.

Agribusiness TNCs have already influenced and conducted operations at a larger scale in low-income countries. This expansion may be expected to continue. So far, there is some evidence that the TNC participation in agriculture in low-income countries through foreign direct investment (FDI) and contract farming can be positive, for example, in Brazil, Kenya and China. The TNCs are ahead in both advanced and general research towards agricultural sustainability, compared to national governments and aid donors. In contrast to aid donors, TNCs are profit-oriented and therefore careful to follow-up the results of their financial investments. Finally, they have both technical competence and practical experience from many countries. They could be a potential actor, albeit with the recognition that their past records also show negative issues and consequences. To regulate this will require a well-designed system of political rules acceptable to all partners concerned.

The British multinational company, Unilever, has gained experience after its introduction of its Sustainable Agriculture Code (Poinksi, 2021). Other TNCs have attempted to work with the concept of sustainability by using agriculture's potential to reduce greenhouse gases through the production of crops for biofuels. Since 1997, companies, such as Archer Daniels Midland and its subsidiaries, have used the UN-approved methodology to trade according to the Kyoto Protocol and benefit from carbon credits through production of agrofuels and waste biomass. In the area of biotechnology TNCs have used genetically engineered microorganisms and enzymes to break down biomass into agrofuels, reduce emissions through livestock manure management, and increase cellulose biomass from crop waste and residues. This advanced research can be important in a global context.

Another area where TNCs have been active is carbon sequestration by using chemical no-till agriculture, sometimes called conservation agriculture by the corporations. In 2021, Unilever added a new layer of practices in five priority areas, planning to incorporate regenerative agriculture throughout their supply chain. Nestlé announced a US$1.8 billion investment in regenerative agriculture to reduce their emissions by 95% (Mellor, 2021). In 2023, the PepsiCo Positive provided support to nearly 5500 farmers adopting regenerative farming techniques (PepsiCo, 2024). It had

doubled its regenerative farming footprint year-over-year from more than 365,000 ha to more than 725,000 ha globally It had also exceeded its agricultural water use efficiency target compared to 2015, and more than 57,000 people had been reached in the company's agricultural supply chain. They have been assisted with programming that aims to support economic prosperity and women's empowerment initiatives. The VF Corporation (formerly Vanity Fair Mills) has entered a partnership with Terra Genesis International to create a supply chain for the fashion industry's first regenerative rubber supply system (Terra Genesis, 2021).

The TNCs have one great advantage over other actors with their access to advanced agricultural research, for example, biotechnology, nanotechnology and genetically engineered crops that are adapted to the climate extremes of drought, heat, flooding and salt. Another research area is nitrogen inoculation of legumes and, possibly, cereals. If so, that may reduce the use of nitrogen fertilizers. Soon, new crop seeds may be combined with new chemical ingredients without negative environmental effects but able to control insects, diseases and weeds. The TNCs are aware of the need to develop new products that can be classified as sustainable – and be profitable in new markets, instead of continuing with current chemicals that have negative effects.

The current solutions are, however, not of the highest relevance to most low-income countries; modern seeds are expensive, they threaten smallholder farmers' rights to local seed agrobiodiversity and they introduce complications with patent claims at the national level. It would be advantageous for interested countries to attract TNC expertise by engaging them in national research agreements on some important tropical and subtropical crops that may become profitable both to the TNCs and the concerned countries. This ought to include food crops, where future research cooperation would be expected between agribusiness TNCs and individual Consultative Group for International Agricultural Research (One CGIAR) centres, and their network of partners, for both basic and applied agricultural research.

The Political and Regulatory Requirements for More Involvement of Agribusiness Transnational Corporations

Political leaders and policy makers face a great challenge in devising strategies for more public and private financial investment into agricultural development. This would include what exact role TNCs could play and how they may generate genuine development benefits for the whole of society. Governments of low-income countries should address the specific obstacles to efficient cooperation between TNCs and local farmers, their organizations and smallholders. In this process, the World Bank can play an advising role in securing that any agreement between a TNC and a government in a low-income country is beneficial to the recipient countries.

A future strategy must include rules on social and environmental concerns about any TNC involvement. There are issues such as land rights, local landholders' tenure rights, protection of the rights of Indigenous peoples and how to secure adequate legal instruments for dispute settlement. Other issues may be the provision of government-led extension services and the establishment of standards and certification procedures for new products.

One complex and controversial issue is the transfer pricing usually practised by TNCs. They avoid both market mechanisms and national laws by using internal costing and accounting. This transfer pricing means that any transaction between two entities of the same TNC are priced as if the transaction was conducted between two unrelated parties. This can include raw materials, finished products, management fees, intellectual property royalties, loans, interest on loans, payments for technical assistance, head office overheads, and research and development. The group is composed of five different institutions that can connect global financial resources, knowledge and innovative solutions to the needs of low-income countries. The International Bank for Reconstruction and Development (IBRD) and International Development Association (IDA) form the World Bank, providing financing, policy advice and technical assistance to governments in low-income countries. The IDA focuses on the world's poorest countries, while the IBRD assists middle-income and creditworthy poorer countries. The International Finance Corporation (IFC), Multilateral Investment Guarantee Agency (MIGA) and International Centre for Settlement of Investment Disputes (ICSID) focus on strengthening the private sector in developing countries. The World Bank Group could influence the TNCs by strengthening the negotiation power of the low-income countries for sufficient focus on both poverty reduction and towards agricultural sustainability. It would also be advantageous if low-income governments with experience of collaboration with a TNC can provide advice and support to other interested governments.

Taxes are important both for governments and TNCs and offer a dilemma for governments. They can be minimized by TNCs since they lower prices in countries where tax rates are high and raise them in countries with a lower tax rate. If national governments attempt to impose higher tax obligations upon TNCs they can transfer these prices among their affiliates, shifting funds around the world to avoid taxation. Some time ago, the intrafirm taxes within the TNCs was found to be 'of immediate concern' to low-income countries (Patel, 1981). In one recent study, it was estimated that about US$420 billion in corporate profits were annually shifted out of 79 countries every year (Janský and Palanský, 2019). This would equate to about US$125 billion in lost tax revenues for these countries. There are different schemes to avoid paying taxes in countries where the revenues are high and law firms can provide guidance, exemplified by EY (2024) or RSM (2021). Individual low-income governments have little ability to control the transfer pricing of TNCs which must be managed prior to any agreement on foreign direct investment (FDI) where the World Bank Group can serve as a mediator. Another issue can be overpricing. If a national government has been able to prevent a corporation from setting product retail prices above a specific percentage of prices of imported goods or the cost of production, the TNC can inflate import costs from its subsidiaries and then impose higher retail prices. This is called transfer pricing (Seth, 2024).

Towards Future Global Leadership in Sustainable Agriculture

Previous large powers in agriculture

Throughout history, several great civilizations have developed and been controlled by imperial powers and ruling dynasties for varying time periods (Annex VI). All long-lasting empires have been based on agriculture, except the latter part of the British Empire. Since then, the great economic powers have been primarily based on industrialization, trade and military strength; a build-up that has taken place merely during a few centuries. In contrast, few industrial investments have shown that kind of long-time survival.

The old Hungduan rice terraces in the Philippines are an example demonstrating sustainability still in operation after many centuries (Chapter 8). If the balance is disturbed, the whole complex agricultural system begins to decline since its components are fragile and based on biology. In modernized agriculture of the industrial type there has been a split into very separate components, just like an industrial manufacturing process. Natural processes are bypassed when only economics is prioritized.

One common feature is that all former empires finally lost their power and collapsed. This might signal that modern industrialized civilization may not be an exception to this pattern, considering the threatening global issues and negligence of the biological issues. Experience from previous major cultures shows that the urban societies of those times were not sustainable and suffered from food shortages. This is interesting since many people today want to move to ever-growing cities and urban areas, part of a global trend. This trend justifies an attempt to review the factors, other than warfare, which led to the collapse of former great powers. A few examples may illustrate these developments.

Sumer was an area in southern Mesopotamia (now south-central Iraq), which was first settled by humans from about 5000 BCE, though it is probable that some settlers had arrived much earlier. The fertile area between the Euphrates and Tigris rivers in the Middle East yielded good harvests, when farmers used irrigation in the form of small canals and crane-like water lifts. They also raised cattle leading to an increased population. When the climate became warmer, the cultivated area expanded northwards and the Sumerian culture emerged. Agricultural growth had led to abundance of cereals, trade, the rise of urban centres and prosperity. Over time, drought became more severe, seriously affecting agriculture around 2000 BCE. Reduced rainfall and constant flooding had resulted in the soils becoming salt-enriched, which gradually caused a very fertile area to deteriorate. This led to economic decline since a shortage of food caused urban areas to decline and trade was disrupted. Altogether, this caused political instability, which worsened when several groups invaded the area, especially around the time of the fall of Ur. After the collapse of Sumerian agriculture, this kind of development seems to have been repeated elsewhere in history.

The highly developed and agricultural Indus Valley civilization was based on both copper and bronze around 2600 BCE. It spread south and lasted until about 1500 BCE. Although it is unclear exactly why the Indus culture fell, food shortages led people to leave the urban communities.

In China, agriculture has been especially important for centuries, right up to the present day. Official Chinese sources speak about 13 dynasties. Starting somewhere between 2123–2025 BCE, the Xia dynasty is believed to have developed a technique for flood control. The Qin dynasty (221–206 BCE) marked the beginning of the Chinese Empire, followed by the Han dynasty (202 BCE–220 CE), known as a Golden Age of Chinese history with a period of stability and prosperity (Chao-Fong, 2021).

In 1206, Genghis Khan created the Mongol Empire. His grandson, Kublai Khan, was one of the successors and started the Yuan dynasty in China (1279–1368). The Mongols' reign in China ended after a series of famines, plagues, floods and peasant uprisings. The Qing dynasty ruled from 1644 until perishing in the revolution of 1911 after the rulers had been weakened by rural unrest and attacks by foreign powers during the 1800s (Chao-Fong, 2021).

Climate change contributed to the fall of a long dynasty in Egypt. Initially, the Nile's annual floods and a well-developed irrigation system enabled the cultivation of many crops in the fertile Nile Valley. Several dynasties flourished for three millennia. The decline started around the time of the murder of Ramses III in 1155 BCE, and continued gradually through century-long droughts, economic crises and several foreign invasions until the Romans conquered Egypt in 30 BCE.

The Roman Empire became a great power after the wars against the North African city of Carthage in the 3rd and 2nd centuries BCE and the incorporation of several areas bordering the Mediterranean Sea. The Roman Empire lasted over 500 years with a varying climate. The number of poor people increased and unemployment rose, which led to increasing food shortages in urban areas, causing them to be depopulated. Climate change, in combination with bacteria and viruses, has been mentioned as the main cause of the collapse of the Roman Empire (Harper, 2017).

The historic Mayan civilization or the Mayan Empire existed during three major periods from about 2000 BCE to 1500 CE. The Mayan culture reached its peak between 300–900 CE but a decline had begun prior to the Spanish conquest in 1519. Among the reasons for the decline were crop failures after droughts, the effects of a climate catastrophe in Peru, diseases and internal strife (MesoAmerican Research Center, 2024).

The European colonial empires did not last so long and usually ended after warfare. The British Empire was the largest empire. By 1913 it held over 23% of the world's population on about one-quarter of the Earth's total land area (Maddison, 2001). It started to decline at the beginning of the 20th century, challenged by other emerging industrialized nations. In addition, some of the colonies were gaining control over their internal affairs after being granted the status of dominions in 1907. Nationalist movements fought for freedom and the World Wars weakened the British Empire economically. It ended formally in 1997, when Hong Kong was handed back to the People's Republic of China.

In the early 1900s, the United States had started to develop, learning from British developments in infrastructure, industrialization and education. Emigration from Europe to the United States had a positive effect on a country with plenty of land and the United States was more democratic than in the United Kingdom and other European countries. When gold was discovered in California in the mid-19th century, it attracted hundreds of thousands looking for a quick fortune and the California dream also spread to emigrants. The United States was seen as a land of opportunity. Well-educated Germans who had fled the 1848 revolution found the country more politically free than their homeland. There was no despotism, privileged orders, nobility or constraints on beliefs and conscience (Bogen, 1851). This was different from their German hierarchical and aristocratic society, which set a ceiling for their talents and aspirations, as was the case in many other European countries.

In the 1920s, known as the 'Roaring Twenties', the United States was also seen as a land of hope and opportunity. It was a period of great economic growth, consumerism and cultural change. This led many to believe in the American Dream; anyone who worked hard could achieve success and prosperity (Adams, 1931). But not everyone benefitted equally from this prosperity. More recently, the original emphasis on liberty, democracy and equality has changed towards a focus on achieving material wealth and upward mobility.

By the turn of the 20th century, Germany had developed into a leading industrial country with Europe's largest population but few colonies compared to the United Kingdom. It was attracted to colonize several more since they could supply its industry with cheap raw materials and provide a larger market. Thus, Germany, together with the United States, challenged the United Kingdom's economic power. This took place in an era of growing nationalism, an idea that was in conflict with socialism. Altogether, this led to the World War I (1914–1918). After the United States entered the war (1917), Germany had no reserves to deploy and soon ran into supply difficulties. About 10 million soldiers and 7 million civilians lost their lives, and millions were wounded.

Food shortages and rationing were key issues for all war combatants. The war had a significant impact on agriculture and food distribution. Naval blockades reduced food imports and the cost of food more than doubled. Much food was sent away to feed the fighting soldiers. Fresh fruit, vegetables, meat and bread were in shortage.

In Germany, one focus of Nazi-government war policy in the 1930s was overcoming food deficits by conquering Poland and the fertile chernozem or 'black earth,' region of Ukraine and its neighbouring republics in the Soviet Union. By starving, killing and expelling the native populations, the idea was to resettle German farmers on the conquered land, assuring Germany self-sufficiency in food and thereby becoming a world power alongside United Kingdom and the United States.

According to Collingham (2012), food, and the lack of it, was central to the experience of World War II (1939–1945). During that war, some 20 million people were killed by starvation and associated illnesses. World War II is the deadliest and most destructive war so far in history, claiming between 40–50 million lives, wounding millions more and displacing tens of millions (Chmielewski, 2024). Rationing was introduced in the United Kingdom and Germany, but Japan allowed civilians and soldiers to starve. In the Soviet Union, the authorities destroyed their own food supplies so that invading Germans would starve. In contrast, the United States developed both its industrial and agricultural production, which expanded to the extent that it became a global and dominant agricultural power. It took command of agricultural technology and development. That model has been highly successful for increasing food production and has spread to most other countries. Simultaneously, US-based TNCs expanded their global activities.

During the Cold War period (1947–1991), the main United States rival, the Soviet Union, could not match with its collective agriculture. However, a new dictatorship evolved in Russia in the early 2000s, characterized by stability, growth and increasing incomes from oil production with some early positive effects for many Russians, including an increase in conventional food production.

Towards the future

In the past, greed was evil for both the state and the Church. As of today, this principle is nullified by the need for increased growth, modernity and convenience. People want more, which capitalism has sanctified, and this has proved remarkably successful. The problem is that such a capitalist empire, with rapidly growing global consumerism, is not sustainable in the long term for all countries at the same time. The risks of a global ecological meltdown have been ignored. According to Stephen Hawking, the biggest threats to planet Earth were of our own making, including nuclear war, global warming and genetically engineered viruses (Knapton, 2016).

Researchers at the Global Sustainability Institute at the Anglia Ruskin University in the United Kingdom have analysed the ability of 20 countries to cope with a global collapse of civilization. One is described as 'likely within a couple of decades'. If it did occur, it would be driven by ecological destruction, limited resources and a growing population, as well as being exacerbated by climate change. The study also showed that five countries – for varying reasons – would have the best chances of managing to maintain a stable civilization (King and Jones, 2021). They were New Zealand, Iceland, the United Kingdom, Ireland and Australia. A common factor is that they are all islands with fewer drastic temperature fluctuations.

Power ascension correlates strongly to available resources and economics. According to Kennedy (1987) the rise and decline of a great power during 1500 to 1980 was not only the consequence of fighting by its armed forces but also the consequences of how the state had utilized its productive economic resources in wartime and its actions in peacetime. The relative strengths of leading nations in world affairs never remain constant. The uneven development of technical and economic growth leads to greater advantage for one society compared to another. The decline of the Soviet Union was predicted as well as the rise of China, with its four modernizations, and Japan, but also the relative decline of the United States. The most important reason for the decline of a great power was identified as deficit spending, especially on military build-up, and negligence of its access to and utilization of natural resources (Kennedy, 1987)

In a war situation, the future supply of food is important for both warfare and the population, as demonstrated during the two world wars in the 20th century. The war fleets blocked all imports for most countries, seriously affecting food, feed supplements and fertilizers. When the UN was created in 1945, international rules were established to prevent international wars, but compliance has been marginal. It would be naïve to believe that the world's great powers could be controlled by such an institution (Blix, 2023). Humanity can be willing to collaborate but is also egotistical, aggressive and capable of evil. The hope of growing democracy after the collapse of the Cold War in 1989 has not proved to be realistic. In contrast, there is risk of war, given that there are three military superpowers (the United States, China and Russia) and several other states with tactical nuclear weapons. Warfare will not only be about access to national resources and territories but also about culture between primarily Western ideology, communist China and escalating attacks by terrorist groups. In 2014, there was a new war in Europe when Russia annexed the Crimean Peninsula and later invaded Ukraine in 2022.

Countries have forgotten the old Latin adage *si vis pacem, para bellum* from the 4th or 5th century CE. It means that the conditions of peace are often preserved by a readiness to make war when necessary. This implies a need for permanent defence rather than the dismantling of military forces, which has been taking place in Sweden since the end of the Cold War in 1991. During 1996–2006, Swedish defences were dismantled by half according to the then prime minister, Göran Persson (Persson, 2007). Until very recently, there has been no substantive discussion of how the country would manage its access to food in the event of war nearby, isolation due to international conflict or an international trade stoppage.

In the case of a war or isolation, more food must be produced nationally – and in a sustainable way. This challenge has received little attention by both policy makers and agricultural scientists in discussions about future food security. Most countries will be affected in wartime even if they are not actually involved in battles, as exemplified by Sweden. World War I led to urgent measures, and new laws and authorities. A poor harvest in 1917 led to malnutrition, but famine was avoided because Sweden received food from Denmark in exchange for iron and wood products. There was a shortage of gasoline and housing. Sweden was better prepared at the beginning of World War II with 200 national depots spread across the country of durable foodstuffs. However, price controls,

import and export bans on certain goods, and rationing of all foodstuffs were introduced. Real famine occurred only in Eastern Europe and the Netherlands, though not as severely as in Bengal, India. The dismantling of rationing took time in all countries affected by the war. Coffee was rationed in Sweden between 1947 and 1951, while meat was rationed in the United Kingdom until 1954 and until 1958 in East Germany.

Like many other industrialized countries, Sweden would be worse off than in 1939, at the start of World War II. Today, many food products are imported (50%) and there are no food stocks any longer. This would apply to most countries that are dependent upon imported food, such as China, Saudi Arabia and all countries which have relied on food imports with too little priority on their national agriculture to ensure food security. The dependence on one global food system for business, distribution and transportation is risky.

American global agricultural power in research and development is based on industrialized, highly mechanized agriculture with synthetic fertilizers, pesticides, herbicides and fungicides influencing and degrading the environment. Up to 2023, the total planted area of the eight major field crops is expected to increase over the first years and then decline (Williams *et al.*, 2023). Prices for most crops are expected to remain low or stable with increased nominal prices for cotton, rice and wheat. This can hardly be classified as sustainable food production and in the short term this model looks likely to continue up to 2032.

There has been an increase in ecological farming since 1990 but it is of minor importance. Based on USDA Agricultural Statistics surveys (2011 and 2021) certified organic cropland increased by 79%, to 1.5 million ha, over the 2011–2021 period (USDA, 2024). This may be a way forward towards sustainability and such farming practices differ with farm size, which is considered to improve sustainability (Liebert, 2022). On average, the larger organic farms used fewer agroecological practices and the way they operated more closely resembled conventional farming. Instead of redesigning their farms to use a broader range of sustainable agroecological practices, they frequently substituted synthetic pesticides and fertilizers with inputs permitted in organic production.

Organic farms are smaller (about 135 ha) compared to the average size of conventional farms (180 ha). Pasture or rangeland has decreased by 22%, to 0.5 million ha), and certified farming operations have increased by more than 90% to 17,445 farms. In 2021, organic retail sales were estimated to be about 5.5% of all retail food sales (USDA, 2024).

The United States can continue to be a global power in producing food, but would it be sustainable? In agricultural trade, current projections assume existing trade agreements and domestic policies remain in place. But the China–United States trade war has had negative effects for both economies since 2018. Members of the US Congress have argued that China's Belt and Road Initiative (BRI) will advance China's geopolitical and economic goals, challenging American interests. A long-term goal for agriculture was elaborated at the US Agricultural Forum Outlook in 2021. The plan is to increase agricultural production by 40% while cutting the environmental footprint of American agriculture in half by 2050 (USDA, 2021). This can be a step forward towards sustainability, but its implementation is unclear. This change is unsatisfactory to make the country show global leadership towards sustainability. The result of the 2024 presidential election may even make the United States more protectionist.

The long-term goal for agriculture is an ambitious step towards sustainability, but its implementation is unclear which is why it may not enable the United States to show leadership in agricultural sustainability. Future research strategy is a directional vision for an agricultural innovation strategy (USDA, 2021). Building on specific technical innovations, further improvements of current agricultural operations and continued use of synthetic fertilizers and pesticides, with significant negative environmental consequences, may hardly be regarded as sufficient for sustainable agriculture with societal development in the longer term.

Organic production has been practiced in the United States since the late 1940s, with small garden plots and, more recently, larger farms. This organic food is produced and processed without using synthetic fertilizers or pesticides. But organic farming makes up a very small share of US farmland overall. In 2016, there were about 2 million ha of certified organic farmland in 2016. representing less than 1% of almost 370 million ha of total farmland nationwide (Bialik and Walker, 2019). For a long time, California has been the leading state for organic farming with 21% of all US certified organic land, followed by Wisconsin. A growing market, driven by consumer demand, has led the number of organic farms to increase from 10,903 in 2008 to 17,445 in 2021 (USDA, 2022). The large farms sell their surplus products under a special organic label, being certified by about 40 private organizations and state agencies (EPA, 2024). In 2021, the top commodities sold were milk, chicken, eggs, apples and maize for grain.

China is the main challenger to the United States as a world power in agriculture. The BRI is a long-term economic and political strategy to accelerate economic development in both China and emerging countries, and to increase China's global influence. It is an infrastructure network launched in 2013 and comprises a land-based 'Silk Road Economic Belt' and an ocean-going 'Maritime Silk Road'. On land, it stretches from Xi'an in China through Central Asia to Moscow, Rotterdam and Venice. The BRI has expanded but has met problems since China has used both loans and aid to gain global influence.

The BRI could potentially reshape the world order and circumvent the United States in global trade. The tensions between these two countries, combined with the extensive investments required by the energy transition, represent something like a new world order for both the private sector and politicians. Other countries have increased their scrutiny of China's investments in strategic sectors and infrastructure.

It is questionable whether China will take the leadership in sustainable agriculture. Since the 1980s, its agricultural and rural sector has changed dramatically. China has increased agricultural productivity by using modern inputs and through substantial changes to the structures of agricultural production and food consumption. These are based on Western-type innovations, leading to environmental degradation and are far from sustainable. The use of chemicals on Chinese farms is 2.5 times the global average per acre of land (Si, 2019). The country uses four times more fertilizer per unit area than the global average and it accounts for half of the world's total pesticide consumption (Statista, n.d.).

There has been a substantial reduction in rural poverty, but the Communist Party needs rapid growth to lift many more people out of poverty and get them into other work. Soon, China will no longer have the advantage of low wages. Situated far from the major food importing markets, transportation costs will increase. Moreover, the country itself must import food. In 2023, the gross domestic product (GDP) growth was the lowest in three decades, the COVID-19 pandemic years excluded. Another problem is a declining population due to too few births, which may lead to labour shortages and soaring wages. Moreover, reforms are needed of rural institutions and marketing, and many more technological innovations are required. These reforms are considered essential for a success of the national Rural Revitalization Development Strategy (Huang *et al.*, 2019).

So far, there are few signs of a notable change to the current model of modernizing agriculture with too many ecological footprints. One change is that the total area of certified organic agriculture cultivation increased more than five-fold between 2005 and 2018, meaning 3.1 million ha, according to a 2019 government report (Scott and Si, 2020). Chinese farmers have started to cut down on the use of chemicals. The quadrupling to the total grain output since 1961, with high fertilizer applications per hectare, has had significant environmental consequences. According to a Chinese report on organic agriculture certification and industry development in 2019, Europe, Japan and the United States were the biggest markets for Chinese organic food exports.

In the EU, the share of agricultural land under organic farming had annually increased by 5.7% during 2012–2020 (European Commission, 2023). In 2020, 9.1% of the EU's agricultural area was farmed organically but about 61% of all EU land under organic farming received specific organic support payments from the Common Agricultural Policy (CAP). France, Spain, Italy and Germany had the largest areas of organic farming. The largest shares were permanent grassland (42%), green fodder (17%), cereals (16%) and permanent crops. Organic animal production accounted for a small share of total EU animal production.

Another important agricultural actor is India. Agriculture plays a significant role in the country's growing economy, contributing almost 18% of the country's gross value added (Invest India, 2022). India may be an emerging global agricultural power that could compete for leadership towards sustainability. Some 54% of the total workforce is involved in agriculture and allied sector activities (IBEF, 2023). In 2022–2023, total agricultural exports had increased by 20% as it did during the previous 2 years (IBEF, 2024). The agricultural sector is expected to grow by some 3% annually.

So far, sustainable agriculture is not mainstream agriculture in India. Only five sustainable agriculture practices and systems are on a scale beyond 5% of the net sown area (Gupta *et al.*, 2021). They include crop rotation, agroforestry, rainwater harvesting, mulching and precision. In total, 16 structural adjustment programmes (SAPs) have been identified but existing literature is considered to lack long-term assessments of SAPs across the sustainability dimensions (economic, environmental and social). Although SAPs operate on a small scale nationally it means that about 5 million farmers practice such systems. There are also expectations and recommendations on how sustainable food and farm systems may expand towards 2030 (Chand *et al.*, 2022).

There are many diverse types of agricultural practices still being actively used and based upon long experience of different climate conditions and limited resources for societal development in the Indian context. This experience would be highly relevant for many other countries to consider in their approach to sustainable food production. Thus, India may turn out as an interesting actor and as a future global leader towards sustainability.

In Australia, agriculture accounts for 11% of Australia's goods and services exports. Two-thirds of this export goes to Asia (Hatfield-Dodds *et al.*, 2020). The country is the twelfth largest global agricultural exporter. Australian farmers are using certain sustainable practices such as crop rotation, limited use of pesticides and fertilizers, and no-till farming as the norm. Government agricultural support reinforces this approach since there is 'virtually no direct farm support.' Most Australian farmers manage native vegetation based on environmental considerations.

In 2020 there were 190 countries in the world practicing a form of organic production on about 75 million ha; an increase from 11 million ha in 1999 (Willer *et al.*, 2022). Its share of total agricultural land was 1.6% in 2020. In all, there were 3.4 million producers with 1.6 million in India alone. The top countries were Australia (35 million ha), Argentina and Uruguay.

Major agricultural powers have been slow in changing the food production to ecological or organic farming as a first step towards sustainability. Such an initial movement does not comprise all dimensions of sustainability and is not homogeneous, although it is an alternative to conventional agriculture.

A recent political initiative may stimulate some practical action. It is the Emirates Declaration on Sustainable Agriculture, Resilient Food Systems and Climate Action and it was signed at COP28 in Dubai by many of the countries with the highest food-related greenhouse gas emissions, committing them to integrate food into their climate plans by 2025. As with all political declarations its success will be determined by whether countries follow up the commitments with policy reforms at the national level (World Resources Institute, 2023). Moreover, the declaration does not suggest how sustainable agriculture may be achieved and is primarily confined to how food can be better included in climate plans. This will not suffice.

There is a great urgency for individual governments to initiate discussions on how to achieve sustainable agriculture and to identify what this means in practical terms with regard to the technical, environmental and social dimensions of future agriculture that lead to societal development. These discussions cannot be confined only to innovations or modern technology, although research is important. Governments must provide the political framework and give guidance for the future. They need to agree internationally on the practical meaning of a sustainable agriculture for 2040 and identify the necessary steps to gradually achieve this goal.

A regional approach would be an alternative to avoid large international gatherings like the COP meetings which have more than 75,000 participants. Instead, individual governments ought to organize national discussion to formulate policy briefs for national sustainable agriculture for societal development, that consider the technical, environmental and social dimensions. The

national level is the most fundamental. Prior to their finalization, it may be very useful to invite the donor community and agribusiness TNCs, that are already operating in the country, for consultations on their long-term interest in providing future financial and technical assistance for national capacity building of sustainable agriculture for societal development. These briefs should be forwarded to regional meetings for further discussions and possible conclusions.

The FAO could play a more active and constructive role and initially serve as the coordinator at the regional level. The organization is well quipped and present in more than 130 countries through a decentralized network. It includes regional and subregional offices and several country offices (FAO, 2024). Over many years they have acquired a lot of experience of the countries in the respective regions.

The FAO regional offices work to ensure multidisciplinary representation in projects and programmes, identify priority areas for the FAO and maintain a dialogue with countries in the region. This would serve as an excellent basis for initially inviting some 50–60 interested member countries, representing all continents and major climatic regions, to prepare a country brief on sustainable agriculture for societal development. This means, however, that the briefs should not only deal with agricultural production, but also its productivity, research and innovations, and agricultural trade. They are some of the components of societal development. Thus, they must be considered in relation to relevant social aspects, local organizations, informal institutions and overall society, where farming is one important actor.

In a first round, the FAO offices should select countries that have showed some progress in the transformation process towards achieving the SDGs and/or tackled specific environmental concerns related to agriculture. These briefs should not exceed 60 pages, including the summarized experiences of projects by the FAO, IFAD and agribusiness TNC activities after 2010. After receiving these briefs, it would be useful to discuss and debate the findings at regional meetings organized by the FAO, for instance in 2027. This approach will facilitate the exchange of views in a frank dialogue, allowing for more detailed discussion than at a large global conference.

References

Abegaz, B. (2015) Aid, accountability and institution building in ethiopia: the self-limiting nature of technocratic aid. *Third World Quarterly* 36(7), 1382–1403. Available at: https://doi.org/10.1080/01436597 .2015.1047447 (accessed 17 September 2024).

Adams, J.T. (1931) *The Epic of America*, 1st edition. Routledge, London.

Besson, G.A. (2018) Sweet sorrow: The timeline of sugar in Trinidad and Tobago. The Caribbean History Archives. Available at: https://caribbeanhistoryarchives.blogspot.com/2018/12/sweet-sorrow-timeline -of-sugar-in.html (accessed 17 September 2024).

Bialik, K. and Walker, K. (2019) Organic farming is on the rise in the U.S. Pew Research Center. Available at: https://www.pewresearch.org/short-reads/2019/01/10/organic-farming-is-on-the-rise-in-the-u-s/ (accessed 17 September 2024).

Blix, H. (2023) *A Farewell to Wars: The Growing Restrains on the Interstate Use of Forces*. Cambridge University Press, Cambridge, UK.

Bogen, F.W. (1851) The German in America. Quoted in Ozment, S.E (2005). In: *A Mighty Fortress: A New History of the German People*. Harper, New York, pp. 170–171.

Brown, J. (2005) Indian rebellion of 1857: Two years of massacre and reprisal. *Warfare History Network*. Available at: https://warfarehistorynetwork.com/article/indian-rebellion-of-1857-two-years-of-mass acre-and-reprisal/ (accessed 17 September 2024).

Burman, M. (2024) What does Sida's annual report say about the results of development assistance? The example of employment (in Swedish). Background report May 2024: Report commissioned by the Expert Group for Aid Studies (EBA), Stockholm. Available at: https://eba.se/wp-content/uploads /2024/06/Vad-sager-Sidas-arsredovisning-om-bistandets-resultat.pdf (accessed 17 September 2024).

Chafer, T. (2002) *The End of Empire in French West Africa: France's Successful Decolonization*. Berg, Oxford, UK.

Chand, R., Joshi, P. and Khadka, S. (eds) (2022) *Indian Agriculture Towards 2030: Pathways for Enhancing Farmers' Income, Nutritional Security and Sustainable Food and Farm Systems*. Springer, Singapore.

Chao-Fong, L. (2021) The 13 dynasties that ruled China in order. *HistoryHit*. Available at: https://www. historyhit.com/the-dynasties-that-ruled-china-in-order/ (accessed 13 August 2024).

Chmielewski, K. (2024) Casualties of World War II. *Encyclopedia Britannica*. Available at: https://www.brit annica.com/event/casualties-of-World-War-II-2231003 (accessed 7 September 2024).

Collingham, L. (2012) *The Taste of War: World War II and the Battle for Food*. Penguin Press, New York.

Dasgupta, S, 2024. BRICS' de-dollarization has a long way to go. Voice of America, October 26, 2024 at https://www.voanews.com/a/brics-de-dollarization-agenda-has-a-long-way-to-go/7840686.html

Dollar, D. and Pritchett, L. (1998) *Assessing Aid: What Works, What Doesn't, and Why*. World Bank, Washington, D.C.

Easterly, W. (ed.) (2016) *The Economics of International Development: Foreign Aid versus Freedom for the World's Poor*. Institute of Economic Affairs, London.

European Commission (2023) Organic farming in the EU: A decade of growth. European Commission, Brussels. Available at: https://agriculture.ec.europa.eu/news/organic-farming-eu-decade-growth-2 023-01-18_en#more (accessed 17 September 2024).

EY (2024) Worldwide Transfer Pricing Reference Guide 2022-23. Ernst & Young Global Limited (EY), United Kingdom. Available at: https://www.ey.com/en_gl/tax-guides/worldwide-transfer-pricing-ref erence-guide (accessed 17 September 2024).

Ferro, M.V. and Nishio, A. (2021) From aid recipient to donor: Korea's inspirational development path. *The Korea Herald*. Available at: https://blogs.worldbank.org/en/eastasiapacific/aid-recipient-donor-kore a-inspirational-development-path (accessed 17 September 2024).

Food and Agriculture Organization (FAO) (2022) *The State of Food and Agriculture 2022. Leveraging Agricultural Automation for Transforming Agrifood Systems*. FAO, Rome.

Food and Agriculture Organization (FAO) (2024) FAO, Rome. Available at: https://www.fao.org/home/en/ (accessed 17 September 2024).

Frühling, P. (ed.) (1986) Swedish development aid in perspective: Policies, problems & results since 1952. Värnamo, Sweden.

Gehring, K., Michaelowa, K., Dreher, A. and Spörri, F. (2017) Aid fragmentation and effectiveness: What do we really know?. *World Development* 9, 320–334. Available at: https://doi.org/10.1016/j.worldde v.2017.05.019 (accessed 17 September 2024).

Göteborgs-Posten (2021) New minister: High aid budget gives Sweden weight (in Swedish). Göteborgs-Posten. Available at: https://www.gp.se/nyheter/sverige/nye-ministern-hogt-bistand-ger-sverige-ty ngd.4edbfe44-f246-5421-855d-1465ed9f86d2 (accessed 17 September 2024).

Gupta, N., Pradhan, S., Jain, A. and Patel, N. (2021) Sustainable agriculture in India 2021: What we know and how to scale up. Council on Energy, Environment and Water, New Delhi, India.

Harper, K. (2017) *The Fate of Rome. Climate, Disease, and the End of an Empire*. Princeton University Press, Princeton, NJ.

Hatfield-Dodds, S., Greenville, J., Burns, K. and Downham, R. (2020) *Pathways to Sustainable and Productive Agriculture – An Australian Perspective*. Notes prepared for webinar on EU–Australia Farming: One goal but many paths to sustainability on 28 October 2020, ABARES, Canberra, Australia.

Huang, J., Rozelle, R., Zhu, X., Zhao, S. and Sheng, Y. (2019) Agricultural and rural development in China during the past four decades: An introduction. Special Issue for the 40th Anniversary of China's Rural Reforms, *The Australian Journal of Agricultural and Resource Economics (AARES)* 64(1), 1–13. Available at: https://doi.org/10.1111/1467-8489.12352 (accessed 21 August 2024).

India Brand Equity Foundation (IBEF) (2023) Agriculture 4.0: Future of Indian agriculture. IBEF, New Delhi. Available at: https://www.ibef.org/blogs/agriculture-4-0-future-of-indian-agriculture (accessed 21 August 2024).

India Brand Equity Foundation (IBEF) (2024) Agriculture and food industry and exports. IBEF, Ministry of commerce and industry, Government of India. Available at: https://www.ibef.org/exports/agriculture-and-food-industry-india (accessed 17 September 2024).

International Fund for Agricultural Development (IFAD) (2022) IFAD Evaluation Manual. IFAD, Rome. Available at: https://ioe.ifad.org/en/w/the-2022-ifad-evaluation-manual (accessed 17 September 2024).

International Rice Research Institute (IRRI) (1959) Legal Status of the International Rice Research Institute, Annex 1, Memorandum of Understanding between Ford and Rockefeller Foundations and the Government of the Republic of the Philippines. Available at: https://cgspace.cgiar.org/items/2280bf7c-186a-4b60-93ad-d003e7dee6fc (accessed 17 September 2024).

Invest India (2022) Indian Agriculture: Investments and Achievements. Strategic Investment Research Unit (SIRU), New Delhi. Available at: https://www.investindia.gov.in/team-india-blogs/indian-agriculture-i nvestments-and-achievements (accessed 17 September 2024).

Janský, P. and Palanský, M. (2019) Estimating the scale of profit shifting and tax revenue losses related to foreign direct investment. *International Tax and Public Finance* 26, 1048–1103. Available at: https://doi.org/10.1007/s10797-019-09547-8 (accessed 17 September 2024).

King, N and Jones, A (2021) An analysis of the potential for the formation of 'nodes of persisting complexity'. *Sustainability* 13(15), 8161. Global Sustainability Institute, Anglia Ruskin University, Cambridge CB1 1PT, UK. https://doi.org/10.3390/su1315861

Kennedy, P. (1987) *The Rise and Fall of Great Powers*. Random House.

Knack, S. and Rahman, A. (2007) Donor fragmentation and bureaucratic quality in aid recipients. *Journal of Development Economics* 83(1), 176–197. Available at: https://doi.org/10.1016/j.jdeveco.2006.02.002 (accessed 17 September 2024).

Knapton, S. (2016) Prof Stephen Hawking: Disaster on planet Earth is a near certainty. *The Telegraph*. Available at: https://www.telegraph.co.uk/news/science/science-news/12107623/Prof-Stephen-Ha wking-disaster-on-planet-Earth-is-a-near-certainty.html (accessed 17 September 2024).

Lawi, Y.Q. (2007) Tanzania's operation Vijiji and local ecological consciousness: The case of eastern iraqwland, 1974–1976. *The Journal of African History* 48(1), 69–93. Available at: https://doi.org/10.1 017/S0021853707002526 (accessed 17 September 2024).

Liebert, J. (2022) Sustainable management practices vary with farm size in US organic crop production. *Nature Plants* 8, 871–872. Available at: https://doi.org/10.1038/s41477-022-01194-y (accessed 21 August 2024).

Lundberg, U., Tydén, M. and Berg, A. (2021) *A Dizzying Task: Sweden and Development Assistance 1945–1975* (in Swedish). Ordfront, Stockholm.

Maddison, A. (2001) *The world economy: A millennial perspective*. Organisation for Economic Co-operation and Development (OECD).

Makwega, R.M. (2024) Assessment of china's relations to tanzania: A journey of cooperation for mutual success. *Modern Diplomacy*. Available at: https://moderndiplomacy.eu/2024/08/03/assessment -of-chinas-relations-to-tanzania-a-journey-of-cooperation-for-mutual-success/ (accessed 17 September 2024).

Manning, P. (1982) Back matter. *The Journal of African History* 23(2), i–ii. Available at: http://www.jstor.or g/stable/182095 (accessed 17 September 2024).

McGillivray, M., Carpenter, D., Morrissey, O. and Thaarup, J. (2016) *Swedish Development Cooperation with Tanzania: Has it helped the poor? Expertgruppen för biståndsanalys (EBA)*. Elanders Sverige AB, Stockholm.

Mellor, S. (2021) Nestlé trumpets its green credentials as shareholders approve $3.5 billion net-zero plan. *Fortune*. Available at: https://fortune.com/2021/04/15/nestle-net-zero-plan-shareholders-approve-3 -5-billion/ (accessed 21 August 2024).

MesoAmerican Research Center (2024) Ancient maya civilization. *MesoAmerican Research Center*. Available at: https://www.marc.ucsb.edu/research/maya/ancient-maya-civilization (accessed 17 September 2024).

Narayan, T., Usmani, F., Faye, A., Felix, E., Chen, H. *et al.* (2022) Sustainable agriculture decision support tool: application to dual-purpose crops in Senegal. USAID Bureau for Africa Implemented by the Institut Sénégalais de Recherches Agricoles (ISRA) and the Feed the Future Sustainable Intensification Innovation Lab (SIIL). Available at: https://www.climatelinks.org/sites/default/files/a sset/document/2022-05/Sustainable%20Agriculture%20DST%20Final%20Report_04-29-22.pdf (accessed 17 September 2024).

Openaid (2024) Universidade Eduardo Mondlane (UEM) Mozambique 2017-2024. Available at: https:// openaid.se/en/contributions/SE-0-SE-6-51140073 (accessed 17 September 2024).

Owino, E. (2022) Chinese lending to east africa: what the numbers tell us. *Development Initiatives*. Available at: https://devinit.org/resources/chinese-lending-to-east-africa-what-the-numbers-tell-us/ (accessed 17 September 2024).

Patel, M. (1981) A note on transfer pricing by transnational corporations. *Indian Economic Review* 16(1/2), 139–152. Available at: http://www.jstor.org/stable/29793313 (accessed 17 September 2024).

PepsiCo (2024) Positive agriculture. PepsiCo. Available at: https://www.pepsico.com/our-impact/sustai nability/esg-summary/pepsico-positive-pillars/positive-agriculture (accessed 17 September 2024).

Persson, G. (2007) *My Way, My Choices (in Swedish)*. Bonnier, Stockholm.

PhilRice (2024) About PhilRice. Philippine Rice Research Institute, Nueva Ecija, Philippines. Available at: https://www.philrice.gov.ph/about-us/philrice/ (accessed 17 September 2024).

Poinksi, M. (2021) How Unilever is turning sustainability into opportunity. *Food Dive*. Available at: https://www.fooddive.com/news/how-unilever-is-turning-sustainability-into-opportunity/595399/ (accessed 17 September 2024).

Reimers, C.-V. and Luth, E. (2018) Then Tanzania became Sweden's pet pig and when aid became social-ist. Del 1 & 2, (in Swedish). *Smedjan*. Available at: https://timbro.se/smedjan/sa-blev-tanzania-sveri ges-kelgris/ (accessed 17 September 2024).

Roberts, A.D. (1982) Notes towards a financial history of copper mining in Northern Rhodesia. *Canadian Journal of African Studies/Revue Canadienne Des Études Africaines* 16(2), 347–359. Available at: https://doi.org/10.2307/484302 (accessed 17 September 2024).

Ronen, D. (1971) Review: constitutions, republics and social reality: the legacy of french rule in dahomey, by maurice A. Glele. *Africa Today* 18(2), 77–81. Available at: http://www.jstor.org/stable/4185159 (accessed 17 September 2024).

RSM (2021) Six key transfer pricing considerations for multinational companies. RSM, United States. Available at: https://rsmus.com/insights/services/business-tax/six-key-transfer-pricing-consideratio ns-for-multinational-compan.html (accessed 17 September 2024).

Rushe, D. (2020) Big agriculture warns farming must change or risk destroying the planet. *The Guardian*. Available at: https://www.theguardian.com/environment/2022/nov/03/big-agriculture-climate-crisis -cop27 (accessed 21 August 2024).

Rushe, D. (2022) Sustainable Markets Initiative: Agribusiness Task Force Report. Sustainable Markets Initiative. Available at: https://a.storyblok.com/f/109506/x/1eb7531ee2/smi_agritaskforce_2023-final .pdf (accessed 17 September 2024).

Scott, S. and Si, Z. (2020) Why China is emerging as a leader in sustainable and organic agriculture. *The Conversation*. Available at: https://theconversation.com/why-china-is-emerging-as-a-leader-in-sust ainable-and-organic-agriculture-132407 (accessed 17 September 2024).

Seong, J. (2024) The global economy is resetting. China is repositioning itself to export innovative technologies, and its trading partners are more diverse. McKinsey Global Institute. *Hong Kong Economic
Times*. Available at: https://www.mckinsey.com/mgi/overview/in-the-news/the-global-economy-is-
resetting-china-is-repositioning-itself-to-export-innovative-technologies-and-its-trading-partners
-are-more-diverse (accessed 17 September 2024).

Seth, S. (2024) *Transfer Pricing: What It Is and How It Works, With Examples*. Investopedia. Available at:
https://www.investopedia.com/terms/t/transfer-pricing.asp (accessed 17 September 2024).

Si, Z. (2019) *Shifting from Industrial Agriculture to Diversified Agroecological Systems in China*. The Centre
for Agroecology, Water and Resilience (CAWR). Coventry University, Coventry, UK.

Statista (n.d.) Consumption of fertilizer per area or arable land worldwide from 1961 to 2021, by leading
country. Statista. Available at: https://www.statista.com/statistics/1288229/global-consumption-of-f
ertilizer-per-area-by-country/ (accessed 13 August 2024).

Tadei, F. (2020) Measuring extractive institutions: colonial trade and price gaps in french africa. *aEuro-
pean Review of Economic History* 24(1), 1–23. Available at: https://doi.org/10.1093/ereh/hey027
(accessed 17 September 2024).

Terra Genesis (2021) Regenerative rubber. Terra Genesis International. Available at: https://terra-genesis.
com/project/regenerative-rubber (accessed 17 September 2024).

United Nations (UN) (2017) United Nations Development Assistance Framework Guidance. UN Sustainable
Development Group, New York. Available at: https://unsdg.un.org/resources/united-nations-develo
pment-assistance-framework-guidance (accessed 17 September 2024).

United Nations (UN) (2017) Development Group (2017) Monitoring and Evaluation. UNDAF Companion
Guidance, United Nations, New York. Available at: https://unsdg.un.org/sites/default/files/UNDG-U
NDAF-Companion-Pieces-Monotoring-and-Evaluation-PDF (accessed 17 September 2024).

United Nations (UN) (2021) World Public Sector Report 2021. National Institutional Arrangements for
Implementation of the Sustainable Development Goals. A Five-year Stocktaking. United Nations
Department of Economic and Social Affairs (UN DESA), New York. Available at: https://publicadmin-
istration.desa.un.org/publications/world-public-sector-report-2021 (accessed 17 September 2024).

United States Agency for International Development (USAID) (2024) USAID. Washington DC. Available at:
https://www.usaid.gov/ (accessed 17 September 2024).

United States Department of Agriculture (USDA) (2021) U.S. Agriculture Innovation Strategy: A Directional
Vision for Research. USDA, Washington DC. Available at: https://www.usda.gov/sites/default/files/d
ocuments/AIS.508-01.06.2021.pdf (accessed 17 September 2024).

United States Department of Agriculture (USDA) (2022) Certified organic survey 2021. Summary December
2022. USDA, National Agricultural Statistics Service. Washington, DC. Available at: https://www.
nass.usda.gov/Publications/Highlights/2022/2022_Organic_Highlights.pdf (accessed 17 September
2024).

United States Department of Agriculture (USDA) (2024) Organic Agriculture. Overview. Economic Research
Service, USDA, Washington DC. Available at: https://www.ers.usda.gov/topics/natural-resources-en
vironment/organic-agriculture/ (accessed 17 September 2024).

United States Environmental Protection Agency (EPA) (2024) Organic Farming. EPA, Washington DC.
Available at: https://www.epa.gov/agriculture/organic-farming (accessed 17 September 2024).

Walsh, T. (2021) How ethnic and religious divides in Afghanistan are contributing to violence against
minorities. *The Conversation*. Available at: https://theconversation.com/how-ethnic-and-religio
us-divides-in-afghanistan-are-contributing-to-violence-against-minorities-168059 (accessed 17
September 2024).

Willer, H., Trávníček, J., Meier, C. and Schlatter, B. (2022) *The World of Organic Agriculture, 2022*, 23rd
edn. FiBL and IFOAM - Organics International, Frick, Switzerland.

Williams, B., Dohlman, E. and Miller, M. (2023) *Declining Crop Prices, Rising Production and Exports
Highlight U.S. Agricultural Projections to 2032*. Economic Research Service, U.S. Department of
Agriculture, Washington DC. Available at: https://www.ers.usda.gov/amber-waves/2023/february/
declining-crop-prices-rising-production-and-exports-highlight-u-s-agricultural-projections
-to-2032/ (accessed 17 September 2024).

World Bank Group (2017) Trinidad and Tobago: Caroni Sugar Project (English). World Bank, Washington,
DC. Available at: http://documents.worldbank.org/curated/en/774701468914357839/Trinidad-and-T
obago-Caroni-Sugar-Project (accessed 17 September 2024).

World Resources Institute (2023) 134 countries sign the Emirates Declaration on Sustainable Agriculture
and put food high on the climate agenda at COP28. World Resources Institute. Available at: https://
www.wri.org/news/statement-134-countries-sign-emirates-declaration-sustainable-agriculture-and
-put-food-high-0 (accessed 13 August 2024).

Annexes

Annex I

Acronyms and Abbreviations

AAMI	Adama Agricultural Machinery Industry, Ethiopia
ACIAR	Australian Centre for International Agricultural Research
ADM	Archer Daniels Midland Company, US
AfCFTA	African Continental Free Trade Area
Africa Rice	Africa Rice Center, Côte d'Ivoire
AI	Artificial Intelligence
AMOC	Atlantic meridional overturning circulation
AMR	Antimicrobial Resistance
ASEAN	Association of Southeast Asian Nations
ATI	Agricultural Transformation Institute, Ethiopia
AU	African Union
Bioversity	Alliance Bioversity International and CIAT, Italy
BRI	China's Belt and Road Initiative
BRICS	Brazil, Russia, India, China and South Africa
CADU	Chilalo Agricultural Development Project, Ethiopia
CAP	Common Agricultural Policy
Cas9	CRISPR Associated System number 9
CDM	Clean Development Mechanism
CER	certified emission reduction
CFA	Comprehensive Framework for Action
CFC	Chlorofluorocarbon
CFS	Committee on World Food Security
CGIAR	Consultative Group of International Agricultural Research (One CGIAR)
CGN	China General Nuclear Power Group
CHEC	China Harbour Engineering Company
CHF	Swiss Franc
CIFOR	Center for International Forestry Research, Indonesia
CIMMYT	International Maize and Wheat Improvement Center, Mexico
CIP	International Potato Center, Peru

CRISPR	Clustered Regularly Interspaced Short Palindromic Repeats
DFID	Department for International Development, United Kingdom
DNA	Deoxyribonucleic Acid
EABC	Ethiopian Agricultural Business Corporation
EAEU	Eurasian Economic Union, Russia/Belarus
EBA	Expert Group for Aid Studies, Sweden
ECDC	European Centre for Disease Prevention and Control, Sweden
ECHA	European Chemicals Agency, Finland
EFSA	European Food Safety Authority, Italy
EIAR	Ethiopian Institute of Agricultural Research
EIB	Excellence in Breeding Platform
EIC	British East India Company
EIP	European Innovation Partnership
EIP-AGRI	The European Innovation Partnership for Agricultural Productivity and Sustainability
ENSO	El Nino Southern Oscillation
EPA	US Environmental Protection Agency
EU	European Union, Belgium
FAO	Food and Agriculture Organization of the United Nations, Italy
FAS	Foreign Agricultural Service, US
FDA	United States Food and Drug Administration
FDI	Foreign Direct Investment
FEANTSA	European Federation of National Organisations Working with the Homeless
FSANZ	Food Standards Australia New Zealand
GABA	Gamma Aminobutyric Acid
GAFSP	Global Agriculture and Food Security Program
GATT	General Agreement on Tariffs and Trade
GDP	Gross Domestic Product
GEP	Gross Ecosystem Product
GHG	Greenhouse Gases
GM	Genetic Modification
GMO	Genetically Modified Organism
GRIIS	Global Register of Introduced and Invasive Species
GSR	Green Super Rice
HBM4EU	European Human Biomonitoring Initiative
HCFC	Hydrochlorofluorocarbon
HDI	Human Development Index
HLTF	UN High-level Task Force on the Global Food Security Crisis
HVC	High-value Crop
IBRD	International Bank for Reconstruction and Development, US
ICARDA	International Center for Agricultural Research in the Dry Areas, Lebanon
ICC	International Criminal Court, Netherlands

ICCPR	International Covenant on Civil and Political Rights
ICERD	International Convention on the Elimination of all forms of Racial Discrimination
ICESCR	International Covenant on Economic, Social and Cultural Rights
ICIPE	International Centre of Insect Physiology and Ecology, Kenya
ICRAF	World Agroforestry Centre, Kenya
ICRISAT	International Crops Research Institute for the Semi-arid Tropics, India
ICSID	International Centre for Settlement of Investment Disputes, US
IDA	International Development Association, US
IDFIP	ICARDA's Integrated Desert Farming Innovation Program
IDRC	International Development Research Centre, Canada
IEA	International Energy Agency, France
IFA	CGIAR Integration Framework Agreement
IFAD	International Fund for Agricultural Development, Italy
IFC	International Finance Corporation, US
IFPRI	International Food Policy Research Institute, US
IHME	Institute of Health Metrics and Evaluation, US
IITA	International Institute for Tropical Agriculture, Nigeria
ILO	International Labour Organization, Switzerland
IMF	International Monetary Fund, US
INIBAP	International Network for the Improvement of Banana and Plantain (merged with IPGRI to Bioversity International)
IPBES	Intergovernmental Science-Policy Platform on Biodiversity and Ecosystem Services, Germany
IPCC	Intergovernmental Panel on Climate Change, Switzerland
IRA	Inflation Reduction Act
IRRI	International Rice Research Institute, Philippines
IRS	Islamic Relief in Sweden
ISA	International Seabed Authority, Jamaica
ISAAA	International Service for the Acquisition of Ari-biotech Applications, US
ISDC	Independent Science for Development Council
IT	information technology
ITPGRFA	International Treaty on Plant Genetic Resources for Food and Agriculture
ITT	International Telephone and Telegraph, US
ITU	International Telecommunication Union, Switzerland
IUCN	International Union for Conservation of Nature, Switzerland
IWMI	International Water Management Institute, Sri Lanka
JICA	Japan International Cooperation Agency
KTDA	Kenya Tea Development Authority
LCIPP	Local Communities and Indigenous Peoples Platform
LKAB	Luossavaara-Kiirunavaara Aktiebolag (company), Sweden
LMC	Lancang–Mekong Cooperation

LNG	Liquefied Natural Gas
MDGs	United Nations Millennium Development Goals
Mercosur	Southern Common Market, Uruguay
MIGA	Multilateral Investment Guarantee Agency, US
MRSA	Methicillin-Resistant *Staphylococcus Aureus*
NAFTA	North American Free Trade Agreement
NASA	National Aeronautics and Space Administration, US
NFPAP	National Food Production Action Plan
NGO	Nongovernmental Organization
NGT	New Genomic Techniques
NMBU	Norwegian University of Life Sciences
NNR	Nordic Nutrition Recommendation
OECD	Organization for Economic Cooperation and Development, France
PCB	Polychlorinated Biphenyl
PFAS	Per- and Polyfluorinated Alkyl Acids
PISA	Programme for International Student Assessment
PVR	Productivity, Versatility, and Resiliency
RCEP	Regional Comprehensive Economic Partnership
RISE	Research Institutes of Sweden
RNA	Ribonucleic Acid
SADC	Southern African Development Community, Botswana SAP
SAPO	Swedish Security Service
SAREC	Swedish Agency for Research Cooperation with Developing Countries
SARS-CoV-2	Severe Acute Respiratory Syndrome Coronavirus 2
SCB	Statistics Sweden
SDGs	Sustainable Development Goals
SEED	Promoting Entrepreneurship for Sustainable Development
SIANI	Swedish International Agriculture Network Initiative
SIDA	Swedish International Development Agency
Sida	Swedish International Development Cooperation Agency
SLU	Swedish University of Agricultural Sciences
SMI	Sustainable Markets Initiative
SSE	Stockholm School of Economics, Sweden
TCDD	2, 3, 7, 8-Tetrachlorodibenzo-p-dioxin
THC	Tetrahydrocannabinol
TNC	Transnational Corporation
TPP	Trans-Pacific Partnership
TRIMS	Agreement on Trade-related Investment Measures
TRIPS	Agreement on Trade-related Aspects of Intellectual Property Rights
TTIP	Transatlantic Trade and Investment Partnership

UK	United Kingdom
UN	United Nations
UNASUR	Union of South American Countries, Bolivia
UNCCD	United Nations Convention to Combat Desertification
UNCED	United Nations Conference on Environment and Development
UNDAF	United Nations Development Assistance Framework
UNDESA	United Nations Department of Economic and Social Affairs, US
UNDP	United Nations Development Programme, US
UNEP	United Nations Environment Programme, Kenya
UNHCR	United Nations High Commissioner for Refugees, Switzerland
UNOPS	United Nations Office for Project Services, Denmark
US	United States
USAID	United States Agency for International Development
USDA	United States Department of Agriculture
USMCA	United States–Mexico–Canada Agreement
VOC	Dutch East India Company
WFP	World Food Programme, Italy
WHO	World Health Organization, Switzerland
WIC	Dutch West India Company
WMO	World Meteorological Organization, Switzerland
World Fish	World Fish, Malaysia
WRI	World Resources Institute, US
WTO	World Trade Organization, Switzerland
WWF	Worldwide Fund for Nature, Switzerland

Annex II

Sustainable Development Goals

GOAL 1   End poverty in all its forms everywhere.

GOAL 2   End hunger, achieve food security and improved nutrition and promote sustainable agriculture.

GOAL 3   Ensure healthy lives and promote well-being for all at all ages.

GOAL 4   Ensure inclusive and equitable quality education and promote lifelong learning opportunities for all.

GOAL 5   Achieve gender equality and empower all women and girls.

GOAL 6   Ensure availability and sustainable management of water and sanitation for all.

GOAL 7   Ensure access to affordable, reliable, sustainable and modern energy for all.

GOAL 8   Promote sustained, inclusive and sustainable economic growth, full and productive employment, and decent work for all.

GOAL 9   Build resilient infrastructure, promote inclusive and sustainable industrialization, and foster innovation.

GOAL 10   Reduce inequality within and among countries.

GOAL 11   Make cities and human settlements inclusive, safe, resilient and sustainable.

GOAL 12   Ensure sustainable consumption and production patterns.

GOAL 13   Take urgent action to combat climate change and its impacts.

GOAL 14   Conserve and sustainably use the oceans, seas and marine resources for sustainable development.

GOAL 15   Protect, restore and promote sustainable use of terrestrial ecosystems, sustainably manage forests, combat desertification, halt and reverse land degradation, and halt biodiversity loss.

GOAL 16   Promote peaceful and inclusive societies for sustainable development, provide access to justice for all and build effective, accountable and inclusive institutions at all levels.

GOAL 17   Strengthen the means of implementation and revitalize the global partnership for sustainable development.

Source

United Nations (2015) Transforming our World: The 2030 Agenda for Sustainable Development. A/RES/70/1, United Nations, https://sustainabledevelopment.un.org/index.php?page=view&type=111&nr=8496&menu=35.

Annex III

Five Mass Extinctions on Earth with Loss of Species

Extinction event	Major cause	Groups affected	Percentage of species	Millions of years ago
Cretaceous–Paleogene	Asteroid	Dinosaurs	75	66
Triassic–Jurassic	Volcano eruptions, climate change, gas hydrate release	Large amphibians, crurotarsans	70–75	201
Permian–Triassic	Unstable climate, ocean oxygen reduction or asteroid/comet	Marine invertebrates, land plants, insects and plankton	70–96	252
Late Devonian	Global anoxia, forest evolution, reduced carbon dioxide	Tropical coral–sponge reefs, fish and plankton	70	375
Ordovician–Silurian	Massive glaciation and sea-level drop	Brachiopods, trilobites, graptolites and moss animals	60–70	450

Sources

Bond, D.P.G. and Grasby, S.E. (2017) On the causes of mass extinctions. *Palaeogeography, Palaeoclimatology, Palaeoecology. Mass Extinction Causality: Records of Anoxia, Acidification, and Global Warming during Earth's Greatest Crises* 478, 3–29. DOI: 10.1016/j.palaeo.2016.11.005 (accessed 31 July 2027).

Jablonski, D. (1986) Mass and background extinctions: the alternation of macroevolutionary regimes. *Science* 231, 129–133. Available at: https://doi.org/10.1126/science.231.4734.129 (accessed 31 July 2024).

McCallum, M.L. (2015) Vertebrate biodiversity losses point to a sixth mass extinction. *Biodiversity and Conservation* 24(10), 2497–2519.

Ritchie, H. (2022) 'There have been five mass extinctions in Earth's history.' OurWorldInData.org. Available at: https://ourworldindata.org/mass-extinctions (accessed 31 July 2024).

Annex IV

Centres of Diversity and Domestication of Animals and Crops During Early Civilizations: A Summary

A centre of diversity is a region of the world where domestication started. Originally, the famous Russian botanist, Nikolai Vavilov, identified three crop centres in 1924, five in 1926, six in 1929, seven in 1931, eight in 1935 and then reduced them down to seven in 1940. His theory was first published in 1926. This concept influenced thinking about the origins and spread of agriculture for many years. It has been updated and the Mediterranean centre has been replaced by Papua New Guinea (Vavilov, 2009).

Some of the earliest known domestications were of animals. Domestic pigs had multiple centres of origin in Eurasia, including Europe, East Asia and South-west Asia (Larson *et al.*, 2005). Wild boar were first domesticated about 10,500 years ago (Larson *et al.*, 2007). Sheep were domesticated in Mesopotamia between 11,000–9000 BCE (Ensminger and Parker, 1986). Cattle were domesticated from the wild aurochs in the areas of modern Turkey, India and Pakistan around 8500 BCE (McTavish *et al.*, 2013). Bees were kept for honey in the Middle East around 7000 BCE.

Mesopotamia

Sumerian farmers grew barley and wheat and had started to live in villages by about 8000 BCE (van der Crabben, 2023). Agriculture relied on water from the Tigris and Euphrates rivers which led to the construction of irrigation canals. This permitted greater wheat production, sufficient to support cities. In addition, they grew other crops, such as onions, garlic, lettuce, leeks and mustard. Lentils and chickpeas were domesticated around 7000 BCE in this fertile crescent (De Ron, 2015). Faba beans are among the oldest crops spread across Eurasia during the Neolithic, Bronze and Iron ages. Archaeological evidence dates the existence of peas back to 10,000 BCE. Fruit included dates, grapes, apples, melons and figs. Nomadic animal husbandry provided the meat of sheep, goats, cows and poultry (Dong, 2016). Sumerians caught fish and hunted fowl and gazelle. Fish was preserved by drying, salting and smoking.

Ancient Egypt

Dependable seasonal flooding of the Nile River creating fertile soil was the basis for the development of the Egyptian civilization. Along the Nile, the people could practice agriculture with irrigation on a large scale, starting around 10,000–4000 BCE (Janick, 2002). Wheat and barley were major staple food crops, whereas papyrus and flax were used for industrial works (FAO, 2020). Cannabis was a crop used for ropemaking and as a medicine in Egypt by about 2350 BCE (Vergara, 2014). Its origin was much older; some 12,000 years ago in Central Asia. Since then, cannabis seeds have accompanied the migration of nomadic peoples (Crocq, 2020). Records of the medicinal use of its seed are old.

Ancient China

In northern China, millet was domesticated between 8000–6000 BCE and became the main crop of the Yellow River basin by 5500 BCE (Lu *et al.*, 2009). Other crops were mung, soy and azuki beans. Cannabis was in use in China in Neolithic times and may have been domesticated there.

In southern China, rice was domesticated 13,500–8200 years ago in the Pearl River valley region from a single genetic origin of the wild rice *Oryza rufipogon* (Molina *et al.*, 2011), with its earliest known cultivation from 5700 BCE. Gradually, rice cultivation spread to maritime South-east Asia between 3500–2000 BCE, then mainland South-east Asia and later reached the Americas as part of the Columbian exchange. The nowadays less common rice variety, *Oryza glaberrima*, was independently domesticated in Africa, probably around 3000 years ago (Choi, 2019; Choi *et al.*, 2019). Other crops followed later such as water chestnut and mung, soy and azuki beans. Mung beans were an ancient crop in India and spread to southern China, Indochina and Java. Soybeans were cultivated in China before the times of the first written records in 2823 BCE and were a food crop in Korea, Japan and Manchuria (Purseglove, 1974). They were taken to the United States in 1804 but only became a commercial crop in the 20th century. Domestication of both the azuki bean and soybean has enabled them to develop nondormant seeds, nonshattering pods and a larger seed size. Seed remains from the Jomon period recently discovered at archaeological sites in the Central Highlands of Japan (6000–4000 BCE) suggest that the use of azuki beans and soybeans and their increase in seed size began earlier in Japan than in China and Korea. Molecular phylogenetic studies indicate that azuki beans and soybeans originated in Japan (Takahashi *et al.*, 2023).

Contacts with Sri Lanka and southern India led to exchanges of food plants which later became the origin of the spice trade (Gilboa and Namdar, 2016). Between the 5th century BCE and the 2nd century CE sericulture was practiced in China and described in a Chinese book on agriculture, according to Needham (1986). It provides details of various agricultural practices and the culinary uses for crops. Silk was believed to have first been produced in China as early as the Neolithic period. In a Confucian text, the discovery of silk production is dated to about 2700 BCE, although Barber (1991) noted archaeological records on silk cultivation as early as the Yangshao period (5000–3000 BCE).

South Asia

Jujube, also called red date or Chinese date, is one of the oldest cultivated fruit trees in the world, though it is now on International Union for Conservation of Nature (IUCN)'s Red List of Threatened Species. It is a very small deciduous tree or shrub, kept for its fruit, originating from the middle and lower reaches of the Yellow River in China (Liu *et al.*, 2020). Its domestication in the Indian subcontinent took place around 9000 BCE (Gupta, 2004). From the eighth millennium BCE onwards, 2-row and 6-row barley were cultivated, along with einkorn, emmer and durum wheats, and dates (Harris and Gosden, 1996). Game hunting was successively replaced by domesticated sheep, goats and humped zebu cattle by the fifth millennium BCE (Pérez-Pardal *et al.*, 2018). Cotton was cultivated by the 5th–4th millennium BCE (Stein, 1998). Harappa was the centre of one of the core regions of the Indus Valley civilization in central Punjab (3300–1300 BCE). The people of the Harappan civilization not only planted wheat, barley and dates but also peas, lentils, linseed, mustard millet, sesame and rice (Weber and Kashyap, 2016). Other crops included melons and *Brassica* species. This approach allowed people to supply food for themselves and others rather than rely on the ancient nomadic ways of hunters and gatherers living in small isolated groups, relying on their knowledge of water holes, plant locations and game habits. This was in contrast to pastoral nomads, who migrated within established territories to find pastures for their domesticated livestock. Most of such groups have selected sites, where they would stay for longer periods and may practice some kind of agriculture, hunt or even trade with agricultural peoples. Agricultural communities expanded as a result of the use of irrigation, leading to better planned settlements with both drainage and sewers (Rodda and Ubertini, 2004). The inhabitants of the Indus Valley valued agriculture highly.

Sub-Saharan Africa

Sorghum was domesticated in the Sahel region by 3000 BCE and pearl millet by 2500 BCE in Mali (Winchell, 2018). Teff, and most likely finger millet, were domesticated in Ethiopia (D'Andrea, 2008). The country has always had a rich biodiversity. Wild coffee is believed to have been native to the Ethiopian Kefa plateau though its origin and domestication remains unclear (Myhrvold, 2024). Also, noog (Daniel and Beldados, 2017) and ensete have their origin in Ethiopia (Zerihun, 2022). Cowpeas originated in sub-Saharan Africa and wild varieties have been traced to southern Africa (Herniter *et al.*, 2020). Due to their high shade tolerance, they are still important in intercropping systems on the dry savannahs of sub-Saharan Africa. Other domesticated plant foods in Africa included watermelon, okra, tamarind, black-eyed peas, kola nuts and oil palms.

Ancient Greece

The major cereals of the ancient Mediterranean region were wheat, emmer and barley, while common vegetables included peas, beans, faba and olives (Mefleh, 2021). Dairy products came mostly from sheep and goats. Meat came from pigs, cattle and sheep. It was consumed occasionally by most people (Koester, 1995). Due to the topography of mainland Greece, grain had to be imported from its colonies, whereas oil and wine was exported. During the Hellenistic period (323–30 BCE), the Ptolemaic empire controlled grain-producing Egypt, Cyprus, Phoenicia and Cyrenaica (Ferguson, 2023). That production played an important role in the rise of the Roman Republic.

The Roman Empire

Roman agriculture was built on techniques developed by the Sumerians. After the Iron Age, Rome expanded during the Republic and the Empire across the Mediterranean and Western Europe. Roman agriculture was based on an old manorial economic system with serfdom according to Oxford Reference (2023). This system lasted throughout medieval agriculture. There were four types of farm management: work by the owner and his family; slaves under the supervision of slave managers; tenant farming or sharecropping; and a farm leased to a tenant (White, 1970). Farms were categorized into small farms (less than 20 ha), medium-sized farms (20–132 ha) and large estates/latifundia (more than 132 ha).

The Americas

The earliest known areas of possible agriculture in the Americas, dating to about 9000 BCE, are in Colombia, near present-day Pereira, and in Ecuador, on the Santa Elena peninsula, from the Las Vegas culture (Piperno, 2011). These areas have been identified as one of the four oldest places of origin for agriculture, along with the Fertile Crescent, Mesoamerica and China (Larson, 2014). Cultivated plants included arrowroot, squash and bottle gourd, which probably came from more humid climates. Early American agriculture usually consisted of intensive polyculture with hand labour in several different climatic zones close to each other, often in mountainous regions (Drake,

2012). This gave rise to the domestication of many different plants (Diamond, 2002). However, arriving Europeans found the polyculture of less interest as a gardening activity, and more as a commercial activity over which they wanted full control.

In the Andes region, the major crop was the potato, domesticated between 8000–5000 BCE (Spooner *et al.*, 2005). Other crops were coca, still a major crop, and peanuts, tomatoes, beans, cocoa and pineapple (Murphy, 2011). The genus *Phaseolus* of beans originated approximately seven million years ago (De Ron, 2015). Wild forms of common beans can be found from north-western Argentina to northern Mexico. Several Andean species were cultivated in early times and later became minor crops, such as quinoa, kiwicha and cañihua (Atchison *et al.*, 2016). It was noted that the Andean lupin had originated in the highlands of northern Peru, whereas the white lupin was domesticated in the Old World (Hufnagel *et al.*, 2021). Cotton was domesticated in Peru by 4200 BCE (Rajpal, 2016), but another species of cotton was domesticated in Mesoamerica, and this became the most important species. Farm animals were also domesticated such as llamas, alpacas and guinea pigs.

Over several thousand years to about 7000 BCE, Mesoamerican farmers transformed wild teosinte, through human selection, into the ancestor of modern maize (Kistler *et al.*, 2018). It was a very important crop to indigenous Americans at the time of the arrival of European explorers in the 16th century. Maize was more productive than wheat and barley from the Old World. The turkey was probably also domesticated in Central America before the arrival of Europeans (Speller *et al.*, 2010).

Cassava was domesticated in the Amazon Basin and tropical lowlands before 7000 BCE (Isendahl, 2011).

North America

The indigenous people of the eastern North America domesticated crops, such as sunflower, tobacco, squash and *Chenopodium* species (Smith, 2011). Wild foods were harvested, including wild rice and maple sugar (Smith, 2013). Pecans and Concord grapes were used but are considered to have been domesticated in the 19th century. The indigenous people in modern California and the Pacific Northwest, practiced several types of forest gardening and fire-stick farming. They used low-intensity fire ecology and managed to prevent larger fires; a sort of 'wild' permaculture (Blackburn and Anderson, 1993).

'The Three Sisters' (maize, beans and squash) complemented each other when planted together in local communities (Steinberg, 2020). The maize gave the beans a structure to climb, eliminating the need for poles. The beans provided the nitrogen for the soil and the squash could spread along the ground, preventing weeds. This system involves species/cultivars that benefit one another in the same area and can adapt to adverse weather conditions (Misra and Ghosh, 2024). This ancient agricultural practice has recently been highlighted by the USDA Agricultural Research Service (2021).

Australia

Indigenous Australians were predominately nomadic hunter-gatherers and current consensus suggests various agricultural methods were employed by the indigenous people. They are also reported to have used systematic burning or fire-stick farming to enhance natural productivity (Gerritsen, 2020). Certain groups are reported to have developed sophisticated eel farming and fish trapping systems which have been used for nearly 5000 years (Williams, 1988).

In the central west coast and eastern central regions of Australia, forms of agriculture were practiced. People living in permanent settlements of over 200 residents sowed or planted on a large scale and stored the harvested food. Yams was grown in eastern central Australia, together with bush onions and native millet, according to Gerritsen (2008).

New Guinea

Sugarcane and taro were domesticated in New Guinea around 7000 BCE (Murphy, 2011). Also, bananas (*Musa* spp.), native to the South-east Asia/Oceania region were domesticated in New Guinea some 7000 years ago (Bower, 2003). Initially, the crop was not the seedless, fleshy fruit of today but had hard black seeds, almost inedible. Thus, cuttings were taken to find seedless mutants and reproduce them asexually, leading to genetically identical populations. Such cloned populations became vulnerable to pests and diseases, requiring cross-breeding over the years, which has resulted in a complicated domestication process with several wild ancestors, for example the predominant *Musa acuminata*. It seems to have evolved in the northern borderlands between India and Myanmar, having existed across Australasia for approximately 10 million years before it was first domesticated (Sardos *et al.*, 2022). A recent extensive genetic DNA analysis of wild and cultivated bananas revealed the existence of three previously unknown – and possibly still living – ancestors, including *Musa balbisina*. Currently, there are over 1000 varieties of banana which has become a globally popular commodity for consumption only within the last 150 years (Pennisi, 2022).

References

Agricultural Research Service (2021) USDA Agricultural Research Service Sets Stage for Next-Generation Researchers. USDA Agricultural Research Service. Available at: https://www.ars.usda.gov/news-events/news/research-news/2021/usda-agricultural-research-service-sets-stage-for-next-generation-researchers (accessed 5 August 2024).

Atchison, G.A., Nevado, B., Eastwood, R.J., Contreras-Ortiz, N., Reynel, C. *et al.* (2016) Lost crops of the Incas: Origins of domestication of the Andean pulse crop tarwi, *Lupinus mutabilis*. *Botany* 103(9), 1592–1606. Available at: https://doi.org/10.3732/ajb.1600171 (accessed 5 August 2024).

Barber, E.W. (1991) *Prehistoric Textiles: The Development of Cloth in the Neolithic and Bronze Ages with Special Reference to the Aegean*. Princeton University Press, Princeton, NJ.

Blackburn, T.C. and Anderson, K. (eds) (1993) *Before the Wilderness: Environmental Management by Native Californians*. Ballena Press, US.

Bower, B. (2003) New Guinea went bananas: Agriculture's roots get a South Pacific twist. *News Anthropology, Science News*. Available at: https://www.sciencenews.org/article/new-guinea-went-bananas-agricultures-roots-get-south-pacific-twist (accessed 5 August 2024).

Choi, J.Y. (2019) The complex geography of domestication of the African rice *Oryza glaberrima*. *PLOS Genetics* 15(3), e1007414. Available at: https://doi.org/10.1371/journal.pgen.1007414 (accessed 24 August 2024).

Choi, J.-Y., Zaidem, M., Gutaker, R., Dorph, K., Singh, R.K. *et al.* (2019) The complex geography of domestication of the African rice *Oryza glaberrima*. *PLOS Genetics* 15(3), e1007414. Available at: https://doi.org/10.1371/journal.pgen.1007414 (accessed 5 August 2024).

Crocq, M.A. (2020) History of cannabis and the endocannabinoid system. *Dialogues in Clinical Neuroscience* 22(3), 223–228. Available at: https://doi.org/10.31887/DCNS.2020.22.3/mcrocq (accessed 5 August 2024).

D'Andrea, A.C. (2008) T'ef (*Eragrostis tef*) in ancient agricultural systems of highland Ethiopia. *Economic Botany* 62, 547–566.

Daniel, H. and Beldados, A. (2017) Ethnoarchaeology study of *Noog* (*Guizotia abyssinica* (L.f.) Cass., Compositae) in Ethiopia. *Ethnoarchaeology* 10(1), 16–33. Available at: https://doi.org/10.1080/19442890.2017.1364513 (accessed 5 August 2024).

De Ron, A.M. (ed.) (2015) *Grain Legumes, Handbook of Plant Breeding 10*. Springer Science Business Media, New York.

Diamond, J. (2002) Evolution, consequences and future of plant and animal domestication. *Nature* 418, 700–707. Available at: https://doi.org/10.1038/nature01019 (accessed 5 August 2024).

Dong, S. (2016) Overview: Pastoralism in the World. In: Dong, S., Kassam, K.A., Tourrand, J. and Boone, R. (eds) *Building Resilience of Human-Natural Systems of Pastoralism in the Developing World*. Springer, Cham, Switzerland.

Drake, S.G. (2012) *History of the Early Discovery of America and Landing of the Pilgrims: With a Biography of the North American Indians (1854)*. Ulan Press.

Ensminger, M.E. and Parker, R.O. (1986) *Sheep and Goat Science*, 5th ed. Interstate Printers and Publishers.

Ferguson, J. (2023) Hellenistic age. *Encyclopedia Britannica*. Available at: https://www.britannica.com /event/Hellenistic-Age (accessed 13 June 2023).

Food and Agriculture Organization (FAO) (2020) Ancient Egyptian Agriculture: Agricultural System. Food and Agriculture Organization of the United Nations (FAO). Available at: https://www.fao.org/country -showcase/item-detail/en/c/1287824/ (accessed 5 August 2024).

Gerritsen, R. (2008) *Australia and the Origins of Agriculture*. Archaeopress, Oxford, UK.

Gerritsen, R. (2020) Australia and the origins of agriculture. In: Smith, C. (ed.) *Encyclopedia of Global Archaeology*. Springer, Cham, Switzerland, pp. 1118–1126.

Gilboa, A. and Namdar, D. (2016) On the beginnings of South Asian spice trade with the Mediterranean region: A review. *Radiocarbon* 57(2), 265–283. Available at: https://doi.org/10.2458/azu_rc.57.18562 (accessed 5 August 2024).

Gupta, A.K. (2004) Origin of agriculture and domestication of plants and animals linked to early holocene climate amelioration. *Current Science* 87(1), 58–59.

Harris, D.R. and Gosden, C. (1996) *The Origins and Spread of Agriculture and Pastoralism in Eurasia: Crops, Fields, Flocks and Herds*. Routledge.

Herniter, I.A., Muñoz-Amatriaín, M. and Close, T.J. (2020) Genetic, textual, and archaeological evidence of the historical global spread of cowpea (*Vigna unguiculata* [L.] Walp.). *Legume Science* 24(4), e57. Available at: https://doi.org/10.1002/leg3.57 (accessed 5 August 2024).

Hufnagel, B., Soriano, A., Taylor, J., Divol, F., Kroc, M. *et al.* (2021) Pangenome of white lupin provides insights into the diversity of the species. *Plant Biotechnology Journal* 19(12), 2532–2543. Available at: https://doi.org/10.1111/pbi.13678 (accessed 5 August 2024).

Isendahl, C. (2011) The domestication and early spread of manioc: A brief synthesis. *Latin American Antiquity* 22(4), 452–468. Available at: https://doi.org/10.7183/1045-6635.22.4.452 (accessed 5 August 2024).

Janick, J. (2002) Ancient egyptian agriculture and the origins of horticulture. *Acta Horticulturae* 583(582), 23–39. Available at: https://doi.org/1017660/ActaHortic.2002.582.1 (accessed 5 August 2024).

Kistler, L., Yoshi Maezumi, S., Souza, J., Przelomska, N., Malaquious Costa, F. *et al.* (2018) Multiproxy evidence highlights a complex evolutionary legacy of maize in South America. *Science* 362(6420), 1309–1313. Available at: https://doi.org/10.1126/science.aav0207 (accessed 5 August 2024).

Koester, H. (1995) *History, Culture, and Religion of the Hellenistic Age*, 2nd ed. Walter de Gruyter, New York.

Larson, G. (2014) Current perspectives and the future of domestication studies. *Proceedings of the National Academy of Sciences* 111(17), 6139–6146. Available at: https://doi.org/10.1073/pnas.1323 964111 (accessed 5 August 2024).

Larson, G., Dobney, K., Albarella, U., Fang, M., Matisoo-Smith, E. *et al.* (2005) Worldwide phylogeography of wild boar reveals multiple centers of pig domestication. *Science* 307(5715), 1618–1621. Available at: https://doi.org/10.1126/science.1106927 (accessed 5 August 2024).

Larson, G., Albarella, U., Dobney, K., Rowley-Conwy, P., Schibler, J. *et al.* (2007) Ancient DNA, pig domestication, and the spread of the Neolithic into Europe. *Proceedings of the National Academy of Sciences* 104(39), 15276–15281. Available at: https://doi.org/10.1073/pnas.0703411104 (accessed 5 August 2024).

Liu, M., Wang, J., Wang, L., Liu, P., Zhao, J. *et al.* (2020) The historical and current research progress on jujube: A superfruit for the future. *Horticulture Research* 7, 119. Available at: https://doi.org/10.1038 /s41438-020-00346-5 (accessed 5 August 2024).

Lu, H., Zhang, J., Liu, K.B., Wu, N., Li, Y. *et al.* (2009) Earliestdomestication of common millet (*Panicum miliaceum*) in East Asia extendedto 10,000 years ago. *Proceedings of the National Academy of Sciences of the United States of America* 106(18), 7367–7372. Available at: https://doi.org/10.1073 /pnas.0900158106 (accessed 5 August 2028).

McTavish, E.J., Decker, J.E., Schnabel, R.D., Taylor, J.F. and Hillis, D.M. (2013) New world cattle show ancestry from multiple independent domestication events. *Proceedings of the National Academy of Sciences* 110(15), 1398–1406. Available at: https://doi.org/10.1073/pnas.1303367110 (accessed 5 August 2024).

Mefleh, M. (2021) Cereals of the Mediterranean region: their origin, breeding history and grain quality traits. In: Boukid, F. (ed.) *Cereal-Based Foodstuffs: The Backbone of Mediterranean Cuisine*. Springer, Cham, Switzerland, pp. 1–18.

Misra, S. and Ghosh, A. (2024) Agricultural paradigm shift; a journey from traditional to modern agriculture. In: Singh, K., Ribeiro, M.C. and Calicioglu, O. (eds) *Biodiversity and Bioeconomy: Status Quo, Challenges, and Opportunities*. Elsevier, Amsterdam, pp. 113–142.

Molina, J., Sikora, M., Garud, N., Flowers, J.M., Rubinstein, S. *et al.* (2011) Molecular evidence for a single evolutionary origin of domesticated rice. *Proceedings of the National Academy of Sciences* 108(20), 8351–8356. Available at: https://doi.org/10.1073/pnas.1104686108 (accessed 5 August 2024).

Murphy, D. (2011) *Plants, Biotechnology and Agriculture*. CABI, Wallingford, UK.

Myhrvold, N. (2024) History of coffee. *History of Coffee*. Available at: https://www.britannica.com/topic /history-of-coffee (accessed 28 June 2024).

Needham, J. (1986) Science and civilisation in China, Vol. 6(2). Caves Books Ltd, Taipei, pp. 55–56.

Oxford Reference (2023) Roman agriculture: Overview. Available at: https://www.oxfordreference.com /view/10.1093/oi/authority.20110803095356559 (accessed 5 August 2024).

Pennisi, E. (2022) Researchers have gone bananas over this fruit's complex ancestry: genomic analyses suggest three mysterious ancestors contributed DNA to the modern-day banana. *Science Adviser*. Available at: https://doi.org/10.1126/science.adf3420 (accessed 24 August 2024).

Pérez-Pardal, L., Sánchez-Gracia, A., Álvarez, I., Traoré, A., Bento, S, *et al.* (2018) Legacies of domestication, trade and herder mobilityshape extant male zebu cattle diversity in South Asia and Africa. *Scientific Reports* 8, 18027. Available at: https://doi.org/10.1038/s41598-018-36444-7 (accessed 5 August 2024).

Piperno, D.R. (2011) The origins of plant cultivation and domestication in the New World Tropics: Pattern, process, and new developments. *Current Anthropology* 52(S-4), S453–S470. Available at: https:// doi.org/10.1086/659998 (accessed 5 August 2024).

Purseglove, J.P. (1974) *Tropical Crops.Dicotyledons. Volume 1*. Longman, London.

Rajpal, V.R. (2016) *Gene Pool Diversity and Crop Improvement, Volume 1*. Springer, Cham, Switzerland.

Rodda, J.C. and Ubertini, L. (2004) *The Basis of Civilization: Water Science? Volume* 286. International Association of Hydrological Science, Wallingford, UK.

Sardos, J., Breton, C., Perrier, X., Houwe, I., Carpentier, S. *et al.* (2022) Hybridization, missing wild ancestors and the domestication of cultivated diploid bananas. *Frontiers in Plant Science* 13, 969220. Available at: https://doi.org/10.3389/fpls.2022.969220 (accessed 24 August 2024).

Smith, A. (2013) *The Oxford Encyclopaedia of Food and Drink in America*. Oxford University Press, New York.

Smith, B. (2011) The cultural context of plant domestication in Eastern North America. *Current Anthropology* 52(54). Available at: https://doi.org/10.1086/659645 (accessed 5 August 2024).

Speller, C.F., Kemp, B.M., Wyatt, S.D., Monroe, C., Lipe, W.D. *et al.* (2010) Ancient mitochondrial DNA analysis reveals complexity of indigenous North American turkey domestication. *Proceedings of the National Academy of Sciences* 107(7), 2807–2812. Available at: https://doi.org/10.1073/pnas.09097 24107 (accessed 5 August 2024).

Spooner, D.M., McLean, K., Ramsay, G., Waugh, R. and Bryan, G. (2005) A single domestication for potato based on multilocus amplified fragment length polymorphism genotyping. *Proceedings of the National Academy of Sciences* 102(41), 14694–14699. Available at: https://doi.org/10.1073/pnas .0507400102 (accessed 5 August 2024).

Stein, B. (1998) *A History of India*. Blackwell Publishing, Oxford, UK.

Steinberg, I. (2020,7 October) The Three Sisters: What an ancient agricultural technique can teach us about community. *Resilient Palisades*. Available at: https:/resilientpalisades.org/the-three-sisters/ (accessed 5 August 2024).

Takahashi, Y., Nasu, H., Nakayama, S. and Tomooka, N. (2023) Domestication of azuki bean and soybean in Japan: From the insight of archaeological and molecular evidence. *Breeding Science* 73(2), 117–131. Available at: https://doi.org/10.1270/jsbbs.22074 (accessed 5 August 2024).

van der Crabben, J. (2023) Agriculture in the fertile crescent and mesopotamia. *World History Encyclopedia*. Available at: https://www.worldhistory.org/article/9/agriculture-in-the-fertile-crescent--mesopotamia/ (accessed 24 August 2024).

Vavilov, N.I. (2009) Origin and geography of cultivated plants. In: *In Cooperation with and Translated by Doris Love*. Cambridge University Press, Cambridge, UK.

Vergara, D. (2014) Cannabis: Marijuana, hemp, and its culturalhistory. Cannabis Genomics Conference, December 2014.

Weber, S. and Kashyap, A. (2016) The vanishing millets of the Indus civilization. *Archaeological and Anthropological Sciences* 8, 9–15. Available at: https://doi.org/10.1007/s12520-013-0143-6 (accessed 5 August 2024).

White, K.D. (1970) *Roman Farming: Aspects of Greek and Roman Life.* Cornell University Press, Ithaca, NY.

Williams, E. (1988) *Complex Hunter-Gatherers: A Late Holocene Example from Temperate Australia.* British Archaeological Reports, BAR Publishing, Oxford, UK.

Winchell, F., Brass, M., Manzo, A., Beldados, A., Perna, V. *et al.* (2018) On the origins and dissemination of domesticated sorghum and pearl millet across africa and into india: A view from the butana group of the far eastern sahel. *The African Archaeological Review* 35(4), 483–505. Available at: https://doi.org/10.1007/s10437-018-9314-2 (accessed 5 August 2024).

Zerihun, Y. (2022) *Annotated Bibliography of Enset. Southern Agricultural Research Institute (SARI).* Hawassa, Southern Nation Nationality People Regional State, Ethiopia.

Annex V

Underutilized Minor Crops with Potential for a Sustainable Future

The Andean region has many indigenous crop species in a biodiversity that has been maintained and cultivated for generations in variable environments, variable rainfalls and changing soil types.

Amaranthus is a cosmopolitan genus. Species are cultivated as leaf vegetables, pseudocereals and ornamental plants. Both amaranth (*Amaranthus* spp.) and quinoa are considered pseudocereals. They have similarities to cereals in flavour and cooking. Several species are grown for 'grain' production in Asia and the Americas. Amaranth is a gluten-free grain that has been cultivated for several thousand years. It was already part of the agriculture of the Aztec civilization in the 1400s. The seeds are rich in protein and minerals such as calcium, magnesium, phosphorus and potassium.

Quinoa (*Chenopodium quinoa*) has been a staple crop for thousands of years in the Andean region of South America. It is still a staple crop grown by the Quechua people in Peru. Its seeds are high in protein and fibre, and the young leaves can be eaten as a vegetable similar to spinach. The seeds are a good source of iron, magnesium, phosphorus, potassium, calcium, zinc, copper, vitamin E and several antioxidants. Because of its health benefits, quinoa is now grown in many countries, including the United States.

Cañihua (*Chenopodium pallidicaule*) is native to the Andean region with more than 200 varieties. For thousands of years, it has been farmed in high mountain conditions in the Andes. It is drought resistant, has a high protein content (15–19%), fibre and the polyunsaturated fatty acids Omega-3, 6 and 9. It is rich in calcium, phosphorus and iron. It is a gluten-free alternative in cooking, similar in character and uses to quinoa, but smaller in size.

Kiwicha or guihuicha (*Amaranthus caudatus*) is a fast-growing plant, adapted to different environments at altitudes between 1400–2400 m above sea level in Peru and other Andean regions. Estimates indicate some 1200 varieties remain. The plant contains a more balanced composition than conventional cereals and better-quality proteins (13–18%). It is rich in calcium, phosphorus, iron, potassium, zinc and vitamin E.

Coca (*Erythroxylum coca*) is one species of cultivated coca. It is a tropical shrub whose leaves are the source of the alkaloid cocaine, and several other alkaloids. Coca is cultivated from seeds in Africa and South-east Asia. It requires very acidic soil and water conditions. It has a greater resistance to the use of glyphosate than its relative, the Amazonian coca. In contrast, the Amazonian coca (*Erythroxylum coca* var. *ipadu*) is vegetatively propagated which is why plantations may consist of the same clone. It is specially adapted to the agriculture of Amazonian peoples.

Oca (*Oxalis tuberosa*) is a perennial herbaceous plant that overwinters as underground stem tubers. It is a staple crop of the Andean highlands, mainly in Peru and Bolivia. It is easy to propagate, and tolerates poor soil, high altitudes and a harsh climate. It is cultivated primarily for its edible stem tuber. In addition, its leaves and shoots are eaten as green vegetables. It has a high nutritional content and is second only to the potato in the area over which it is planted within the central Andean region. It was introduced into Europe in 1830s to compete with the potato and later to New Zealand.

Olluco (*Ullucus tuberosus*) has many different names. Its tuber is the part that is primarily eaten, but its leaf can also be used, like spinach. The tuber is brightly coloured, is grown at higher altitudes (2500–4000 m) and has few pest and disease problems. Olluco contains high levels of protein, calcium and carotene.

Mashua (*Tropaeolum tuberosum*) is particularly grown in Peru and Bolivia, but also to some extent in Ecuador and Colombia, for its edible high-yielding tubers. It is a herbaceous perennial climber, growing vigorously in marginal soils and a good competitor for weeds. Mashua is well-adapted for high-altitude subsistence agriculture and it is resistant to insect, nematode and bacterial pests, which is attributed to high levels of isothiocyanates. Its strong flavour may prevent popularization, together with a reputation as an anaphrodisiac. However, traditionally roasted tubers are considered a delicacy.

Yacon or the Peruvian ground apple (*Smallanthus sonchifolius*) is vigorous grower, cultivated for its crisp, sweet-tasting, tuberous roots. It is composed mostly of water and fructooligosaccharide. It is grown in the northern and central Andes from Colombia to northern Argentina on the mid-elevation slopes of the range. The plant is easily infected by bacteria, nematodes, fungi, viruses and insects. Its use has spread since 2000 to several Asian countries and it was recently introduced into farmers' markets in the United States.

Khorasan wheat or Asian wheat (*Triticum turgidum* ssp. *turanicum*) is commercially known as Kamut. It is a tetraploid, drought-tolerant wheat species with a grain twice the size of modern-day wheat. Its high-fibre grain contains gluten and has a chewy, nutty texture with more proteins, lipids, amino acids, vitamins and minerals than modern-day wheat.

Freekeh or farik is a staple in Middle Eastern cuisine made from roasted and rubbed green durum wheat (*Triticum turgidum* var. *durum*). It contains gluten and a variety of nutrients and powerful carotenoid compounds such as lutein and zeaxanthin.

Farro is an ethnobotanical term for three species of hulled wheat: spelt (*Triticum spelta*), emmer (*Triticum dicoccum*) and einkorn (*Triticum monococcum*). In Italian cuisine, the three species are sometimes distinguished as *farro grande*, *farro medio* and *farro piccolo*. Farro contains gluten, is high in protein and fibre, and has antioxidants such as polyphenols, carotenoids and phytosterols.

Bulgur or Persian wheat, sometimes called cracked wheat, is durum wheat with a high fibre content. It contains gluten. Cracked wheat is crushed wheat grain that has not been parboiled, unlike bulgur.

Further reading

Bazile, D., Jacobsen, S.E.. and Verniau, A. (2016) The global expansion of quinoa: Trends and limits. *Frontiers in Plant Science* 7, 622. Available at: https://doi.org/10.3389/fpls.2016.00622 (accessed 5 August 2024).

Choque Delgado, G.T., da Silva Cunha Tamashiro, W.M., Maróstica Junior, M.R. and Pastore, G.M. (2013) Yacon (*Smallanthus sonchifolius*): A functional food. *Plant Foods for Human Nutrition* 68, 222–228. Available at: https://doi.org/10.1007/s11130-013-0362-0 (accessed 5 August 2024).

Corke, H., Cai, Y.Z. and Wu, H.X. (2016) Amaranth: Overview. *Reference Module in Food Science*. Available at: https://doi.org/10.1016/B978-0-08-100596-5.00032-9 (accessed 5 August 2024).

Delaney, M. (2023) How to Grow Oca in Australia. August 22, 2023, Regeno organic gardening in Australia.

Geisslitz, S. and Scherf, K.A. (2020) *Rediscovering Ancient Wheats*. Institute of Technology (KIT), Karlsruhe, Germany, Cereals & Grains Association.

Leidi, E.O., Altamirano, A.M., Mercado, G., Rodriguez, J.P., Ramos, A. *et al.* (2018) Andean roots and tubers crop as sources of functional foods. *Journal of Functional Foods* 51, 86–93. Available at: https://doi.org/10.1016/j.jff.2018.10.007 (accessed 5 August 2024).

Purseglove, J.W. (1974) Tropical Crops: Dicotyledons, Volumes 1 and 2. Longman, London.

Purseglove, J.W. (1974) Tropical Crops: Monocotyledons, Volumes 1 and 2. Longman, London.

Yadav, R., Gore, P.G., Gupta, V. and Saurabh and Siddique, K.H.M. (2023) Quinoa (*Chenopodium quinoa* Willd.): A smart crop for food and nutritional security. In: Farooq, M. and Siddique, K.H.M. (eds) *Neglected and Underutilized Crops*. Academic Press, Cambridge, MA, pp. 23–43.

Annex VI

Some Ancient Empires and Dynasties

Empire	Starting year	Ending year
Egyptian Empire	3150 BCE	30 BCE
Marharsi (Iranian Plateau)	2550 BCE	1900 BCE
Assyria	2025 BCE	609 BCE
Babylonian Empire	1792 BCE	626 BCE
Shang Dynasty (China)	1600 BCE	1046 BCE
Kingdom of Kush	1070 BCE	550
Kingdom of Israel	1047 BCE	930 BCE
Zhou Dynasty (China)	1046 BCE	256 BCE
D´mt (Geez in Eritrea)	980 BCE	400 BCE
Pandyan Dynasty (India)	500 BCE	1618
Ghana Empire	300	1240
Byzantine Empire	395	1453
Gurja-Pratihara Dynasty (Hindu India)	600	1136
Abbasid Caliphate (Iraq)	750	1258
Khmer Empire (Cambodia)	802	1431
Abyssinia	1137	1974
Mongol Empire	1206	1368
Ahom Dynasty (Assam, India)	1228	1838
Mali Empire	1235	1610
Habsburg Empire (Germany/Austria)	1282	1867/1918
Ottoman Empire	1299	1922
Portuguese Empire	1415	1999
Spanish Empire	1479	1975
French Colonial Empire	1534	1980
Dutch Empire	1568	1975
British Empire	1603	1997

Index

www.ingramcontent.com/pod-product-compliance
Lightning Source LLC
Chambersburg PA
CBHW040137200326
41458CB00025B/6294